Teaching Children Science

Teaching Children Science

Hands-On Nature Study in North America, 1890–1930

SALLY GREGORY KOHLSTEDT

The University of Chicago Press

CHICAGO AND LONDON

SALLY GREGORY KOHLSTEDT is professor in the Program of History of Science, Technology and Medicine at the University of Minnesota. She is author of *The Formation of the American Scientific Community: The American Association for the Advancement of Science, 1848–1860*, coauthor of *The Establishment of Science in America: 150 Years of the American Association for the Advancement of Science* (with Michael M. Sokal and Bruce V. Lewenstein), and coeditor of *Historical Writing on American Science* (with Margaret Rossiter), *International Science and National Scientific Identity* (with Roderick Homes), and *Gender and Scientific Authority* (with Helen Longino), among others.

The University of Chicago Press, Chicago 60637
The University of Chicago Press, Ltd., London
© 2010 by The University of Chicago
All rights reserved. Published 2010
Printed in the United States of America

19 18 17 16 15 14 13 12 11 10   1 2 3 4 5

ISBN-13: 978-0-226-44990-6 (cloth)
ISBN-10: 0-226-44990-4 (cloth)

Library of Congress Cataloging-in-Publication Data
Kohlstedt, Sally Gregory, 1943-
Teaching children science : hands-on nature study
in North America, 1890–1930 / Sally Gregory Kohlstedt.
p. cm.
Includes bibliographical references and index.
ISBN-13: 978-0-226-44990-6 (cloth : alk. paper)
ISBN-10: 0-226-44990-4 (cloth : alk. paper)
1. Nature study—United States—History—19th century. 2. Nature study—
United States—History—20th century. 3. Science—Study and teaching—
United States—History—19th century. 4. Science—Study and teaching—United
States—History—20th century. I. Title.
LB1585.3.K64 2010
372.3'5097309041—dc22
2009044289

For David, Kris, and Kurt

Contents

Illustrations

Acknowledgments

BECAUSE THIS BOOK took over two decades, weaving sometimes imperceptibly among other projects, a long list of colleagues, librarians, graduate students, family members, and others encouraged my efforts, tolerated my impatience with the process, and offered cogent advice as I gathered materials and speculated about the process of educational innovation.[1] My first public presentation of nature study came while I taught at Syracuse University, where an audience of library friends and academic colleagues included a woman who remembered her excitement at being taught nature study in her upstate New York school. In later years, professional colleagues in the history of science asked about why it was important to study the history of such early education in science and sharpened my research and arguments by their queries. Others offered useful suggestions and even invited me to present my findings at venues as varied as those of the History of Science Society annual meetings, the Special Collections Library at the University of Wisconsin at Madison, and, while I was a visiting Fulbright Fellow, at the Department of History seminar at the University of Auckland. Writing, however, is ultimately a solitary task. I am particularly grateful to those closest to me for their sustenance, sometimes intangible but often quite material, while I pursued an elusive topic that took me on research odysseys in search of the sources far from home. Such trips were essential to document the lives of teachers, scientists, and others engaged in nature study education in years before the Internet made many such travels, perhaps unfortunately, less justifiable.

Identifying relevant primary and even secondary sources was harder when I started, before the age of the World Wide Web. This book would not have been written without the many, many librarians and archivists at the long list of archives and special collections at the rear of the book who responded to queries, pointed me to sources, listened to my frustrations about missing volumes, and sometimes came up with small treasures that would otherwise have escaped my notice. Some library staff members worked with me for days and weeks; others responded during a half-day visit and had already put together materials for a harried investigator. Often the research was tedious as I combed through dusty volumes of city and state educational reports, diaries with little on my topic, and unpublished organizational minutes in hopes of finding data on the teaching of nature study in places as remote as Saskatoon or as centrally located as Chicago. From the librarian who mailed to me without charge copies of a few manuscript letters that I had misplaced, to the archivist who took me back into stacks to scan folders that had not been processed, to the manuscripts librarian who shared recently acquired materials not yet in the catalogue, I was remarkably lucky in finding professional staff helpful and responsive. As the list of libraries visited suggests, they proved too many to name here.

The University of Minnesota librarians, especially those in archives and special collections as well as those who expedite interlibrary loans, have been unfailingly responsive; special thanks go to Elaine Challacombe at the Wangensteen Historical Library of Biology and Medicine for her help with illustrations. We historians must count ourselves fortunate in the dedication of those specialists who value one-of-a-kind resources and who retain an avid intellectual curiosity about their holdings. In a few libraries I was lucky enough to be just ahead of the reorganization in academic and public libraries that is relegating "old" books to storage or even disposal—and thus I was able to enjoy the serendipity of shelf browsing that turned up the kinds of ephemeral course binders and pocket-sized guidebooks that make this a richer account (and deepened those footnotes). By the end of this project, I was able to rely on search engines to help me finalize and document significant points in the history of nature study. It is time to publish this book, but I am quite aware that there are many more accounts of educators, programs, individual schools and school systems, and programs in the states, provinces, cities and country schools that are yet to be explored. I expect that regional studies, biographies of key figures, and attention to literary and conservation aspects of nature study will open out and undoubtedly challenge some of my conclusions. I look forward to subsequent studies that will enlarge, correct, and bring to light the multifaceted ways that nature study was shaped by and, in turn, helped inform contemporary social and intellectual life at the turn into the twentieth century.

Along the way and over a long period of time, students, colleagues, and friends have found useful bits of ephemeral remnants of nature study teaching, heard me lecture, provided critical commentary, and read sections of this work. Roots of this project go back to my graduate student days at the University of Illinois where Winton Solberg introduced me to the history of education and to the extension work of Liberty Hyde Bailey; Raymond P. Stearns encouraged my interest in science in North America; and Wilbur Applebaum grounded me in the early history of modern science. More immediately, I want to thank friends and fellow scholars who assisted along the way: Sari Biklen, Janet Bogdan, Judith Wellman, Lester Stephens, Pamela Henson, Jennifer Gunn, Nancy Beadie, Jane Maienschein, Karen M. Reed, Keith Benson, Rob Kohler, Lynn Nyhart, Ruth Barton, and several others mentioned in notes. An extended line of graduate students at the University of Minnesota have offered wonderful inspiration, many of them providing me with references, handwritten notes on a relevant source they had uncovered, and ideas that advanced my thinking about nature study. At this point, I specifically recall assistance, comments, or recommendations from Donald Opitz, Mark Largent, Tom Haakenson, Joe Cain, Mark Jorgenson, Horace Taft-Ferguson, Erik Conway, Erika Dirkse, Paul Brinkman, Michael Reidy, Michael Buckley, Mary Ann Andrei, Olivia Walling, Juan Ilerbaig, Karin Matchett, Juliet Burba, Georgina Montgomery, Susan Rensing, Margot Iverson, and Gina Rumore—indeed, many more graduate students have helped me become ever more aware of the ways in which science is embedded in American culture. Their energy and ideas have influenced and enriched my life in multiple ways. Colleagues Michele Aldrich, David Sloane, and Rima Apple read and commented in detail on the entire manuscript, for which I am very grateful. Margaret Rossiter offered a sustaining friendship and pointed out useful materials as she found them in her own extensive research efforts. Over the course of the project, I also had editorial help from Dianna Kinney and Jan Zita Grover; most significantly near the end of the project, the deft hand of Frances Kohler improved and clarified my prose. Karen Marikangas Darling, Abby Collier, Maia Rigas, and Elissa Park, with other staff at the University of Chicago, as well as anonymous referees, offered encouragement and helped shape the manuscript into the book in hand.

Writing this book pushed me to think more comprehensively about the ways in which children learn about the natural world and the importance of studying how they are taught and by whom. Although this book did not explore the role of families in such education—a subject that is elusive but deserves further attention—it did give me an opportunity to reflect on the importance of family and to acknowledge that my debt goes back to my maternal grandparents, no longer alive, who provided my earliest introduction to the natural world. On a farm in the fertile "thumb" of Michigan, George and Amanda Stoll Bitzer gave me a home for most of the first three years of my life and welcomed me back each summer to "help" on a property that in immediate post-Depression America boasted self-sufficiency, with its orchards, chicken coop, pig pen, large pantry garden, small dairy herd, feed crops, and commercial sugar beets. Over subsequent summers, I regularly went back for extended visits, learning from my grandparents the rigors and satisfactions of working with, and sometimes against, nature. My parents offered me different and complementary encouragement. My mother, Tula Bitzer Gregory, raised five children to be observant, independent, hard working, concerned for neighbors, and thorough in all we undertook. My father, Frederick Gregory, a veteran of World War II, encouraged us to read avidly, think critically, and attend to issues in the larger world. I like to think that their generation, most no longer living, would have found ideas that resonated with their experience in this account of education provided in rural, small town, and urban America in the early twentieth century. One family member of that generation remains, my Aunt Eileen Trott, herself a teacher for several decades; I hope she will enjoy this book.

The immediate family members who live with a book project, however, deserve particular thanks. They are the people who inspire, distract, and in many ways sustain me. I started collecting materials for this project in the 1980s when my children were relatively young and our family lived in century-old home in upstate New York, surrounded by the challenging farms and returning woodlands that nurtured the vision of nature study educators Anna Botsford Comstock and Liberty Hyde Bailey. Just a few miles away lived John Gustafson, a man who had long been active in the surviving Nature-Study Society and served as its treasurer; he spent a few hours one afternoon providing fond recollection of friends who had maintained it through the last half of the twentieth century. In the semirural countryside of Cortland County, my sons, Kris and Kurt, accompanied me on short hikes through farm fields to the Tioughnihoga River, rich with aquatic and shoreline life, where their curiosity about insects and enthusiasm for capturing tadpoles reminded me of the importance of hands-on learning. Nature study leaflets remained good guides to that terrain. In the last decade, Minnesota's Arrowhead region has provided additional dimensions of natural life for our family, now happily including Courtney Mollan, also a teacher, to investigate. My husband, David, has been steadfast in taking my work and me seriously since we met as undergraduates; his encouragement, confidence, and love sustain me on a daily basis.

Historians are indebted to those who make and record the history we write about. My final gratitude is for the teachers, supervisors, and advocates, women and men, who left traces of their educational practice and insight into their purposes in public documents, professional publications, and private correspondence sufficient to make this history possible. Their accounts, in important ways, prove particularly resonant in our own day. Many contemporary educators, working formally and informally with children, are today reviving nature study techniques in their classrooms, making extensive use of local nature reserves, and teaching about the complex interdependency of the natural environment surrounding us. If the details of such educational innovation vary from a century ago, the intention of teaching children to understand and appreciate the substance and dynamics of nature, locally and around the globe, remains surprisingly similar and of continuing importance.

INTRODUCTION

THE NATURE STUDY MOVEMENT introduced science into public schools of North America. The innovative curriculum established that even very young pupils could and should learn about the natural world. The widespread acceptance of nature study, as teachers taught with natural objects from their local environments, proved to be emblematic of public attitudes toward both nature and the human condition in the anxious and yet progressive years of social and intellectual change at the turn into the twentieth century. The nature study movement was deeply rooted in the American enthusiasm for the natural sciences and was spurred by a deepening commitment to education for all children. Elements of its agenda can be traced to the antebellum decades when several prominent naturalists advocated teaching school pupils botany, zoology, and other scientific subjects, but only a few pupils had access to such studies. The explicitly titled nature study program at the end of the nineteenth century added to its subject matter an ambitious and imaginative pedagogy directly shaped by ideas about how children learn and what impact learning should have on their lives.

The nature study outlook attracted a wide range of supporters critical of contemporary methods and curricula in the public schools who wanted change to start with the elementary and grammar grades. Reforming and progressive schoolmen sought to transform, standardize, and establish strong public support for educational programs that had demonstrable results. They found ready collaborators among scientists keen to enlarge the ranks of science and disseminate scientific ideas. This book is fundamentally an institutional account of the circumstances that brought the idea of nature study into prominence, some of the key advocates who framed its fundamental principles, the complicated threads of preparation by teachers and supervisors who implemented it, and the multiple ways that the concept continued to resound long after the term had receded from school usage.[1] It is also an account of the way in which civic science presented itself in the early decades of the twentieth century.

One goal of this book is to represent the teachers themselves, joining historians of education who seek to elaborate on the often elusive working lives of teachers, many of whom taught only a few years and whose papers were seldom kept. School teachers, primarily women by the end of the nineteenth century, were key participants in nature study.[2] As an emergent professional group, they pursued the credentialing and certification of specialized training that could open opportunities and provide career direction and security. Although some were ambivalent about the standardizing forces that imposed nature study on public school curricula and continuously constrained its expression, establishing a permanent place for the subject in the public schools as far down as the elementary grades required teachers' cooperation and enthusiasm.[3] Many of them embraced the opportunity to broaden the curriculum, excite pupils' imaginations with hands-on experiences, and extend their own creative capacity for teaching.

I first became familiar with nature study education while studying natural history in nineteenth-century Worcester, Massachusetts, at the American Antiquarian Society. The Worcester Natural History Society archival materials recorded the enthusiasm of local teachers who brought their classes to its museum and the efforts of a full-time staff member who worked with them and with individual pupils after school.[4] Curious about the evident emphasis on teaching children about nature at the end of the century, I was intrigued to find that most twentieth-century histories of education dismissed the courses of nature study referenced by these teachers as an insignificant fad that had been promoted in textbooks and by a few activist school administrators.[5] Gradually I discovered miscellaneous remnants of nature study education in local, state, and college libraries that suggested the pervasiveness of this curriculum across the country and among a wide spectrum of pupils and teachers, even as my casual search revealed how little remained of the content and practice of elementary education.[6]

Educational materials and accounts by teachers made me skeptical about the hint of denigration in scholarly accounts of historians of education, the environment, and science. I was impressed by the ambitious goals expressed in textbooks and educational journals, the controversies swirling about nature study in magazines at the turn of the century, the implementation of a curriculum so situated in local sites, and its impact on the progressive agendas in public schooling. I gathered information gradually while working on other research projects and checking what were then card catalogues of public and academic libraries as I attended professional meetings and went on family trips. Later I collected evidence more systematically using travel funds awarded by the National Science Foundation and enlisting the help of numerous librarians and archivists who helped me locate rare ephemera. I also identified several important repositories with specialized collections of books and curriculum materials (see appendix A).[7]

Although I was searching for a simple and comprehensive definition of nature study, I slowly accepted the fact that its rich and varied expressions helped explain its success. It was the flexibility of nature study that made it particularly useful at the turn of the century, when progressive, vocational, and scientific educational curricula were being negotiated in North America. No single or simply defined program served to dominate or universalize nature study as educational theorists and grade-school teachers expressed their particular interpretations and worked within their own natural environments. Their differences of opinion and contentions were genuine, like those among religious denominations, yet proponents usually shared a general agreement on certain fundamentals and persisted in their efforts to disseminate these core ideas.

Nature study built on the naturalist tradition with its emphasis on material objects and visual as well as textual representation, while turning away from a focus on the organization of nature, or taxonomy, in order to understand animal and plant life in environmental context. At the core of nature study was a pragmatic insistence on using local objects for study, emphasizing the connection between those objects and human experience. Nature study was an educational movement intent on enhancing individual learning, and it aligned readily with other contemporary reform movements that saw community life in relationship to the natural environment, especially those involved in preservation and conservation. Advocates combined a sometimes nostalgic outlook on rural life with an appreciation for the cultural possibilities of urban life, well aware of the importance of technological innovation while seeking a counterweight to it.

Acknowledging its complexity was not sufficient as I sought to answer other historical questions. How widespread was the nature study movement? How did its purposes and practices differ depending on where it took root? What was its influence? Why has it been ignored or dismissed in historical literature? Because nature study emphasized the use of plants, animals, and observations of other aspects of nature readily accessible to pupils, I anticipated that curricula would differ based on geographical and ecological settings. But other significant variations emerged. For example, because of economic and racial factors, most public schools in the post–Civil War South lagged behind schools elsewhere. Only scattered instances of nature study appeared in that region. Perhaps ironically, nature study in the South was most notably found in programs for African American students subsidized by northern philanthropists; however, even here racial attitudes meant nature study had a distinctly vocational intention. Occasionally, too, nature study could be found in Southern college and university towns where public schools enjoyed better support. Nature study flourished in the Northeast, the upper Midwest, and the Far West, most often in regions where it was established in normal schools or taught through extension programs at land grant universities. Major urban areas, from New York City to Chicago to Los Angeles, offered comprehensive school programs, typically about the time that their school systems were consolidated under reforming city administrators. It seemed important to capture the range of this genuinely national movement, often recorded in scattered materials or tucked inside archived school board reports.

Nature study evoked multiple possibilities in part because it attracted such a wide range of participants. Reforming educational philosophers, several of them trained in German psychology, saw in its use of familiar materials a potential for the child-centered curriculum that took the developmental stage of the child into consideration. A generation of academic educators under ambitious presidents at new research universities like the University of Chicago and Teachers College at Columbia University sought to promote theory and methods based on an emerging child psychology. Normal school faculty, determined to relate theory and practice, integrated nature study into their model schools on campus and formulated parallel courses that taught content and method side-by-side or sequentially. Advocates in rural areas were motivated by a range of expectations, with some believing that nature study would teach children sympathy for nature and others believing that it would enhance farming skills. Leading urban reformers wanted to introduce inner-city children to living nature, most with the thought that through school gardens and nature walks these residents would learn community and moral values. All these supporters argued that knowing nature through science was increasingly important in the modern world and thus fundamental for those coming of age in the twentieth century.

Although goals were diverse and sometimes improbable, documentation of nature study practice provides evidence that its fundamental goals were advanced by teachers who concentrated on materials close at hand and taught children to observe carefully. Nonetheless, application of nature study principles left considerable room for variation: Should nature study be an opportunity for children to learn about nature in their own way and at their own pace? Should their lessons have practical goals, perhaps relating to agriculture or furthering advanced science studies? How scientific should the study be? Would using nature study in correlation with art or literature enhance or detract from its potential? Who were the appropriate educators to guide pupils' open-ended studies? Should nature study be understood and introduced as having religious or ethical meaning? These issues make nature study a useful site to investigate the ways in which nature and the natural sciences were understood in an era when the expectations framing science and educational structures were being transformed.

Teaching children about their natural world was not without precedent. Nineteenth-century enthusiasm for public lectures on science, illustrated books on birds and plants, popular field guides, personal collections, and public museums in Britain and Europe had counterparts in North America. The transatlantic transfer was spurred by immigrants like Louis Agassiz, whose students and public audiences were drawn by his principle that it was best to "study nature, not books." Teachers in private schools and academies introduced botany, chemistry, and astronomy to their pupils, and, as public education became normative, scientists were keen to make such subjects available to everyone. Important experiments in introducing science into grammar school grades in Boston and St. Louis reflected the aspirations of prominent educators in the 1870s, but their short-lived efforts also demonstrated the difficulties entailed when too much depended on individual leaders, often in the absence of material infrastructure, well-prepared teachers, or committed administrators.

An expansion of public schooling and the advocacy of progressive reformers dramatically changed the educational landscape, especially in the last decade of the century. The reforming educators promoted an ambitious "new education" intended to overthrow outmoded teaching methods, to introduce a broader curriculum, to enhance the preparation of teachers, to standardize classrooms by grades, and to make public education a local and national priority. Assisted by a rapid expansion of state normal schools and the introduction of colleges of education at research universities, influential educators moved surprisingly quickly. A specific curriculum for teaching the sciences to students from the start of their education was established under the rubric of "nature study" at Cook County Normal School, where the principal was well-respected educator Francis W. Parker. His observation that "the school is society shaping itself"[8] resonated with other educators who shared his interest in science and the new social sciences. Parker's young faculty member Wilbur Jackman provided a succinct name, the singular phrase "nature study," for his curriculum for elementary and grammar schools. Jackman then played a pivotal role in designing, publicizing, and establishing credibility for this program, especially in urban school systems.

Despite its familiar name, nature study posed very distinct challenges for teachers and administrators in cities, small towns, and rural areas who implemented it. The emphasis on using natural objects found in the pupils' own out-of-door environment required teachers to adapt their lessons based on the availability of specimens and interests of students. Jackman developed a manual and recommended teaching strategies, including school gardens, which were in the first instance intended for classrooms in Chicago. In New York, his general program was readily adapted at the new Teachers College loosely affiliated with Columbia University, in part because his basic formulation fit the ambitions of its faculty. Here, as in other urban areas where teachers and administrators had limited access to nature in the vicinity of the schools, schools collaborated with museums, botanical gardens, and park boards to provide the requisite specimens and outdoor spaces. Other urban programs, including the decidedly applied nature study effort introduced by Clark University faculty in Worcester, approached nature study from the standpoint of an industrializing society; they introduced projects that could ameliorate problems of pollution and disease.

In country schools, teachers encouraged pupils to take a fresh look at the abundant nature around them. Cornell University's College of Agriculture, with rapidly expanding farm extension programs, provided leadership for rural nature study that also addressed the problem of rural out-migration. Under the leadership of Liberty Hyde Bailey and Anna Botsford Comstock, Cornell became widely known for its emphasis on building observational skills and an appreciation of wildlife as well as domestic plants and animals. This well-funded program published leaflets, taught institute and summer school courses, and produced textbooks that were widely distributed throughout the country and even abroad. Nature study educators were very serious about science and accuracy even as they insisted that pupils should be taught using materials attractive to them and at a level readily understood.

Journal articles, teachers' manuals, and textbooks proliferated as nature was introduced into classrooms. The authors were variously committed to child-centered pedagogy, to imported German theories of apperception and correlation, and to inculcation of moral and spiritual values. Chicago remained a center that provided practical advice about nature study content for teachers through its College of Education, while the faculty at Illinois State Normal concentrated on adapting the theories of Johann Herbart for the American setting. President G. Stanley Hall at Clark University encouraged faculty member Clifton Hodge to use educational psychology to test the outcomes of nature study activities. The result was a rich mix of educational theory and practice.

Although strong advocacy and resources were important in promoting the nature study movement, teachers were essential for introducing its activities into their classrooms. New teachers, most of them women, learned nature study fundamentals through normal school and institute courses, but established teachers needed to do some remediation either through self-study with books and leaflets or by participating in special summer or weekend classes. Nature study typically required more effort to implement than many other school subjects as teachers acquired specimens, planned excursions, or coordinated school gardens. Even those eager to take up nature study often requested help implementing it. As a result, larger school systems hired nature study supervisors to provide advice and material resources, opening a niche of opportunities for women trained in science and interested in administration and thus another extension of what Margaret Rossiter identified as "women's work in science."[9] Women's advancement did not go unnoticed, however, and some social scientists questioned whether nature study was becoming too "sentimental" and feminized.

In an effort to both stem criticism and provide a forum to discuss the issues facing those who taught nature study, whether in colleges, normal schools, or public classrooms, Maurice A. Bigelow inaugurated *Nature-Study Review* in 1905. With the Nature-Study Society, formed just two years later, the *Review* invited philosophical debate about the goals and methods of nature study even as it provided dedicated space for teachers to describe successful practices and projects. Its pages were full of definitions and the boundary work of establishing both connections to and distinctions from agricultural education, social and sexual hygiene, applied science, literature, geography, and other subjects. By this time, too, nature study had attracted international attention, moving particularly fast into the schools of Canada, Australia, and New Zealand.

Curricular innovations often last only a generation, and nature study proved no exception. Criticism came from teachers already overburdened with other responsibilities, social scientists worried that the field was not sufficiently systematic, academics fearful of the feminization of education, and science educators seeking to enhance their status and that of their work. Tension also existed because nature study seemed a contrast to the growing masculinist and positivist insistence on disciplinary science training.[10] An assertive group of women advocates, whose independence was gradually eroded, nonetheless boldly integrated literature into their teaching about nature and worked well alongside a public interested in applications of nature study education to conservation and preservation efforts, collaborative gardens, and other public programs.

For many teachers, the excitement of exploring new pedagogy and their students' responses encouraged them to incorporate the ambitious curriculum even though it required extra time and creative attention. A few exceptional teachers found that their new skills opened further opportunities for them as nature study textbook writers or administrators. The lives of these innovative teachers and administrators also make clear the nature of their classroom and school activities.[11] Particularly notable is that many collegiate-level instructors and systemwide supervisors were women, in parallel with others in the early twentieth century whose expertise in the natural sciences allowed them to pursue such unanticipated career paths as home economics and museum work. But the innovations aroused some criticism.

A generation of Ph.D. educators in the 1920s began to argue that the term "elementary science" would be most appropriate for a curriculum that covered explicitly scientific topics. They wanted standardized materials and sought to use social science tools to evaluate educational outcomes. During and after World War I, nature study as a distinctive curriculum was gradually displaced in public schools. By that time, however, the term nature study and its hands-on approach had gained a new foothold in extracurricular venues. Teachers trained in nature study were among the leaders who introduced their well-honed approaches to nature observation into scouting, summer camping and park activities, 4-H, and other out-of-school programs for young people. Some of them joined the ranger systems in national, state, and even local parks where they provided lectures, conducted tours, and initiated small museums on site. Colleges shifted their programs to offer degrees appropriate to these emerging careers even as school curriculum moved away from the vocabulary and methods of nature study.

Nature study, like all educational movements, was possible because it resonated with other social, intellectual, and cultural currents in late-nineteenth-century North America. Its broad public appeal echoed the turn-of-the-century enthusiasm for nature evident in amateur clubs in botany and ornithology, in the myriad of fiction and nonfictional nature books, and in conservation and preservation projects concentrated in the West but evident in regional parks and new nature preserves near major cities and across North America. Nature study offered one effective way to combat what many late-nineteenth-century educational reformers viewed as the dry and narrow recitation style characteristic of grammar school teaching. It flourished because prominent philosophers and psychologists found in this curriculum a way to test and implement educational theories they had encountered while studying in Europe. Nature study education, though provided to children, reflected and directed concerns in the larger community.

Nature study illustrates the complex relationship between local control and national movements within American education. National organizations and professional societies, as well as federal and state agencies, advocated for particular elements in the new curriculum, but it was local and regional activists who worked with community leaders and parents to mobilize resources for classrooms. Revisionist historians earlier argued that education in this period was didactic and formulated by those with social and political power, whereas more recent scholars identify the multiple ways that parents and teachers resisted imposed programs, modified them to suit local needs, and encouraged alternatives.[12] Both accounts have concentrated largely on the structures governing education and issues of consolidation, standardization, and requirements.[13] Curriculum, the actual content of schoolwork, opens a useful window onto the negotiations among administrators, teachers, parents, political leaders, community activists, and educational theorists. The practices surrounding nature study in schools across the United States reveal how gender, class, and ethnicity were inevitably woven into them, even as the curricula were caught in crosscurrents of educational theorizing and contestation about content. What is striking is the persistence of the nature study curriculum through these turbulent decades, and its tactile materiality made it distinctive from the rote learning that had preceded it. To this day, the legacy of nature study remains in the special corners, tables, and window ledges of elementary school classrooms devoted to natural objects — birds' nests, rocks, salamanders, shells, leaf collages, and more — brought in by children for show and tell.

This account of nature study reveals that the history of education, down to the primary level, is intimately intertwined with the history of science. Education in modern democratic societies is fundamental, and so public schools become an important reflection of cultural priorities. The introduction of nature study as a required school subject reflected the growing authority of the natural science community and a recognition that knowing more about nature in an analytical way would be fundamentally important for future generations. The coincidence of nature study with such familiar markers as the establishment of research universities, specialized societies, and philanthropies also dedicated to advancing knowledge reflects the intensifying prominence of science. Moreover, as nature study became an integral part of school curricula, it intersected with neighboring cultural institutions, particularly municipal, state, and national parks, natural history museums, botanical gardens, and zoos.[14] All these institutions at the turn of the century promoted the idea that knowing nature directly provided individual and community benefits. As conservation and preservation became common public concerns, advocates sought to use the public schools to advance their causes and equip future active citizens with knowledge to advance their agendas. In this context, nature study played a critical role in shaping certain ways of knowing about the natural world and indeed integrating specific knowledge into the lives of pupils in early-twentieth-century America.

CHAPTER ONE

Educating with Nature's Own Book

*When the forms of animals are as familiar to children as their A, B, C, and the
intelligent study of Natural History from the objects themselves, and not from
textbooks alone, is introduced into all our schools, we shall have popular names
for things that can now only be approached with a certain professional stateliness
on account of their technical nomenclature.*

Louis Agassiz[1]

LOUIS AGASSIZ, the prominent Harvard naturalist, addressed his remarks
to receptive, educated readers of the *Atlantic Monthly* who undoubtedly
agreed that children should be as acquainted with natural history as with
their alphabet.[2] The popular professor not only drew crowds in the hundreds to his lectures in Boston and on tours as far west as Chicago but also
taught prospective teachers in nearby normal schools how to use natural
objects to stimulate the curiosity of their pupils. Throughout the nineteenth
century most children lived in proximity to living plants and animals. Those
with education were often introduced to the natural sciences through story-
books, family activities, museums, or public lectures. An occasional enterprising teacher would develop a program for teaching the natural sciences,
particularly if she had come from an academy for young women that had a
program in science.[3] In the period after the Civil War, as colleges developed
elective courses and technical schools intensified their science curriculum,
discussion also turned toward providing science preparation during earlier
years of schooling. Despite a few systemwide efforts, however, few schools
and school systems had consistent or persistent commitment to teach the

natural and physical sciences in the public school classrooms. Informal projects for young people in and beyond the schools, including Agassiz Clubs, increased through the 1870s and 1880s, as did amateur interest in local ornithology and other natural history subjects.

From the early days of the republic, interest in the natural sciences was evident and children as well as educated adults pursued them. Polite audiences in Philadelphia and Boston provided a market for large, beautifully illustrated books like those by John James Audubon on birds, while local and regional societies collected cabinets filled with insects, shells, and other natural specimens.[4] Itinerant lecturers and local amateurs attracted audiences to lyceum presentations on mineralogy, chemistry, and natural history in small towns as well as in major cities. Their informal activity inspired some well-trained teachers to bring the natural history subjects into their classrooms, most often those in better-equipped private academies and schools, but also some public ones.[5] Not all interest was academic. Agassiz and many of his contemporaries held a natural theology, which stressed the wisdom and power of God in creation, an outlook that gave additional meaning to their investigations.[6] Transcendentalists and their contemporaries linked American identity to nature in the antebellum period, suggesting the country was "nature's nation" and that it had produced a culture of "nature addicts."[7]

Agassiz was only the most visible among a cadre of educational enthusiasts whose efforts created rich and complex, if unsustained opportunities for teachers and their pupils to learn about nature in the last half of the nineteenth century.[8] Well known on the public lecture circuit, Agassiz was an educator at heart and volunteered to talk at spring and fall teachers institutes throughout Massachusetts, and he worked with teachers at Framingham Normal School in the 1850s (fig. 1). Beyond such encouragement, innovative teachers relied primarily on their own interests, local materials, and in-class techniques to acquaint their pupils with natural science subjects. Gradually, public and private school administrators expanded the breadth of their curriculum, starting with geography and incorporating history and the sciences. Advanced pupils could nurture personal expertise in such studies through popular books and juvenile periodicals with nature content.[9] How often teachers used additional materials is difficult to estimate; such efforts were limited, since relatively few teachers had much advanced education—or discretionary funds for equipment and books. Many elementary teachers simply went from being pupils to teaching, but a privileged few attended an array of pedagogical seminaries, academies, teachers' institutes, high schools, normal schools, preparatory departments,

FIGURE 1. Louis Agassiz, shown here examining a sea urchin, was a dynamic instructor who brought natural specimens to class and was eager to encourage teachers to bring nature into their classrooms. Radcliffe College Archives, Schlesinger Library, Radcliffe Institute, Harvard University.

and colleges. In the latter decades of the century, more widespread introduction of public and private teacher training schools, special courses to assist practicing teachers, and curricular programs developed at the local and state level that emphasized methods for teaching dramatically changed this situation.[10] In the process, classrooms became more standardized, teachers were under greater scrutiny, and the distinction between those who taught in schools and those who administered them grew wider.[11]

CHILDREN AND THE AMERICAN NATURALIST TRADITION

Ongoing waves of immigrants to North America were highly aware of the climate, geography, and natural history that characterized the "new world" they inhabited and to some extent incorporated the environment into their identity.[12] Seventeenth-century explorers had stimulated a fascination for the natural landscape in Europe and eighteenth-century colonists could make a name for themselves by "discovering" new specimens.[13] Culture and commoditization both derived from describing nature's artifacts — minerals, rocks, plants, insects, shells, and even larger animals — and by the nineteenth century public audiences supported ornamental gardens, collections of curiosities, and popular lectures. Books and natural objects in museums remain as evidence of this fascination with the natural world in what some identified as an empirically minded culture.[14]

Advocates for teaching geography, botany, zoology and other natural sciences appeared in the early days of the republic. The eminent Philadelphia physician Benjamin Rush, following the example of dissenting schools in England, encouraged teaching science as an alternative to the classics in private academies.[15] A generation later the radical utopianist William Maclure argued that science should be at the core of school curriculum. He had visited the renowned German educator Heinrich Pestalozzi and saw firsthand the engagement of children as they collected specimens on field trips and used objects for drawing and model building.[16] Maclure imported Joseph Nef and Madame Fretegot, teachers experienced in Pestalozzian methods, to teach in his experimental community of New Harmony, Indiana, in the 1830s. He intended the natural history collections and emphasis on science as a model for other schools, but his innovations do not seem to have been much noted or emulated beyond his short-lived utopian community.

More persistent advocacy came from naturalists whose work in taxonomy was rapidly improving America's reputation abroad. Their justifications for incorporating natural history into even elementary education included patriotism, moral uplift, religion, and practical applications. They appealed to friends interested in educational reform and published their ideas in educational and literary journals.[17] The entomologist Augustus A. Gould, for example, argued in 1835 that the study of his subject by youth could improve virtue and morality, enhance appreciation for natural life, and provide for mental exercise and physical health.[18] Insects, he declared, provided interesting and varied species, readily accessible. In contrast, he gently derided his conchology colleagues who collected only the "houses" or shells of species rather than the processes of development of the soft creatures inside and those who collected bird skins, a practice that could foster a disregard for life.

What is notable about Gould's argument is that he obviously intended students to do fieldwork. Gould's dictum "I would have children study no books but nature's own book" echoed a tenet of his contemporary Ralph Waldo Emerson. This outlook would resonate later in elementary nature study programs.[19] Implementing that ideal proved difficult, but particularly enterprising teachers did take their pupils into the outdoors, and others brought natural specimens into the schoolroom for study. Popular textbook writers like Almira Hart Phelps encouraged outdoor study for girls as well as boys, starting when they were young, with the goal of establishing a lifelong pursuit. When field trips were not possible, she instructed readers to take a flower in hand in order to follow her botanical lesson.[20] Her well-illustrated and widely published textbooks, including the frequently reprinted *Familiar Lectures on Botany*, were addressed to parents on the presumption that such studies were undertaken by families.[21]

As Margaret Welsh has pointed out, readily accessible publications on zoology in the middle of the nineteenth century typically belonged to a distinctive "life history" genre, with virtual biographies that went well beyond taxonomy and classification to document the activities, communication, habitat, life stages, family connections, and geographical distribution of particular animals.[22] By using terms like biography and portrait in animal descriptions, authors could offer moral analogies between readers' experience and those of animals. When etched and hand-colored illustrations gave way to lithographed illustrations that were cheaper and readily available, more readers were able to see firsthand the specimens previously described in texts. By the 1850s, an audience had developed for specialized, if avocational, handbooks for studying birds, flowers, insects, amphibians, and other natural subjects.[23]

At midcentury, children's interest in nature books and the cultural significance attached to knowledge of nature led a few educators to raise the issue about whether such studies belonged in the early years of public schooling and, if so, what should be taught and how? A. Constantine Barry, the Superintendent of Public Instruction for Wisconsin, wrote to ask the advice of Spencer F. Baird, assistant secretary in charge of the museum of the Smithsonian Institution, on how to teach natural history to school pupils; he apparently did not receive a response.[24] In Waltham, Massachusetts, the school committee instructed staff to lecture on the natural sciences and use local examples to demonstrate the interrelatedness of vegetation, topographical features, and the soil in the region. However, the teachers, who knew little about such matters, were left to their own devices when it came to implementing these general instructions.[25] Evidence of administrators' interest did not insure the teaching of nature in even advanced school systems.

Until very recently, historians have focused more on schools and exceptional educators in urban areas than on the one- and two-room district schools prevalent until well into the twentieth century.[26] Reconstructing the actual instruction in rural or urban settings—methods, content, and pedagogy in general—remains difficult.[27] Evidence of intent is easier to find, such as the advertisements and catalogues of proprietary schools like that of Emma Willard in Troy, New York, which provide detailed information on a proposed curriculum in botany and geology.[28] The reminiscences of students like Thomas Cushing provide some insight into classroom organization. Cushing remembered his teacher Moses Mandell as "an extraordinary man and very skillful teacher," one of those who "make their own preparation and their own methods, which will always work successfully in their hands, even if not strictly and theoretically the best."[29] Mary Peabody (later Mann) described her teaching techniques in print, acknowledging the role her own personal fascination with natural history had in building her pupils' interests in botany and conchology. An innovative educator like her sister Elizabeth Peabody, Mary described taking her pupils on field trips and on visits to the small but systematic collections of the Boston Society of Natural History, then housed in a lawyer's office downtown.[30] Such rare but specific accounts of nineteenth-century teachers in private schools taking class excursions suggest that families with resources valued having their children familiar with birds and minerals, as well as with music and art, as supplements to the standard basic subjects.[31]

Although teachers were expected to take initiative in bringing resources to their classrooms, they were not necessarily encouraged to be innovative. In one of several handbooks written for teachers in the 1830s, the prolific textbook author Jacob Abbott, a former teacher, cautioned that pupils' enthusiasm for new techniques might reflect the novelty, rather than the results, of a genuinely effective teaching method. He also warned teachers that they were responsible to their constituents (presumably the local school board and parents) and offered a cautionary example:

> A young lady, I will imagine, wishes to introduce the study of Botany into her school. The parents of the committee object; they say that they wish the children to confine their attention exclusively to the elementary branches of education. . . . "We want them to read well, to write well, and to calculate well, and not to waste their time in studying about pistils and stamens, and nonsense."[32]

Abbott's choice of botany was probably not arbitrary, for indeed some school sponsors were uncertain about "useless" scientific subjects like botany. It is revealing that his advice books devoted more pages to providing methods to manage unruly pupils than to encouraging teachers to broaden their curriculum.[33]

Despite Abbott's warning and certainly reflecting a widespread interest in natural sciences, popular children's books and magazine articles on scientific subjects proliferated, many of them intent on demonstrating simple techniques for identifying and classifying objects. This juvenile scientific literature adapted language and complex ideas from science and advocated industry, self-reliance, and hard work as a requirement for successful scientific study.[34] As Patricia Pond has pointed out, the first scientific books for children appeared "at the beginning of an era in which biology and geology were increasingly being used to discover new facts about the universe and its origins."[35] A few were directed at very young children, such as *Lessons for Children from Four to Five Years Old* by English author Anna Letitia Barbauld; this and several of her other books were modified slightly and republished in Philadelphia for decades.[36] Texts for children used straightforward descriptions, questions and answers, or conversational techniques as they linked specific and sometimes technical ideas about nature to the local environment of their readers.[37]

Starting children young was the lesson of the story of *Rollo's Museum*, one in an innovative and didactic series of fourteen fictional books by Jacob Abbott. In this particular story, when Rollo learns that his eyes have been strained from too much reading—that problem itself a commentary about classroom education—he goes outside and begins to learn nature lore from the family's hired man, Jonas. A plainspoken but observant farmhand, Jonas encourages his curious pupil to gather objects of interest—rocks from the brook, a hornet's nest, raspberry seeds, and leaves representing all the trees in the neighborhood—and he converts a large wooden furniture box into a cabinet for curiosities. Their interest sparked by Rollo's efforts, several young friends and relatives join the enterprise. Cousin Lucy, choosing what was portrayed as a particularly female-appropriate science, botany, presses a "Collection of Common Flowers" for the children's museum. Jonas provides a collection of wood samples from his shop, and Rollo pursues the mystery of a chrysalis that he has initially identified as a "hemp seed." The Rollo stories were a reminder that family life often initiated contact with tame and untamed nature as parents and children maintained gardens for profit and pleasure, attended public lectures and museums, and made pets of traditional breeds of cats and dogs, as well as partially domesticated species of birds, snakes, and rabbits.[38]

Study of life cycles and the dynamics of groups were part of the moral lessons taught indirectly and sometimes explicitly by natural history authors. Abbott's widely read stories instructed youthful readers not only about nature but also about human nature. When strong personality differences threaten the little band in *Rollo's Museum*, the children decide to eliminate competition by forming a society and delegating responsibility. Rollo had presumed a proprietary role as initiator, but his sister reminds her petulant sibling, "when you invite us all to come and form a society, you give up your claim to it [the museum collection] and it comes to belong to the society."[39] The author made it clear that Rollo's behavior did not fit the egalitarian notions implicit in the study of natural history (fig. 2), and when Rollo tried to reassert his authority as the founder of the museum, the other children simply went outdoors to play. Rollo found it "very dull amusement to work there alone." The situation resolved in a democratic fashion. Rollo's museum project pleased his parents and other adults not only because participants found it entertaining and instructive but also because it provided a collaborative model for middle-class readers. Similarly popular were the Peter Parley books of Samuel G. Goodrich, who published compendia with text and illustrations borrowed from naturalists in England as well as the United States under various titles, including *The Naturalist's Library* and *Peter Parley's Tales of Animals*.[40] Such authors advocated natural history studies for children as a means to gain knowledge and also to build social skills.

Books specifically on natural history could be for self-instruction, for parental and teacher guidance, for leisure activities of adults, or for serious self-study.[41] Most were written by naturalists or by practicing educators. Harvard's Asa Gray's *Botany for Young People and Common Schools: How Plants Grow* went through numerous editions.[42] Worthington Hooker's *Child's Book of Nature* made it clear that the intended audience included "the mother and the teacher" who wanted to go beyond school curriculum.[43] These texts were intended to prepare children for later advanced study in botany and zoology by teaching them nomenclature and definitions, although the authors typically included anecdotes to maintain the interest of younger pupils. A significant number of women, like Phelps, wrote books and thus extended the broad social acceptability of such studies for women.[44] Women's botanical interests also led to highly detailed natural history in their needlework, fabrics, and interior design in the mid-Victorian period. Nature study would repeat and extend elements of these efforts to interest children in the natural sciences.

"NO. IT IS NOT YOUR CABINET," SAID HENRY.—Page 107.

FIGURE 2. The popular series of Rollo books by Jacob Abbot, particularly this one, *Rollo's Museum*, introduced children to natural history collecting and to the lesson of cooperative learning as well. Author's copy.

Mary Peabody's students had been privileged to visit the museum of the Boston Society of Natural History through their teacher's social contacts, but after mid-century natural history collections were available to the public more generally, through access to previously private museums, natural history collections of state academies of science, and collections established by colleges as part of their scientific apparatus that built on specimens acquired by students and teachers year-by-year.[45] These private and public institutions made natural history part of public life and extended their goals to provide formal access to children through their teachers and parents.

INDIVIDUAL INITIATIVES
AND ENTERPRISING PROGRAMS

Advocates of teaching natural history in the schools relied on individual initiatives and special projects. In the 1870s and 1880s, programs in Boston and St. Louis introduced the natural sciences, with the former largely dependent on private and individual initiatives and the latter on an effort to impose a curriculum systemwide. The results were limited, but they gained considerable public attention.

In Massachusetts, the French-influenced normal schools inaugurated by Horace Mann in the 1840s were the mechanism to prepare graduates for teaching, and they quickly surpassed women's colleges in the number of graduates who went into education.[46] Because Louis Agassiz recognized that teachers were the critical link between specialized knowledge and children during a time when many pupils were surpassing the educational level of their parents, he lectured occasionally at West Newton (later Framingham) normal school near Boston. His famous dictum "Study nature, not books" paraphrased Ralph Waldo Emerson, but his own example was equally influential, whether supervising advanced students at the Harvard Museum of Comparative Zoology or teaching the younger pupils in his wife Elizabeth Agassiz's school for girls in Cambridge.[47]

One of his students at West Newton, Lucretia Crocker, went on to a distinguished career advocating science in the schools.[48] Agassiz's influence would ripple into the 1880s in other innovative teachers' training schools, such as Bridgewater in Massachusetts, where Arthur G. Boyden, who had attended classes with Agassiz, introduced natural history and physical science into a solidly packed curriculum. Elsewhere, innovators like James Johonnot and his son Marian, both at the state Normal School in Warrensburg, Missouri, shaped their curriculum around the natural sciences.[49] Enthusiastic graduates of these normal schools might incorporate such topics into their own classes, and they often pursued advanced skills in natural science through voluntary participation in summer institutes and at seaside laboratories.[50]

In the early 1870s, Louis Agassiz established the Anderson School of Natural History on Penikese Island in the Long Island Sound, thus creating yet another innovative program for ambitious teachers—those who had financial resources to spend several weeks. Often identified as a forerunner of the present-day Marine Biological Laboratory at Woods Hole, the original school was fundamentally dedicated to education.[51] Agassiz's stated goal was to equip the fifty attendees, who ranged from elementary teachers to aspiring academic faculty, to "introduce natural history into our schools."[52] For three years experienced and prospective teachers, men and women studying side-by-side, took out boats and waded along the shore to capture specimens for careful on-site and laboratory analysis. The graduates remained committed to Agassiz's view that science was based on keen observation.[53] Participant David Starr Jordan's summer class notebook recorded Agassiz's outlook, and he advocated its extension into all levels of education: "There is no part of the country where in the summer you cannot get a sufficient supply of the best specimens. Teach your children to bring them in for themselves. Take your text from the brooks, not from the booksellers. It is better to have a few forms well known than to teach a little about many hundred species."[54]

Thus the ebullient Louis Agassiz established a model for teaching natural sciences that embedded his own famous technique of teaching from objects. The summer school created an extended field experience where living sea forms and the environment were simultaneously in view. In addition, he introduced microscopic and comparative techniques of study even as he acquainted students with current scientific thinking. Jordan later would recall this as the "school of all schools in America which has had the greatest influence on American scientific education."[55] Hyperbole aside, Penikese drew an impressive and influential group of men and women who went on to careers in science education at every level, and the seaside program did inspire related efforts across the country. Agassiz's vision was gender specific as he integrated two groups that he thought would profit from their interaction. The oceanside summer school was designed to "afford the scientific men an opportunity to carry on original researches in natural history" and at the same time provide prospective teachers (mostly women) the "facilities to improve and prepare to teach well."[56]

The Penikese Island program did not outlive Agassiz but coincided with similar laboratory developments in Europe and stimulated others in the United States.[57] One was established in Illinois in 1875 (and taught again in 1877 and 1878) by Stephen A. Forbes, who oversaw the state Natural History Museum and was then teaching at the Illinois State Normal School. He used fresh supplies of marine organisms acquired from the eastern seaboard and through exchange with colleagues in Naples, Italy, along with freshwater specimens from Midwest lakes and streams for the study of comparative anatomy and histology.[58] A former Agassiz student, Alpheus S. Packard, promoted a summer program under auspices of the Peabody Academy of Science in Salem, Massachusetts. Just a few years later, in 1880, the curator of the museum of the Boston Society of Natural History, Alpheus Hyatt, organized a summer school near his summer home in Annisquam, Massachusetts, on the shore of the Gloucester Peninsula. Another zoologist noted with evident approval, "There seems to be considerable interest in summer schools throughout the country."[59] Hyatt turned much of the teaching over to his assistant, who apparently grew impatient with the "raw recruits" among the participants, and this school, too, was short-lived.[60] Although the initiative for most of the seaside programs came from scientists, Homer B. Sprague, principal of Girls' High School in Boston, established a summer school for city teachers on Martha's Vineyard.[61] All these projects were intended to supplement the training received in colleges and normal schools by providing advanced students, as well as practicing educators, with opportunities to pursue outdoor studies of nature.

Hyatt, as curator for the Boston Society of Natural History, was deeply interested in public education in the city. He arranged the exhibits in its new museum building, which had been completed in the early 1860s, with an eye to instruction, and he created a Teachers' School of Science that offered evening and weekend classes to public school teachers.[62] Like Agassiz, he believed that teachers and scholars "should work as companions learning from each other's observations."[63] He also oversaw the production of thirteen "Guides for Science Teaching" between 1876 and 1896.[64] Elizabeth Agassiz prepared *A First Lesson in Natural History* as number four in the series (fig. 3), written in a familiar style as though to two young women companions, perhaps loosely representing her own stepdaughters, about the relatively complex seashore life in eastern Massachusetts (fig. 4). Her engaging story of hydroids is fact filled but rich in metaphor:

> In many of the Hydroid communities, the work is curiously divided between different individuals. Some are the sportsmen and feeders of the community. It is their business to catch the prey, and they are furnished with the lasso cells. . . . Next, there are the swimmers; for this community is not attached, but floats freely in the water; their office is to move the whole establishment; and one may see such a Hydroid community moving along like one individual, though all the motion is performed by these swimming members alone.[65]

Elizabeth Agassiz, an avid naturalist in her own right, included the natural sciences in the curriculum of her school for girls and at Radcliffe College when it opened as a coordinate college to Harvard.[66]

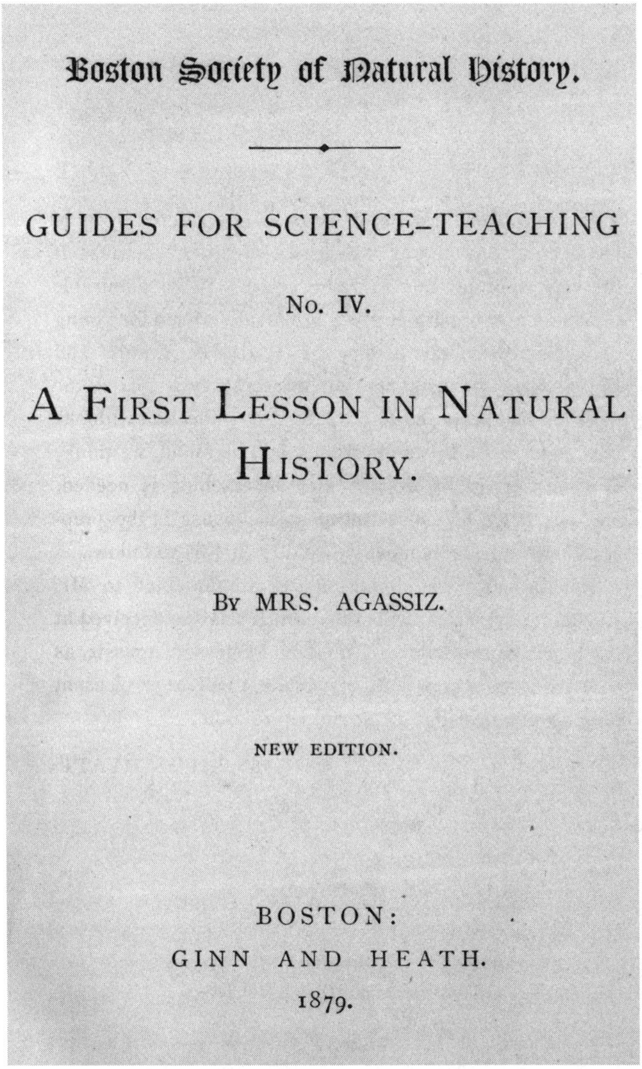

Boston Society of Natural History.

GUIDES FOR SCIENCE–TEACHING

No. IV.

A FIRST LESSON IN NATURAL HISTORY.

By MRS. AGASSIZ.

NEW EDITION.

BOSTON:
GINN AND HEATH.
1879.

FIGURE 3. Elizabeth Cary Agassiz's guide for science teaching, *A First Lesson in Natural History*, written in a familiar style and produced under the auspices of the Boston Society of Natural History, could easily be read by a parent or teacher to younger pupils or used for classroom work by those more advanced. Author's copy.

FIGURE 4. Elizabeth Cary Agassiz for a time conducted school in her own home, eager to help young women like her stepdaughters, Pauline and Ida, pictured here with her; her initiatives helped to found Radcliffe College. Radcliffe College Archives, Schlesinger Library, Radcliffe Institute, Harvard University.

Actively involved in improving science education, among other philanthropic projects, was the Women's Educational Association (WEA). It founded a famous summer seaside program at Woods Hole after Agassiz's death, but the oceanographic center quickly transformed into a research facility.[67] The new director at Woods Hole, Charles Whitman, a faculty member at the University of Chicago, lacked the charisma and enthusiasm for general education that had been evident at Penikese and Annisquam. Under his leadership, the facility was largely dominated by ambitious biologists whose goals emphasized research and community among an inner group of academic scientists.[68] As a result, the cadre of largely women school teachers was virtually excluded from the main research program at the summer school run by Whitman. Faculty who taught at women's colleges managed to maintain a significant presence there, most notably Cornelia Clapp of Mount Holyoke, who voluntarily ran the library.[69]

In Boston the WEA had more successfully encouraged the educational work of Lucretia Crocker and Alpheus Hyatt as they coordinated school programs with the museum of the Boston Society of Natural History. Crocker, a graduate of the State Normal School of West Newton in 1850, was among the first women who moved into school administration. She taught briefly at Antioch College, but then returned to New England, where she worked for a time with children with physical disabilities.[70] Her acquaintance with Louis Agassiz and her position as one of the first women members of the Boston School Committee enabled her to influence the curriculum of the Boston public schools.[71] As the new museum opened in Back Bay, she worked with Hyatt to develop his Teachers' School of Science that in the 1870s provided Saturday and evening classes for teachers. Her position, Superintendent of Science Education, would be renamed "Nature Study Supervisor" in the 1890s. By 1893, the museum's classes were structured into a formal course that provided credentials to regular participants.[72]

Having taught school, Crocker understood that equipment, textbooks, and adequate preparation were essential for effective teaching. In collaboration with Ellen Swallow Richards, who was teaching applied chemistry at Massachusetts Institute of Technology, she drew up and tested a course on mineralogy for teachers. The museum's *Guides for Science Teaching* were small and inexpensive publications that gave teachers and visitors to the museum access to current scientific ideas of people like Hyatt, a neo-Darwinian, and some of his Boston-area colleagues. Their success led the publisher, D. C. Heath, to publish Nathaniel Southgate Shaler's *First Book of Geology* and Cornelia Clapp's *Observations on Common Minerals*.[73] Other publications like the "Nature Reader" series soon followed, including Julia McNair Wright's *Seaside and Wayside*, a series meant to guide outdoor excursions for children.[74]

These initiatives were important, but a Boston commentator pointed out two decades later that, notwithstanding the "painstaking and faithful labor" of Miss Crocker, the normal school, and the teachers' school of science at the museum, elementary science was in fact a "comparative failure" in the Boston school system.[75] As the School Committee's annual report for 1891 lamented, only a few teachers were doing satisfactory work with natural sciences, and "the majority were making no attempts whatsoever."[76] Teaching teachers and implementing systemwide programs would take more than voluntarism and private patronage. The Academy of Sciences working with Philadelphia's Committee on Instruction and Lectures and local philanthropists offered a "Normal School of Science" for local teachers, but it had no comprehensive or long-term impact aside from demonstrating the interest of educators and the aspirations of local scientists.[77]

About the same time, a very ambitious and centralized initiative came from the superintendent of education in the St. Louis public school system, William T. Harris. Harris had built a nationally recognized educational system with kindergartens and graded classrooms, innovations that eventually catapulted him to appointment as United States Commissioner of Education from 1889 to 1906. His enthusiastic and detailed report *How to Teach Natural Science in the Public Schools* in 1871 provided a comprehensive proposal for teaching about plants, animals, and physical phenomena from elementary through secondary schools.[78] His method involved a three-year cycle of material so that children reviewed scientific topics even as they added new materials to their knowledge base. Harris, like many scientific colleagues in the post–Civil War period, emphasized elementary science preparation and relied on recitation techniques to reinforce established scientific principles.[79] Perhaps as a result, Harris would later respond with skepticism to the goals, techniques, and content of nature study.[80] The St. Louis program under Harris, like others introduced by the vision of a strong leader, proved ephemeral, but it marked a growing interest in the possibilities of teaching young children about the natural world.

Most advocates who sought to introduce natural science into public schools believed that a limiting factor was teacher preparation, which remained uneven across the country. Many of the normal schools offered courses on science in the 1870s and 1880s, but they offered little advice about how to integrate such topics into school classrooms.[81] It is important to remember that although stable and acclaimed normal schools continued to grow, they still taught a relatively small percentage of rural school teachers. As late as the 1920s as many as 400,000 teachers in classrooms had not completed a normal school course of two years.[82] Preparation for teaching and access to the leading edge of pedagogical thinking about both methods and practice thus varied considerably. For those able to get to the more innovative normal schools, however, the intellectual and career prospects were exciting.

EUROPEAN PEDAGOGY AND TEACHING ABOUT NATURE

Even as the goal of educating all future citizens provided a motivation for American educators, ideas about what constituted an appropriate education were debated. Americans looked to Europe for principles and models, and reformers like Horace Mann and Henry Barnard wrote for education journals and popular magazines about German pedagogy and philosophy.[83] Thus, well before the Civil War, German theories and practice were known, and among the first practice to find general acceptance was the movement for preschool education that emphasized the intersection of play and learning, often involving the out-of-doors.[84] The kindergarten movement, which identified the importance of outdoor learning in its name, was particularly popular in cities, where aspiring classes were enthusiastic about giving children a head start on schooling, and philanthropists underwrote opportunities to provide poor children with skills that could make them productive citizens.[85]

Through the enthusiasm of William Maclure and Horace Mann the ideas of the Swiss educational reformer Heinrich Pestalozzi had been introduced, and his ideas were being taught in most normal school curricula by the 1870s. It was specifically Pestalozzi's challenge to rote learning that attracted reform-minded educators, and the core principles were widely available, as suggested in an Iowa course of study in 1878:

Begin with the senses, and never tell a child what he can be led to discover readily for himself.

Reduce every subject to its elements, one difficulty at a time is enough for a child.

Proceed step by step, be thorough, the measure of information is not what a teacher can give but what the pupil can receive.

Let every lesson have a point, either immediate or remote.

Proceed from the known to the unknown, from the particular to the general, from the concrete to the abstract, from the simple to the difficult. . . .

Fasten every principle by frequent repetition.[86]

Historians rightly point out that there was considerable distance between the actual experiences and practices of teachers and the rarified atmosphere where pedagogical controversy raged, jargon developed, and the child development theories of Pestalozzi and his German protégé Johann Friedrich Herbert were debated.[87] Nonetheless, among administrators and reform-minded teachers, methodological and psychological innovation was clearly under discussion at educational meetings and referenced in textbooks for prospective teachers. The establishment of German-language schools, philosophy textbooks, and reprinting of German pedagogical books extended this influence. The publisher Ernest Steiger developed the largest German publishing house in the United States in New York City, and translated versions of German philosophy and pedagogy were on his lists from the 1880s onward.[88] Kindergartens and elementary classes were particularly encouraged to adopt the child-centered education of Pestalozzi and another disciple, Friedrich Froebel, and to take children into the world beyond the classroom:

> Lead your child out into Nature; teach him on the hilltop and in the valleys. There he will listen, and the sense of freedom will give him more strength to overcome difficulties. But in these hours of freedom let him be taught by nature rather than by you. Should a bird sing or an insect hum on a leaf, at once stop your talking: bird and insect are teaching him; you may be silent.[89]

As these ideas circulated in England and the United States, they inspired one teacher to write a small text introducing children to shells in the Pestalozzi manner.[90]

The widely publicized ideas of Charles Darwin and Ernst Haeckel soon stimulated new texts and discussion among educational theorists and practitioners, and Americans were attuned to contemporary pedagogical theory abroad. Friedrich Junge of Kiel was a key figure, and his 1885 volume *Natural History Instruction in the Elementary School: The Village Pond as Community Life* showed the influence of the zoologist Karl Möbius.[91] Their efforts were particularly helpful to teachers who were instructed by the Prussian leadership to teach using local objects and include "the pulsing life of nature" in their work. Educational reforms in 1872 had placed new emphasis on history and the sciences in European teacher-training institutions, emphasizing chemical laboratories and physics instruments, but how much this was translated into the schools is unclear.[92] By the turn of the century, there was sufficient interest for Otto Schmeil to establish a journal, *Nature and the School,* in 1902 and to write a number of textbooks for teachers and pupils.[93]

Before and immediately after the Civil War, quite a number of Americans went abroad, often to Germany or Switzerland, to attain graduate degrees and to absorb something of the culture. The postgraduates were impressed by the communities in which they lived, as well as the universities at which they studied, noting the strong local identity of neighbors who beautified towns, created walking paths through adjoining reforestation projects, and taught their children local natural history.[94] In particular, the ideal of *Heimat*, which linked local people to their natural environment, was extended to children through teachers and ministers as well as parents.[95] Thus a civic sensibility seemed to some observers to be established by linking the study of nature to educational practices.

Many young men (and a few women) going to Europe attended lectures in educational psychology and philosophy and even investigated German schools directly.[96] Several, like the Civil War veteran Colonel Francis Parker, toured central European countries intent on learning enough to change American educational systems on their return. Parker stayed more than two years in the 1870s, visiting schools and attending university lectures before returning to Massachusetts to teach. By the 1880s, such study became more highly focused. G. Stanley Hall studied psychology with Wilhelm Wundt and returned to join the faculty at Johns Hopkins University, recently established in 1876 to emphasize research and offer graduate degrees. Later, as president of Clark University, Hall directed his attention to the issue of child development, becoming best known for conceptualizing the period of adolescence.[97] Another group interested in German education, many of its members from the Midwest, received Ph.D. degrees from the University at Jena and elsewhere in philosophy, psychology, and other aspects of education. On their return, they quickly disseminated their formulation of German educational ideas through conferences, publications, and lecture tours, as well as through focused study groups like the Pedagogical Club.[98] In subsequent chapters, we will see how these university-based men who were interested in child psychology and educational philosophy promoted and influenced nature study. Their education in Germany would be reflected in an empirical approach, often using quantification as well as emphasizing psychology. The extent to which German educational practices on the secondary level, which emphasized experimental biology (as well as hygiene and physiology) and discouraged loosely constructed natural history experiences, had influence in the United States deserves further attention.[99] At the elementary level, the clearest evidence for German impact was a short-lived but intense enthusiasm for Johann Friedrich Herbart's concepts of apperception (encouraging children to learn new things in relationship to things experienced earlier) and correlation (teaching two subjects not intrinsically connected, like mathematics and art, in intersecting ways).

Historians of American education have traced some ways in which European educational ideas and practices were implemented. Much attention has been paid to Oswego, New York, where in the 1860s the superintendent of schools, Edward Sheldon, introduced objects—blocks, pictures, and samples of minerals and seeds—borrowed from Pestalozzian models and taught classes to show elementary teachers how to devise their own projects. Soon the New York State Normal School at Oswego became known for its innovative use of the "object method."[100] Sheldon found natural history specimens ideally suited as familiar objects to teach concepts of color and shape, self-expression through painting and writing, and generalizations about form and categories of experience.[101] Sheldon's object lessons became part of the vocabulary of reforming administrators as they shifted away from reproducing knowledge in young minds through recitation toward engaging those minds on topics appropriate to their developmental level, relating them to other sensory experiences. Later critics, however, would find that the method, like any method of education, could easily become mechanical and rely on standardized and inert objects. Oswego Normal School also pioneered in having its students actually practice teaching under supervision, a technique that quickly proved useful in teacher training.

Multiple and intersecting theories of education were part of the curriculum in many normal schools and in new graduate programs in education in universities. Emphasizing theory and philosophy in the curriculum allowed normal school faculty to validate the intellectual significance of their work. Reliance on theory also formed part of the professionalization of education promoted at regional and national professional meetings. Nature study advocates drew on strands of these theories, not always with complete consistency but with a conviction that the theories would, in fact, fit well with teaching the natural sciences. Continental ideas more attentive to the learning styles and capacities of young children slowly displaced Scottish commonsense philosophy that assumed the study of mathematics and science would improve habits of mind. Although the ideas of Pestalozzi and Froebel differed considerably in detail, both stressed children's stages of development, their readiness for instruction, and methods for presenting materials that would relate well to their interests. Moreover, both philosophers reinforced the concepts of learning with and from nature. In the 1890s, educational psychology and the writings of Herbart would become particularly influential among a group of Illinois educators who were also interested in nature study. It was an exciting and challenging period for educators who took their inspiration not only from Europe but also from the widely evident enthusiasm for the natural sciences in the United States.

AGASSIZ CLUBS

In 1875 Harlan H. Ballard, Berkshire Hills teacher and head of the Lenox Academy, initiated an ambitious and widespread program to promote natural history. Ballard, inspired by reading Jacob Abbot's book *Rollo's Museum* as a child and by accounts of Swiss children going on excursions in fields and mountains, assembled a group of his own students who liked to "collect, study, and preserve natural objects" into an organization named for the well-known naturalist Louis Agassiz.[102] Five years later, he issued a general invitation through the children's magazine *St. Nicholas* for other groups to create local chapters. This relatively new magazine for boys and girls established by Scribner and Company was flourishing under the editorship of the well-known author Mary Mapes Dodge, with a subscription list of about seventy thousand in the 1880s.[103] Dodge promoted the contemporary enthusiasm for nature through her formula for publishing intimate, accurate essays about it, often touched by a poetic sensibility. Her authors were free to use nature to present moralistic, fantastic, romantic or even mechanistic accounts of animals.[104] Once *St. Nicholas* provided space for a regular column describing group activities, the Agassiz Association attracted over six hundred chapters that reported seven thousand individual members.[105] The clubs might involve extended families and neighbors, or clusters of five or six, to sometimes twenty to thirty young men (and some women) who banded together for a few years to present talks to one another and go on collecting excursions, sometimes for several days.[106]

The Agassiz Association provided a symbol—a badge modeled after the Swiss cross, in honor of Agassiz—and a point of affiliation within and beyond the local community. Ballard produced the first of four comprehensive *Handbooks* in 1882 to guide local activities. He maintained a casual organizational structure at the center, recommended reference materials, sketched elementary lessons on how to collect and preserve natural specimens, and identified expert consultants from a voluntary pool of naturalists. In 1884, the Association convened what sponsors hoped would be the first of ongoing annual conventions of Agassiz Association members in Philadelphia; they published what proved to be a single issue of *The Naturalists' Journal*.[107] In 1887 Ballard created *The Swiss Cross: A Monthly Magazine of the Agassiz*

Association, another short-lived independent publication in which the essays emphasized religious themes describing the work of a creator in nature and drawing related moral lessons about the right way to live.[108]

In fact, however, the clubs themselves were quite autonomous under the Agassiz Association umbrella organization, with each taking the name of another well-known naturalist such as Asa Gray or Isaac Lea or a place name such as Acadia. A small group of boys in St. Paul, Minnesota, for example, met together weekly for several years simply using the title Agassiz Club.[109] Such amateur groups might specialize in a particular subject such as fossils, birds, or mollusks that were particularly common in their area; or they might build expertise making collections of pressed flowers, arranging a cabinet for collections, or skinning birds. Massachusetts, which had a strong tradition of amateur natural history and numerous clubs, held regional assemblies in 1888, 1889, and 1890.[110] In other parts of the country, clubs might occasionally join together for regular meetings to present natural history specimens and discuss topics of common interest. Ballard encouraged teachers, community leaders, and parents to help form chapters—only four members were needed—and he directed his materials toward young people of all ages.[111]

The emphasis was on going out of doors to observe nature.[112] Agassiz's birthday on May 28 became a designated date for daylong excursions, but clubs were encouraged to spend one or two hours together, perhaps once every week or two. Ballard thought of the activity as avocational and extracurricular, noting that the club proved useful for Lenox Academy teachers who "have not been able to find a place for Natural Science in the ordinary school curriculum." They banded together once a fortnight to study assigned topics. A roster of the Agassiz Association chapters indicates that the groups might be led by women as well as men. Clubs were established in virtually every North American state and province, with others listed from as far away as Ireland, Chile, Japan, and Persia.[113] This loose confederation had grown significantly in the 1880s, and the association incorporated in 1892. When the correspondence seemed to overwhelm the *St. Nicholas* editor, various efforts were made to produce another periodical for the Association, none of which survived more than a few years.[114] The longest lived was Edward F. Bigelow's *The Observer,* published from 1890 to 1897 in Portland, Connecticut, where he also worked valiantly to keep the Association active.[115]

Even during its most visible period in the 1880s and 1890s, the Agassiz Association's efforts to provide guidance and create a stable organization by publishing textbooks and offering correspondence courses (a botany course cost $2.00, for example) were not effective. As Bigelow took over from Ballard, he emphasized natural theology alongside the systematic studies for which the clubs were intended, and as a result the program seemed quite distinct from that carried out by schools attentive to the new pedagogy based on object teaching and child-centered psychology.[116] The Agassiz Association persisted but without evident momentum after the turn of the century. Although Bigelow claimed in 1907 that it had over 10,000 members, an effort to establish a permanent facility near Greenwich, Connecticut, was unsuccessful.[117]

Other books and magazines intended for young people similarly reflected a "back to nature" enthusiasm at the end of the century. *Youth's Companion*, a best-selling children's magazine that reached its circulation maximum of half a million subscriptions in the 1890s, regularly featured both stories and nonfictional essays about nature.[118] One of the oldest children's magazines, founded in 1827, it reflected changing public enthusiasms and created a department for nature and science that appeared weekly from 1895 to about 1916.[119] This attention to children's access was congruent with a much larger movement in nature activity marked by a concern for wilderness and enabled by intensive railroad networks, resort development, and recreational hunting.[120] The spectrum of enthusiasts ranged from those sufficiently serious to build a personal collection and organize a local natural history society, to those who found some enjoyment in casual observation and an occasional "find" of a new species or expanded location for a known one, to those who consumed nature through illustrated or didactic books about plants and animals or created a market for paintings of the natural world.[121] Varied expressions reflected an enthusiasm for the sublime of majestic western landscapes as well as more systematic concerns of botanical and ornithological societies that began to displace earlier broad-gauged natural history societies. Books, guides, and periodicals both came out of the movement and spread its influence. Ornithologist Elliot Coues's *Key to North American Birds*, published in 1872, was among the first modern field guides, complete with woodcuts, intended for outdoor identification by local amateurs. Over the following decades, a steadily increasing set of active observers joined the American Ornithological Union, while others interested in bird protection joined the popular American Audubon Society.[122]

Not everyone went into nature, and those who wanted to appreciate nature in their armchairs found much to interest them by authors who detailed the places they knew best. Clarence King and John Muir published in literary magazines and books about California and the West, while the "Sage of Slabs," John Burroughs, wrote about wildlife and outdoor life accessible in the Catskills and Adirondacks.[123] When the teacher Mary Burt decided to use Burroughs's *Papacton* for her sixth-grade class, she found that his literary account with its vivid descriptions held the attention of students much better than school textbooks. She subsequently edited two series of readers based on Burroughs's books for school classrooms. Sales indicated both Burroughs's accessibility and student interest in nature topics.[124]

FERTILE GROUND FOR NATURE STUDY

Although popular culture outlooks and educational theories held that children would learn by observing and reflecting on their natural world, scientists continued to encourage more systematic education that was specifically scientific. A scientific curriculum had already found a fundamental place in colleges and universities at the undergraduate as well as graduate level by the late 1880s. Attention was turning toward precollegiate education, primarily at the high school level but with some consideration about earlier years as well. National organizations like the American Association for the Advancement of Science and its nominal competitor, the Society of American Naturalists, tried to mobilize their membership to advocate for more science in the schools, forming committees and listening to proposals from their peers about how best to pursue that goal.[125] Sometimes the presentations were highly specific, as when William North Rice, a geologist teaching at Wesleyan University in Connecticut, called for a science curriculum based on the object method and published a volume with his arguments.[126] In other cases, committees on education simply reported with evident frustration on the limited nature of what was being studied in the schools but with no recommendations for what should be taught and how. Educators and scientists also promoted a more regular inclusion of natural sciences in educational journals.[127]

It was the successful combination of broad cultural enthusiasm, changing educational theory, and a well-articulated program that catalyzed the surprisingly rapid introduction of nature study into the public schools in the 1890s. The phrase *nature study*, perhaps because it seemed to designate a quite specific curriculum, caught on quickly, building on the advocacy of professors and educational administrators, who had often quite distinctive expectations of what such study would provide to children and the society they would enter. There was clearly an educational path prepared, because, as one superintendent of instruction put it, "we had some teachers by this time who would gladly learn and gladly teach of Nature."[128]

The introduction of "Nature-Study" as a new curriculum proved both a challenge to and an opportunity for individual teachers. It required more initiative and extra effort, but nature study also allowed them to be innovative, particularly attracting those who found perfunctory lessons on entomology and other specialized subjects ineffective.[129] Teachers and administrators liked to recall the limits of rote learning from their own experience: "We were taught botany by using Gray's 'How Plants Grow' as a reading book, without a word of explanation, with absolute silence on the 'teacher's' part, and incomprehension on ours."[130] Nature study offered an attractive alternative with its emphasis on direct contact with nature through projects and outdoor work rather than text memorization.

The following chapters will demonstrate how nature study provided a fresh rationale that distinguished it from many of the efforts before 1890. The possibilities were opened up by a rapidly changing educational philosophy and pedagogy as well as a substantial reorganization of educational systems. The new curriculum challenged teachers and administrators to formulate creative local opportunities for students to study nature and provided techniques and methods that would persist even after the next wave of educational reform pushed aside the term nature study.

Nature study's rapid adoption drew on the enthusiasm of progressives who believed public schools could be the "lever for social improvement" in a new era of civic consciousness.[131] Ellwood P. Cubberley, the highly respected and historically conscious professor of education at Stanford University in the early twentieth century, liked to point out that the period of "discussion and criticism" in the schools at the end of the nineteenth century led to core requirements, new subjects, and attention not only to formal studies within the schools but also to the importance of such studies to preparation for life.[132] When a nature study curriculum rapidly emerged in the schools at the end of the nineteenth century, it combined concerns and capacities that had long existed among naturalists and educators. Changing circumstances also meant that several things were different. As Liberty Hyde Bailey of Cornell pointed out, the theme of evolution had overturned attitudes toward nature: "The living creation is not exclusively man-centered; it is biocentric."[133] Women who had an avocational interest or specialized training in the natural sciences were now positioned to be leaders in the new educational movement.[134] Across North America school enrollments were exploding, bringing challenges of sheer numbers alongside issues of diversity and urbanization. Innovative curricula in the natural sciences could fit into this environment.

Insofar as nature study programs were also perceived to coincide with the goals of various reformers—including the natural scientists who wanted education based on scientific principles, psychologists who were interested in individual differences, educational theorists who argued that environmental influences were critical in shaping children, and political and business interests who wanted productive citizens — the curriculum was a vital, lively one. Educators and the public agreed that exposure to nature was an important part of learning, and there were many ideas about precisely what kind of experience that should be. One program, however, stood out from the rest.

CHAPTER TWO

Devising a Curriculum for Nature Study

Whoever plans wisely and applies in a practical manner a thorough-going course of natural science work for the lower grades of the public school will make a contribution to educational progress in this country. . . . Thousands of teachers have stood both ready and eager to do this thing if only they knew positively what to do; but in the absence of such knowledge they have hesitated to take the risk of failure and so have gone on the old track, or else have taken the risk, and failed for lack of knowledge and skill.

Stephen Forbes to Wilbur Jackman[1]

WILBUR JACKMAN AND STEPHEN FORBES shared a common enthusiasm for integrating science into the public schools. Jackman, an instructor at Cook County Normal School in Chicago, had already devised a course that would help elementary teachers to introduce the natural sciences into their classrooms, as Forbes knew when he wrote the encouraging letter above. Forbes had a degree from the State Normal School in Normal, Illinois, and taught school before he was appointed head of what eventually became the Illinois Natural History Survey and professor at the University of Illinois. The two men took different career routes but were united in their determination to encourage active nature study for even elementary education.[2] Forbes drew on personal experience in noting, "I have long believed that the greatest obstacle to a general introduction and maintenance of this [scientific] work in the schools is the lack of definite courses and outlines of work and of published methods which have been tested by experience and found successful."[3]

Forbes and Jackman recognized that such a significant addition to a curriculum required leadership and close collaboration with the teachers who would implement it. Forbes had helped coordinate a School and College Association of Natural History in Illinois in the 1870s that brought together school teachers and collegiate faculty; he also organized summer schools for teachers in 1875 and 1878.[4] The endorsement of a scientific leader like Forbes helped make Jackman's initiative credible, complementing the imprimatur of educational leaders and theorists like Francis Parker, who linked the curriculum to their particular philosophical approaches. The science-trained advocates came together to promote a nature study curriculum intended for pupils from the earliest primary grades, for whom the natural world was important even if they did not pursue more advanced studies.

By the late nineteenth century, elementary schooling had captured the attention of psychologists and school administrators. G. Stanley Hall argued that, in terms of the range and pace of knowledge acquisition, "at six years child has learned already far more than a student learns in his entire university course."[5] Hall, a self-conscious leader in the emerging social sciences, brought his research expertise in psychology and child development to bear directly on practice in pedagogy. The new educational psychology and philosophy of education found a sometimes uneasy home in new research universities like Clark University and the University of Chicago. These institutions achieved status surprisingly quickly, in part because they were innovative and they produced doctoral graduates who, in turn, became leading academic faculty members across the country.[6] Formulating and articulating ideas about nature study became the prerogative, although not exclusively so, of faculty at the University of Chicago, Cornell University, Clark University, and Teachers College at Columbia University. Under often young and certainly ambitious leadership at these institutions, professors pursued research projects even as they taught prospective and experienced teachers, often in what were known as experimental, laboratory, or model schools on campuses. The correlation of research and practice, however, was not always straightforward.

From the outset, faculty involved with educational research made it their primary goal to prepare graduates destined to become instructors at the rapidly expanding networks of normal schools throughout the nation. These university programs carefully distinguished their work from that of normal schools that prepared elementary and grammar school teachers. As Frank McMurry later recalled of his short stint at the University of Illinois in Urbana, before he left for Teachers College in New York City,

"I was advised by good authority not to let it be known that my chief interest was in the primary school, for fear I might lose my position. It was then beneath the dignity of any university to identify itself with the training for the instruction of young children."[7] Nonetheless, social scientists, especially psychologists, sought to conduct research about children and pedagogy that fit university ambitions. Their colleagues in science at research universities were typically interested in promoting more science study in the public schools, so integrating nature study into programs at model, experimental, or practice schools on campus was supported beyond the educational departments.[8]

Educational reformers relished the intellectual excitement of creating new curricula and identifying efficient systems for standardizing and coordinating an array of public school initiatives. These included kindergartens, manual and vocational training programs, evening schools, college preparatory courses, and more. The reformers were also pushed by changing demographics.[9] In large cities especially, school boards and their administrators were aware of new immigrants from southern and eastern Europe who differed culturally from earlier peoples from Britain and western Europe, and many of whom had limited education. Pockets of urban poverty created by unemployed or poorly paid workers, exacerbated in some places by the in-migration of rural American residents, underscored class differences and heightened the conviction that the schools needed to play a fundamental role in creating a common civic consciousness.[10]

City schools confronted overcrowding, underfunding for books and equipment, and often astonishingly high ratios of pupils to teachers.[11] In extreme instances, as in Brooklyn at the turn of the century, classes averaged between fifty and a hundred pupils, yet a high percentage of school-age children were not even in the system.[12] Moreover, between 1850 and 1900, roughly half the population was under the age of nineteen, creating concern about the future of political decision making in a democratic society.[13] Pressures on the system, with over six hundred thousand pupils and over five hundred public schools, were enormous. Some schools reported that nearly half their pupils were unable to speak English. Frequent student absence was such a serious problem that there was a special school for truants, and many classes were housed in outdated or converted facilities. In progressive thinking, it was nonetheless such schools that needed to play a critical role in improving the social and economic situation, and the curriculum needed to provide more than the standard 3 Rs: reading, writing, and arithmetic.

Critical to bringing about systemwide changes were teachers. Most of those in the classrooms across the nation in the 1880s had limited educations themselves, perhaps a short course or three-month program offered at a normal school, and they simply replicated the instruction they had experienced. Even those with a one- or two-year normal school education had been taught little about effective teaching methods. Not surprisingly, as interest in public school education grew, state legislatures across the country were persuaded to build large, architecturally impressive normal schools to address such problems. The results were dramatic. In just one generation, teacher training through normal schools was transformed as three-month programs expanded to one-, two-, and even four-year courses. Practicing teachers were encouraged or even mandated to attend weekend and summer institutes to keep up with new developments. Normal schools introduced methods courses, often linked to specific topical courses, thus putting theory and practice side by side for the prospective teachers. Summer schools emphasized content and instructional guidance so that attendees could be "considered as [academic] students as well as prospective teachers."[14]

The introduction of nature study as a new and quite specific subject area was implemented across the country within a decade. Although a few critics greeted nature study as another "fad," reforming schoolmen enlisted a number of rationales for it, not always consistent with one another, as they sought to find what David Tyack called "the one best system."[15] Nature study, although presented as a new formulation, seemed quite familiar to many educational leaders who shared with the majority of native-born Americans in the nineteenth century a farm or small-town background; not surprisingly, they responded to the potential "lure of the country in urban education."[16] Justifications for nature study in rural locations echoed an agrarian nostalgia while simultaneously presenting a forward-looking determination to make agriculture more systematic and attractive. The educational reformer Frederick C. Howe wrote simply: "Human life seems to require a ground wire to the sod, a connection with Mother Earth to maintain its virility."[17] Praise of outdoor life was sometimes expressed as a negative commentary on the chaos and dysfunctionality of urban life, but advocates also asserted that nature study had redeeming aspects that could overcome the negatives of city life. In a talk before the National Education Association (NEA), David Starr Jordan, president of Stanford University, stressed nature's capacity to teach morality, especially to otherwise estranged urbanites.[18]

More optimistic urban leaders turned the argument around by noting that progressive cities offered, in fact, significant opportunities for nature study. Open air spaces, as well as public institutions like natural history museums, arboretums, botanical gardens, and zoological parks, offered material resources as well as knowledgeable experts who maintained them. William T. Harris of St. Louis, serving as the second United States Commissioner of Education, was fond of pointing out that cities had been the birthplace for major discoveries in science, art, literature, and history. Harris also thought that cities could operate as classrooms to dissolve "clanship" (his term for ethnic identity) and teach children interpersonal cooperation through subjects of mutual interest like natural history.[19] Others linked nature study to the contemporary importance of learning science and industrial arts, arguing that nature study could incorporate agriculture, vocational education, and personal hygiene. Themes of natural theology, stated casually, were common in nature study textbooks which referenced an undefined God and his creation, a formulation that most Christian denominations could readily accept.

FROM MASSACHUSETTS TO CHICAGO

The long tradition of education in the greater Boston area and throughout Massachusetts influenced the introduction and development of nature study there as a specific curriculum. The normal schools established by Horace Mann, the kindergarten movement under Dr. Maria Zakrzewska, the summer school of Louis Agassiz, the part-time courses in continuing education at Harvard under Nathaniel Southgate Shaler, and the Teachers School of Science under Lucretia Crocker and Alpheus Hyatt—all encouraged the study of nature by children. In the 1880s, Arthur Gardner Boyden, head of Bridgewater State Normal School, introduced summer lectures at Plymouth on natural history. He had added advanced courses in the sciences to prepare high school teachers, building laboratories for the chemical, physical, and natural sciences.[20] A group of teachers asked that he develop a course on elementary science that they could use in their schools.[21] In the spring and summer of 1890, he traveled in Europe and was much taken with the German idea of *Naturkunde* as a way of engaging children in scientific studies.[22] He used the common expression "nature studies" for his summer institutes and, following Agassiz, advocated studying natural objects and processes out-of-doors. He also felt that in "this age of science" teachers and their students needed to have systematic instruction in basic scientific subjccts.[23] Teachers in training, like Sarah Louise Arnold, pursued the kinds of teaching he suggested. She moved to Minnesota in the mid-1890s, where she introduced nature study into Minneapolis schools before returning to Boston as head of nature study there; she eventually became the dean of students when the progressive Simmons College opened on the Fenway.[24] Boyden also influenced a number of educational leaders who later taught at Illinois State Normal School, which, as discussed in chapter 3, became one of the most prominent normal schools in the Midwest and which produced a number of nature study leaders.

About the same time in Chicago in the 1880s a talented and dedicated combination of educators, social reformers, and academic progressives provided the rationale and institutional base for the somewhat inchoate but growing advocacy to teach more about nature, climate, and geography in the public schools. The leader of the group was a former teacher and administrator who had established his reputation in Boston's North End and then in nearby Quincy. Colonel Francis W. Parker (his title derived from Civil War service) was a charismatic educator whose emphasis on enabling teachers to be innovative and pupils to be active in the classroom was publicized through Amos M. Kellogg's widely distributed teachers' journals. By the 1880s, the "Quincy method" was a highly visible experiment and a major topic among educational reformers.[25] Parker was tapped to head the struggling Cook County Normal School, where he turned its practice school for Chicago area teachers into a demonstration program nationally known for its teacher training philosophy and practice methods.[26]

Parker was teaching in a country school when he read Edwin Sheldon's *Lessons on Objects* (1863). That book inspired him to use the bequest from an aunt to travel in Europe from about 1872 to 1875, especially in Germany and Switzerland, and to observe innovative educational practices. Parker returned to the United States convinced that active learning with objects held the key to younger children's social and intellectual growth.[27] Parker later recalled that natural sciences were not very evident in educational programs of the schools he visited because traditional administrators resisted the rational spirit of such study.[28] In fact, he wrote, "Nature study, the study of elementary science in elementary schools, was kept out of the German schools for many years, for fear that the study of nature would lead children to search out the truth themselves. . . . in other words, that the rational or reasoning spirit would enter children's minds through the study of nature; and with the French Revolution behind them, the German leaders saw nothing but danger lying in the path of nature study." Parker, however, believed that studies of the natural world in which children lived should be central in elementary education.

He envisioned the school as a museum, laboratory, and workshop that provided tools for innovative teachers and their pupils. One commentator reported, "compasses, clinometers, levels, magnets, hammers, trowels, acids, spades, knives, boxes, bottles, jars, railroad utensils and company utensils will be quite as necessary to the up-to-date student in Col. Parker's school as . . . the slates, sponges, pencils, and chalks in the day of the Hoosier schoolmaster." This reporter went on to note that the assignments were not without controversy: "The normal school management is so full of out of door spirit that a student here is practically compelled to be a naturalist of the Robinson Crusoe type if he would pass an examination to the satisfaction of Colonel Parker and his staff."[29] A significant part of Parker's success was, as an admirer put it, his "faith in his teachers; and this belief inspired them to accomplishment. It created in them a power which had not previously existed."[30] His positive assumptions about their skill and motivation appealed to teachers who were too often reminded of their inadequacies, both by advertisers who sold them journal subscriptions and textbooks and by administrators who kept expanding their responsibilities.[31]

Parker identified elementary science as the place to begin integrating schoolwork in language, mathematics, and art. In 1883 he hired Henry Straight, a naturalist who had brought living objects into the well-respected curriculum at the renowned Oswego Normal School, insisting that students must understand each object's intimate relationship to the environment near campus.[32] Here, too, there was a Massachusetts connection because Straight had attended Agassiz's innovative summer school at Penikese Island along with young scientists like David Starr Jordan and outstanding educators like Lucretia Crocker. Straight brought Agassiz's technique to upstate New York, where he took his prospective teachers to the shores of Lake Ontario and into local woods, fields, and even swamps to observe plants and animals. His goal was to develop a holistic understanding that related individual natural objects to their immediate and larger surroundings.[33] After just three years in Chicago, however, Straight died at the age of forty. He had by then established a promising way of teaching with tactile objects from the natural world that eluded the growing critique of "sterile" object teaching for its reliance on artificial objects or for examining natural objects out of context.[34]

To fill the position left by Straight, Parker recruited Wilbur S. Jackman, another experienced, successful teacher committed to introducing more

FIGURE 5. Wilbur Jackman initiated nature study at Cook County Normal School and later the University of Chicago through his textbooks, articles, and professional presentations. *Nature-Study Review* 3 (March 1907): 65. Courtesy of the Wangensteen Historical Library of Biology and Medicine, University of Minnesota.

about nature and science into the schools (fig. 5). Jackman had attended normal school in Pennsylvania and then studied at Allegheny College before completing his bachelor's degree at Harvard in 1884, where he studied with Nathaniel S. Shaler and observed his popular extension courses.[35] Jackman returned to Pennsylvania, where he made a name for himself through his plan to introduce a comprehensive science course into the Pittsburgh school system. His science background and educational initiative made him seem ideal to the enterprising Parker.

At Chicago, Jackman found himself challenged to reshape this preliminary course to match the theoretical perspectives of his new colleagues.[36] Cook County Normal School, with its reputation growing under Parker, gave Jackman a platform from which to advertise his course of nature study, once he had given it shape. He also presented it widely at regional and national meetings for teachers.[37] These presentations and the teachers who studied with him helped make the specific phrase "nature study" one of common parlance and associated it with a specific outlook and approach.[38]

To establish the nature study curriculum as a central component in the practice school at Cook County Normal, Jackman wrote a series of mimeographed lessons titled "Outlines in Elementary Science" for the 1890-91 school year. The following year, he revised and combined them into a comprehensive publication, *Nature Study for the Common Schools* (fig. 6), which he pointedly dedicated to "the common school teacher."[39] The straightforward, detailed, and unpretentious clothbound volume drew on both traditional content from natural science teaching and increasingly popular theories of child development. Noting that teaching science had been gradually working downward from colleges toward the lower grades, he argued that the approaches and current efforts suitable for advanced students had proved "poorly fitted to the needs of beginners."[40] As an alternative, Jackman recommended that teachers take "the spontaneous development of the child's mind under the influence of the natural environment" as a guide to effective instruction.[41] Children were, he suggested, naturally curious, so the teacher could stimulate their interest and observation skills by posing questions for them to investigate, much as informal manuals designed for families had done earlier in the century.[42] His "rolling year" curriculum had students doing projects by season, with field trips to study river basins, insect life, flowers, and woods made during the spring and fall and indoor observation and experiments saved for the winter season in his northern

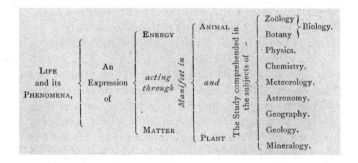

FIGURE 6. Wilbur Jackman's synoptic view of the nature study curriculum attracted the attention of scientists who were eager to have younger children introduced to the broadly defined natural sciences. "Preface," *Nature Study for the Common Schools* (New York: Henry Holt and Co., 1891), iv.

city. He applied the technique of observing nature over the course of a year, an approach familiar through the popular writings of Susan Fenimore Cooper and Henry David Thoreau, to pedagogy, as his text followed plants, animals, and climate changes over time.[43]

Liberty Hyde Bailey of Cornell suggested that Jackman's "rolling year" plan marked the introduction of "elementary science as an organic part of school work, ranking with arithmetic and grammar."[44] This early text and subsequent publications, like his detailed curricular advice in *Nature Study for the Grammar Grades*, gave high visibility to the term and the burgeoning movement.[45] Jackman's principles for this new study, presented to members of the NEA in 1891, never changed: "[A]dapt nature study to the nature of the child, focus on general characteristics and relationship between things, follow the natural cycle of the year in nature study, base nature study on observation, and use nature study as a basis for expression."[46]

Response to Jackman's book was immediate and positive. Stephen Forbes (quoted at the start of this chapter), head of the Illinois State Natural History Survey, wrote a ringing endorsement: "I have examined your book on 'Nature Study,' and have known something personally of your work, and wish to express my hearty approval of them and my very warm personal interest in your success. I trust that the opportunity will be made for you to show, as you can do, what is really practicable with scientific work in the primary grades of the public school, and what is the proper training of the normal school student as a preparation for the teaching of these subjects."[47] John M. Coulter, botanist at the University of Chicago, wrote enthusiastically for the *University Record* that "under the somewhat comprehensive title of 'nature study' there has entered the schools an element of teaching so fresh and vivifying that it has come as a revelation and will certainly work a revolution."[48] The reviewer in *Garden and Forest*, published by Charles Sargent of the Arnold Arboretum, noted Jackman's textbook as "evidence that the study of natural science is making steady progress in our public schools," but regretted the absence of systematics and theory in a book that aimed "to show how one teacher awakened a love of nature study in the pupils entrusted to his care by an intelligent use of materials taken almost at random from nature's own workshop."[49]

Jackman believed that he was guided by educational theory, namely the child-centered approach attributed to the ideas of Swiss and German educators like Johann Pestalozzi and Friedrich Froebel; he also began to incorporate ideas about correlation from Johann Friedrich Herbart.[50] Herbart, among other things, recommended having assignments that integrated various subjects; for example, discussion of natural objects could be, in the term of Herbart, deliberately connected to painting, clay modeling, and written self-expression.[51] Jackman further explored how to correlate nature study with reading, number, history, literature, and language class work. His book on number work, for example, encouraged children to count the number of times a butterfly lighted on a place of protective coloration.[52] Over the next decade, Jackman wrote steadily about nature study, pushing more and more into the educational mainstream through his publications, his participation in professional meetings, and finally his move in 1898 with Parker to the Laboratory School at the University of Chicago.[53] His textbook provided very specific lessons, along with a series of questions and references that could provide answers in geography, meteorology, geology, zoology, botany, mineralogy, and chemistry—in this he drew on a long tradition of literature and children's books that relied on questions to engage pupils in conversation. He positioned nature study as central to the curriculum, an integrative project readily connected to other subjects and related to pupils' lives. As Jackman viewed the situation, "The introduction of nature-study into the common schools proposes, as its final purpose, to establish around neglected childhood the natural environment which the laws of its growth would seem to require."[54] Parker was delighted with the work of his protégé, declaring in his enthusiasm that Jackman was "the recognized leader in the greatest educational movement of the time."[55]

A NATIONAL MOVEMENT

In December 1892, a select group of educators met to discuss how natural history ought to be incorporated into elementary grades and high schools. The goal of this select committee was to create a set of standards and topics relating to science in public education, and their results became reference points for progressive educators. The men who met at the University of Chicago comprised a science subcommittee of the well-known Committee of Ten on Secondary Schools that had been appointed by the NEA to define what many hoped would be a core curriculum, in order to reduce the "crazy quilt" of subjects and methods in high school classes across the country.[56] Like other subcommittees of the so-called Committee of Ten, its members were diverse in terms of geographical location and their professional positions—although none of the subcommittees included women.[57] All of the subcommittee members were science educators: Charles E. Bessey of the University of Nebraska; Arthur C. Boyden of Bridgewater Normal School;

S. F. Clarke of Williams College; D. H. Campbell of Stanford University; John M. Coulter then at Indiana University; S. A. Merritt, a high school principal in Helena, Montana; W. B. Powell, a superintendent of schools in Washington, DC; Charles B. Scott, a high school teacher at St. Paul High School in Minnesota who would later join the faculty at Oswego Normal School; A. J. Tuttle of the University of Virginia; and O. W. Westcott of North Division High School in Chicago.[58] Although assigned primarily to make recommendations on high school curriculum, they also had a mandate to discuss what preparatory curriculum in science ought to be proposed for grades one to eight.

Among other results, the committee endorsed nature study as a curriculum for the early grades. The conference committee report was clear that natural history (defined as botany and zoology) ought to be taught as early as possible, even in kindergarten. Its members recommended that a minimum of at least an hour each week, in two sessions, be devoted to the subject throughout the school year. Their published explanatory notes for the recommendation elaborated a familiar set of pedagogical assumptions:

> One object of such work is to train the children to get knowledge first hand. Experience shows that if these studies begin later in the course, after the habit of depending on authority—teachers and books—has been formed, the results are much less satisfactory. Experience shows also, that if from the beginning, "nature study" is closely correlated with or made the basis of language work, drawing, and other forms of expression, the best results are obtained in all.[59]

Charles Scott provided an outline of a potential course that he simply titled "Nature Study." Its botanical component started with children studying the life cycle of simple plants like beans, peas, and sunflowers. Scott recommended themes that grew in complexity, starting with life growth and development, followed by use or functions, structure, comparison, and with classification as a final step; by the seventh and eighth years, he believed students could be introduced to germination and reproduction.[60] Given the imprimatur of the "feudal barons of the pedagogical realm" who ran the NEA in the 1890s, this report was widely referenced by administrators, reprinted, and discussed at local and regional professional meetings. Its inclusion of the term "nature study" helped circulate the idea that both the subject matter and the approach reflected in Jackman's textbooks were endorsed by the most prominent educational leaders in the country.[61]

Affirmation came as well from interested advocates for progressive education, especially those who were critics of contemporary public schools. In

Roxbury, Massachusetts, the innovative headmaster Henry L. Clapp, who had helped initiate a school garden movement, expressed enthusiasm about the way that nature study illustrated very well the "spirit of the new education."[62] Even best-selling muckraker journalist Joseph M. Rice pulled briefly back from his scathing criticism of education after watching teachers at Parker's practice school at Cook County Normal. He highlighted nature study projects there as a vivid exception to the rote and stultifying classroom exercises that he exposed elsewhere.[63]

Making education work well in the rapidly expanding and socially challenging neighborhoods of the inner city was a major goal of reformers. Working with teachers in training was important, but it could bring only gradual transformation in the public schools. Seeking to make his practice school a model and to assist teachers already in the large and sometimes troubled classrooms of the Chicago public schools, Jackman initiated a Committee of Sixty for Chicago city teachers and worked to promote field work in nature study with them.[64] An elaborate organizational scheme, with subcommittees on maps, syllabi, and libraries (in conjunction with the John Crerar Library) was devised, and the group was to meet monthly to advance their projects and assist teachers. The idea in Jackman's prospectus appealed to the ecologist Stephen Forbes, who observed, "If you succeed in opening up to the schools of Chicago the practically sealed book of their own immediate environment, and to understand its relationship therein, you will have done a great work for the people and for the cause of education in America."[65] For Forbes, as for many natural scientists, a central goal was to provide a better background for students in the "graded schools tributary to the State University."[66] It is not clear just how effectively the area teachers were able to work collectively, but the teachers graduating each year under Parker and Jackman were well prepared for nature study teaching.[67]

THE CHICAGO INSTITUTE

In 1898 Parker seized an extraordinary opportunity.[68] Anita McCormick Blaine offered to establish a private model school and teacher training program based on the "new education" philosophy. Blaine, daughter of wealthy farm implement tycoon Cyrus McCormick, had enrolled in education classes with Parker to assist her in raising her young son after her husband died unexpectedly in 1892 (fig. 7). And, like many other students, she became inspired by Parker's optimism about education and his vision for teacher training.[69] She was also interested in civic affairs in the city, a supporter of

FIGURE 7. Anita McCormick Blaine, shown here with her son, Emmons Blaine, Jr., in 1898, was instrumental in reformulating the University of Chicago's educational program and took special interest in nature study. Wisconsin Historical Society, Madison; WHi-11089.

Jane Addams's Hull House, and eager to put Chicago in the vanguard of educational innovation. She sought advice from other educational leaders and requested detailed information from James Earl Russell at Teachers College in New York about the plans and expenses (salaries, budget, and building costs) required for an experimental school and teachers' training program like the one he was developing in conjunction with Columbia University.[70] Convinced she could help do something similar, Blaine made a substantial investment of her McCormick inheritance to form a new private school to train teachers and develop curriculum, the Chicago Institute.[71] She simultaneously hired Parker away from Cook County Normal School, along with his most outstanding faculty members, who became heads of pedagogic departments. Here, with a closely aligned practice school and teachers' training program, Parker planned to complement his multilayered and interconnecting principles of learning and teaching with a school to test their results. Blaine herself chaired the board of trustees, Parker became president, and Wilbur Jackman was made dean and head of the high school.[72]

While the new building for the Institute was under construction, key faculty were granted support to plan, read, and even travel to study other institutions. Jackman spent several months abroad, visiting schools in England, France, and Germany. Ira B. Meyers, curator of zoology, remained in the United States, visiting normal schools in Philadelphia, New York, and Washington, DC, in preparation for establishing a "school museum" with resources for teachers at the new Institute.[73] The Institute's well-advertised summer school in 1900 enrolled hundreds of aspiring and working teachers, who could choose to attend for one to six weeks. Although a considerable number of the nearly five hundred enrollees in the nature study course were residents of Illinois and Indiana, a significant number came from as far away as New Jersey, Georgia, Texas, and Montana. A few reported themselves as inexperienced teachers, but others had taught for ten, twenty, or more years, and a few were principals and superintendents.[74] The summer program also offered the Institute's faculty members an opportunity to test their teachers' training curriculum for the coming year. The new building for the Institute was not completed on schedule, and classes for teachers were housed in temporary quarters for the first year. Nonetheless, Parker was in his element as he led open-ended discussions about how the faculty could engage their students in ways that modeled child-centered education.

The Institute required that all teachers in training take nature study courses that could be used as a prerequisite for more advanced study of scientific subjects. The preliminary announcement of the Institute explained that in the new practice school, picking up on themes of evolution, method, and sympathy to nature,

> It will be the aim to lead the pupil to habits of thoughtful observation, to cultivate in him the desire and ability to comprehend natural phenomena and to interpret the process of evolution constantly at work in his natural and social environments, and to inspire in him a genuine love for nature. To this end the laboratories in biology, physics, and chemistry will be made as nearly perfect as possible, and the scientific method of study in both laboratory and field will be employed throughout the entire course.[75]

The architectural plans for the new Institute also included a large garden plot on one corner of the practice school grounds, presenting another educational element that became closely associated with nature study.[76]

The entire curriculum was integrated and sequential. The Department of Science's course on natural science, for example, was paired with one on the "pedagogics of Nature Study" so that subject matter was never presented in isolation from method; conversely, simply teaching theory without considering how it could be applied also was avoided. Nothing was to be taken for granted about teaching or learning, and Parker reminded his faculty at regular meetings that they needed to be continually analytical about their goals, methods, and actual results. A series of broad, open-ended questions provided the agenda for one faculty meeting: "What is your purpose in teaching? Is your purpose clear and well-defined or is it vague? Have you more than one purpose in teaching? What is teaching? Do you know when you are teaching?" With such heuristic devices, Parker was a constant prod to his instructors at the Institute and to the students aspiring to be teachers whom he addressed in its courses. At the same time, he left little to chance. He provided an extensive *Course of Study* that began with philosophy but, by its end, specified subjects to be taught in some detail.[77] Thus, although Parker was highly consultative, he was also clearly in charge.

The intensity of the Parker program left some visitors breathless. The former Supervisor of Public Instruction in Iowa wrote to Parker after spending a day in the school: "I do not see what the average teacher in our village and country schools can do with such a course of study as you put up. It is away above their reach. They cannot even see the under side of it. Still, you are working along the right line, and I hope you will keep at it if it takes the present century to work it out."[78] Parker shared the letter with his faculty and observed that its author, Henry Sabin, was a "man with a clear head." The letter, he said, reminded him to provide specific and concrete examples for teachers to follow. Equipping teachers with specimens and other objects along with lesson plans created the connection between experiencing and learning. Another observer quipped that Parker's "school is a museum, a laboratory and a workshop all in one."[79] While press coverage suggested sarcastically that not everyone was enthusiastic about "forcing pupils to take to the woods" and become naturalists "of the Robinson Crusoe type," the Institute's self-selected enrollment meant attendees typically were eager to study with well-known faculty and that nature study was particularly in demand at the turn of the century.[80] Education was both formal and informal, and the faculty even initiated some activity in the history of science. At one of the all-Institute morning assemblies led by the chemistry teacher, Charles W. Carmen, a group of pedagogical pupils reproduced classic experiments of famous scientists, including Luigi Galvani, Alessandro Volta, Hans Christian Oersted, Andre-Marie Ampere, and William Faraday, to trace the history of electricity.[81]

The Institute did not reach financial self-sufficiency or generate the level of income required to supplement Anita Blaine's promised contribution. It was undoubtedly too much to expect in one year as prospective teachers entered the Institute, a new practice school recruited pupils, a library and collection were built, and a subscription journal, *Elementary School Teacher*, was launched. Certainly it did not help that the building was not completed on time and the first year was spent in temporary rented quarters.[82] Although the Institute planned for a museum under Ira B. Meyers, the natural history collections were never fully developed because the curator spent most of his time maintaining an extensive collection of chemical and physical apparatus for use in the classrooms.

Sometime during the Institute's first year, negotiations were opened among Parker, John Dewey, and William Rainey Harper, president of the University of Chicago, about combining the Institute with Dewey's Laboratory (practice) School. In fact, Parker already knew Dewey, who had sent his own children to Parker's school and later called the colonel the "father of progressive education."[83] Dewey had joined the Department of Pedagogy at the University of Chicago in 1894 and there, influenced by his wife, Alice Chipman Dewey, developed the Laboratory School, which was closely modeled on the Parker example.[84] Dewey's program reflected his interest in combining experimental practice with educational psychology, and thus the importance of testing theories.[85] Alice Dewey became principal of the Laboratory School's elementary division, while Ella Flagg Young, later the first woman superintendent of schools in Chicago, supervised overall instruction.[86]

In 1901 the University of Chicago launched its new School of Education, which incorporated several local institutions, including the Chicago Institute founded by Blaine, the Manual Training School, and the South Side Academy. Blaine facilitated this arrangement by donating funds to house the program in new Emmons Blaine Hall, built and named in honor of her deceased husband. Parker was made the Director of the School of Education which oversaw the practice school, and Dewey was named professor in the Department of Education.[87] The merger proved awkward at best, and when Parker died the following year, the combined institutions stumbled forward without his articulate leadership and determined vision.[88] After Parker's death, Jackman tried to position himself to take his mentor's place and to continue consolidation of the two programs. Jackman had been very much in favor of the merger with the University of Chicago because it seemed to offer permanence and university status. [89] Minutes of the Faculties of the School and College of Education, however, reveal early and ongoing tensions over what often seemed to be minor problems among the several departments and divisions dealing with education. Personalities played some role, but the debates also mirrored deep philosophical and professional rifts cutting in various ways among the University's School of Education, the faculty in the teaching facilities, university administrators, and professors in the academic departments of philosophy and psychology. Differences were played out in discussions about very specific issues: basic philosophy of education in both theory and practice, matters of curriculum, methods of practical teacher training, and day-to-day management of the Laboratory School.[90]

In fall 1902, Dewey, by then named Director of the School of Education, pushed to increase the courses in theory and reduce those in nature study, geography, oral speech, and writing. Jackman retained the title of dean of the Laboratory School, managing the day-to-day details until 1904, but he was shaken by Dewey's intervention and the displacement of what Parker had seen as a fundamental part of the curriculum. Tensions between the two men were evident to other staff members. Jackman sought help from Harper, president of the University of Chicago, in administrative matters. He also opposed Alice Dewey's reappointment as head of the elementary school.[91]

By the time Dewey resigned from the University of Chicago in 1904 (some believed because of the treatment of his wife) to go to Columbia University, Jackman seemed to have lost his energy for the new initiatives. He stepped down from the deanship to become principal of what became the university's elementary school and to edit the *Elementary School Teacher*. Although he seemed to be distracted from his leadership role in the nature study movement after the turn of the century, Jackman did produce a *Yearbook* on nature study in 1903 for the NEA that outlined a continuous program from earliest grades to college.[92] Yet another independent, private entity, named the Francis W. Parker School, retained much of the character of the Cook County practice school under the administration of Flora J. Cooke, who had been on the normal school faculty with Parker since 1889. Cooke continued to promote nature study and produced a book, *Nature Myths and Stories*, which sold very well. An administrative leader in her own right, she often served on national panels with educational leaders John Dewey and Frank McMurry.[93] Anita Blaine gradually withdrew her patronage from the private school and worked more closely with Jane Addams and the public schools through the Chicago Board of Education, to which she was appointed in 1905.

The overall curricular program advocated by Parker, including nature study, persisted but without strong leadership or evident innovation in the training school, which remained loosely affiliated with the College of Education at the University of Chicago. Although Jackman's local Committee of Sixty seems not to have been active for very long, Chicago school teachers used the nature study fieldwork curriculum. Their work was encouraged by local women's clubs, some of them interested in linking nature and art, and by the philanthropic support of Norman Harris.[94]

It was the older and reenergized Chicago Academy of Sciences in Lincoln Park, rather than the Field Museum (founded in 1893 after the end of the World's Columbian Exposition), that positioned itself to work with local schools to promote nature study.[95] Its formal program was put in place for the public schools in 1903, when the Chicago Board of Education, raising its standards for teachers, required three years of normal school that included a full semester of nature study. By 1905 the United States Commissioner of Education reported that "Chicago has made nature study the basis for all the work" of summer vacation schools.[96] When W. Moses Willner left a legacy of $100,000 to the Chicago Academy of Sciences in 1907, the directors used much of the endowment income to launch school museum exhibits under Wallace Atwood, its administrative director. Its museum exhibits and programs emphasized Chicago sites and highlighted subjects appropriate for nature study classes. The requests for participation by school classes became so intense that for a time each teacher in the Chicago school system was allowed to send just one pupil "delegate" to the Academy who then returned to report to his or her classmates on the topics presented.[97] In Chicago and elsewhere, nature study relied, in part, on philanthropic and volunteer support to provide requisite training and material.

Several faculty members at the University of Chicago had a sustained interest in nature study. John Merle Coulter, who founded and edited the *Botanical Gazette*, had been drawn to the University of Chicago by Harper in 1896.[98] He helped establish the field of ecology, and he envisioned that nature study could promote similar messages about cycles of life within particular environmental niches.[99] This outlook had influenced Jackman, who pictured in some of his textbooks the encroachment of the sand dunes along Lake Michigan.[100] Coulter also taught Alice Jean Patterson, a leading nature study educator, and collaborated with her on a nature study textbook. The dynamic Patterson, who taught in elementary and secondary schools before joining the faculty at Illinois Normal in 1906, initiated a local and then a state school garden program along with her teaching and textbooks.[101] These fluid connections among university and normal school faculty and exchange of materials among those interested in nature study provided brief but important collaborations.[102]

After the loss of Parker, Dewey, and Jackman (who died in 1907), the University of Chicago's leadership in education faded as attention shifted to the Teachers College in New York City and some highly productive normal schools. Local attention to nature study, however, was rekindled when Elliott Downing joined the faculty and created a nature study series of books at the University of Chicago Press in the 1910s, although he proved a transitional figure in a shifting interest toward elementary science in subsequent decades.

CHICAGO'S LEGACY

Francis W. Parker had been an unstinting supporter of nature study since the 1880s, and his influence spread even after his death through his students and colleagues. Just a few years before he died in 1902, Parker had written,

> Nature Study to-day is pioneering work. The prevailing methods of teaching nature are the old methods dominated by the delusion of logical sequence,—of isolated fact learning. . . . The old method is founded upon a rigid faith in the

book and traditional processes; the new upon the divinity of the child and the influence of God's creation upon his growing mind. One method is fixed; the other is everlasting motion over the infinite line of unrealized possibilities. One method demands accurate imitation; the other original discovery and creation. Under one method the teacher is a pedant; under the other a student.[103]

Parker could argue many sides of pedagogy, offering romantic and theistic justifications for nature study while insisting on a rather structured curriculum with serious content courses for normal school students. Key factors in his success were personal charisma and his frequently articulated faith that dedicated and well-trained teachers would be creative and bring both moral and instructional fervor to their classrooms. The study of nature could be, he thought, at the vital center of such instruction, and he found in William Straight and then Wilbur Jackman trained scientists who caught his vision and worked it out in detail.

Illinois became a national leader in nature study in the 1890s and early 1900s. The University of Chicago was significant, but so were the distinctive and intersecting initiatives coming from the normal schools in Normal and DeKalb, both of which taught the subject in their academic year curriculum and well-attended summer institutes. Charles McMurry, brother of Frank McMurry, of Illinois Normal, for example, taught for a year at the University of Chicago in the late 1890s. Flora Cooke lectured at Illinois Normal and remained friendly with John Dewey even after he went to Columbia. University and normal school faculty worked with teachers in ways that enriched the preparation and practice of nature study and the educational philosophy that undergird it. If Parker had once thought that nature study was the "Cinderella of education," by the turn of the century it had secure footing with important public leaders.[104] The Chicago Academy of Sciences and the City Public Art Society, both privately funded, provided essential support as the program moved into the public schools.[105]

Nature study and early progressive education were mutually reinforcing. The endorsement of the nature study idea in the 1890s by the NEA's Committee of Ten was probably critical in disseminating its rationale and basic content. The rapid expansion of normal schools and the use of nature study in most of their practice schools would equip an emerging generation of teachers to include the subject in their classrooms, and they could reference the widely publicized example of programs in Chicago. Major cities, including the newly combined boroughs of New York City, adopted nature study in a comprehensive way at the turn of the century. Progressive reformers and nature study activists, national and local, framed the discussion of nature study in the 1890s. As urban systems mandated nature study, state boards of education (in some instances state supervisors) added their endorsement, but critical leadership continued to be exercised through the normal schools in an era when local autonomy and school boards remained in control of curricula. Once local political and administrative leaders took initiative, however, the introduction of nature study was quite rapid.

CHAPTER THREE

Framing Nature Study for the Cities

Classroom work on zoology and botany as well as on physiology and hygiene, geography, conservation of birds and plants was enhanced by books, charts and graphs, but the idea of learning directly from nature on a regular basis whether in the classroom or on field excursions remained an important ideal.

Ellen Eddy Shaw[1]

AS ELSEWHERE, nature study in New York City was not accomplished in a single, tidy step nor was it exclusively involved with natural objects. The botanist Ellen Eddy Shaw (quoted above) accurately observed that direct experience with nature was the motivating ideal of nature study educators. The initiation of a formal program in New York City in some ways reflected the Chicago program, and it was formulated at the same time as rural nature study was introduced in upstate New York at Cornell University. The New York City project, however, had local, long-standing elements that came together in a distinct program established by an official school board mandate. Here, and in other cities, it was positioned in a central way by an increasingly coordinated curriculum.

Historians of education have documented the substantial challenges faced by school boards in Manhattan, Brooklyn, and Queens in the latter decades of the nineteenth century. Student-teacher ratios were typically more than fifty to one, some classes were held in rented and sometimes unsanitary facilities, and a myriad of problems caused the majority of students to leave school before their twelfth birthdays.[2] Nature study became one of several programs intended to ameliorate the difficulties facing the pupils, their families, and their teachers in the public schools at the turn of the century. By the time the schools took up the official mantle of nature study, large public parks, natural history museums, botanical gardens, zoos, and other public institutions had created nature havens for New York residents that reflected both rural nostalgia and progressive urban planning. These cultural institutions also provided significant resources for nature study teachers.

NEW YORK CITY AND TEACHER EDUCATION

The pedagogical methods of object teaching developed at Oswego in upstate New York had attracted some downstate teachers and administrators as early as the 1870s. However, the supervising state Board of Education was decidedly lukewarm about such innovations; its annual reports warned city systems that new methods and content should not be substituted for basic instruction. Special subjects should only be used to "supplement common studies of school, and bring the matters of instruction home to the pupil's own experience."[3] But after James Earl Russell was brought to New York City in the 1890s by a promised affiliation between the new Teachers College and Columbia University, he made Teachers' program one of the nation's most highly visible in teacher education based on his innovative curriculum.[4]

Teachers College Record, which Russell started in 1900, documented his intention to develop a professional school to train educational leaders and administrators in the "science of education."[5] As at Chicago, university administrators were interested but uneasy about combining teacher training with liberal arts undergraduate education and a research-oriented graduate program. Yet they were drawn to the possibilities of research in educational philosophy and psychology that might influence educational practice and lead to new insights in these emerging social science disciplines.[6] The compromise at Teachers College was a dual faculty in education: an academic faculty to deal with pedagogical theory and a faculty of practical arts to teach subject matter for practitioners.[7] Within its first decade, Teachers College had established a national reputation, training future teachers in its practice school and preparing advanced students for administrative posts by teaching them educational psychology and philosophy under nationally recognized intellectuals like John Dewey and Edward Thorndike.

Sympathetic to normal schools that "have been knocked about by politicians, starved by legislators, ignored by 'scholars' and despised by 'practical educators,'" Russell positioned himself to strengthen the reputation of teacher education.[8] He envisioned that Teachers College would offer advanced education for outstanding teachers and administrators who wanted to shape education as a science, and whose graduates would, in turn, provide faculty leadership in teacher training schools around the country.[9] Russell, who had briefly taught school in upstate New York after graduating from Cornell, prided himself on encouraging innovation while "resisting pet theory and practice," and he was persuaded that nature study was promising. He supported early initiatives by Francis E. Lloyd, who published outlines of courses for classroom teachers under the title of "Aims of Nature Study" in the first issue of the *Teachers College Record*.[10] Russell subsequently hired Maurice A. Bigelow, first as an instructor and later a full member of the faculty, to develop courses in nature study and in biology.[11] Bigelow, who would become the first editor of *Nature-Study Review*, helped the college develop its summer program for teachers and advanced students.[12] By 1906 Teachers College had put in place a four-year program whose core included biological and physical nature study, taught in both theory and practice courses.[13]

IMPLEMENTING URBAN NATURE STUDY

During the fierce debates described by Diane Ravitch as "the great school wars," New York progressives took on the political machine known as Tammany Hall in the 1890s.[14] The reformers changed political leadership in the New York boroughs and pushed administrators to tackle widespread problems of overcrowding, dilapidated schools, rote learning techniques, and the entrenched and unresponsive teachers who used them. The progressive enthusiasm for efficiency and nonpartisanship had as its underside a disdain for many of the new immigrant poverty-stricken children in the system, and arguments for reform inevitably were slanted toward moral reform and uplift. Efficiency and centralization were the themes of reformers along the lower Hudson River who wrested local control of the wards in Manhattan, Brooklyn, Queens, Staten Island, and the Bronx away from a leadership deemed corrupt and inefficient. The boroughs were merged at the end of the century, and a new superintendent, William Henry Maxwell, took over the combined systems. He candidly identified a reform agenda for the consolidated system, eliminating the politically motivated teacher appointments and standardizing the certification requirements for teachers. His goal was to disestablish a system where "a young woman who is without friends or influence, no matter what her attainments may be, usually receives an appointment, if at all, only after her inferiors who have 'influence' have been provided for."[15] Maxwell embodied the contradictions of progressives, combining autocratic administration, expert elitism, and democratic idealism. He also subscribed to the now common supposition that schooling could be the antidote to social and economic problems.[16] He welcomed new ideas, such as Emily Huntington's "kitchen garden," which opened multiple collaborations between reformers and the schools in the inner-city slums.[17]

New York's borough schools developed their own distinctive nature study programs that took account of developing models in Chicago and at Cornell but required that pedagogical principles fit local interests.[18] Teachers College was particularly influential, undoubtedly because its administration could deal well with what Philip Pauly characterized as the "bureaucratic standardization, metropolitan singularity, and special-interest politics" ruling curricular design and propagation in New York City.[19] In 1899, the New York City superintendent's report indicated courses on nature study at the New York Training School for Teachers and the Normal College of New York City, the two schools that chiefly supplied teachers for Manhattan and the Bronx.[20]

In 1903, the Board of Supervisors of the New York City schools completely restructured the pubic school curriculum to provide "A Correlation of the Pupil's Course of Study with the World in Which He Lives; His Spiritual and Natural Environment."[21] They identified seven key areas of tutelage, one of them nature study, and asserted that the work in these natural history subjects should build skills and subject expertise from year to year for all students—and should coordinate specifically with such subjects as zoology, botany, and geology. A committee of principals, teachers, and district superintendents worked with the supervisors to prepare a *Course of Study in Nature Study, Elementary Science, and Geography.*[22] The goals amalgamated Parker's ideas about elementary science with Bailey's postulate that nature study was to "cultivate a sympathetic acquaintance with nature and to develop the powers of observation" with living animals and actual objects where possible. It had a seasonal arrangement following Jackman's ideas and further stressed "lessons on kindness to animals." The chair of the committee was Dr. Gustave Straubenmueller, then an assistant supervisor, who remained involved with nature and environmental studies in the New York City system for the next three decades.[23] Deeply progressive in his outlook, Straubenmueller took "an active part in developing cooperation between the schools and the museums of the city" in order to support nature study. Over the course of his career, he would be in charge of the expansion and revision of nature study and initial implementation of its extensive school garden program.[24]

Nature study came with standardized instruction for teachers but left considerable room for them to be creative in content and in linking it to other subjects. From the outset, the administrators of the newly combined boroughs of the New York Education Department emphasized nature study and geography as the subjects preliminary to science. They provided elementary teachers with outlines, references, and sometimes even material resources to assist in what was a new area of study in some schools.[25] The initiative seemed successful, as the city proudly highlighted its nature study curriculum in displays at the Louisiana Purchase Exposition in 1904 in St. Louis.[26] Teachers were encouraged to develop their own class projects, always keeping within the general outlines provided, and one, Michael Levine, reportedly structured his fifth grade class at P.S. 64 in Manhattan so that his topics showed the evolutionary development of life.

The city appointed a number of people to assist with nature study instruction. Evangeline Whitney, District Superintendent in charge of Vacation School, Playgrounds, and Recreation, born in Ohio and educated at Oberlin College, taught in Brooklyn and then became associate superintendent of schools before taking responsibility for playgrounds and recreation in 1901, a post she held until her death in 1909. Her annual reports reveal that most school buildings in the boroughs had a nature room dedicated to "growing plants, boxes for seed sowing, caged birds, aquariums, and as many specimens of field and water life as the Teachers could produce."[27] In order to make living plants available to students, she encouraged roof gardens where land was not readily available. She carried this work into the summer, when vacation schools and playgrounds kept city children from dangerous streets and offered them an alternative place for play and learning.[28] Over the next two decades, school gardening would become an extensive program in New York, coordinated by Van Evrie Kilpatrick and a cadre of enthusiastic supporters. Some projects were on school property or on abandoned lots near the schools. One of the most extensive was the DeWitt Clinton Park school garden program that, in conjunction with the Juvenile Agricultural School, provided four-by-twelve foot "farms" for neighborhood pupils (fig. 8).[29] Such programs garnered sponsorship from women's clubs and from philanthropic organizations, as well as from local communities that benefited when children were constructively occupied in projects to improve the quality of life in neighborhoods.

FIGURE 8. Built on a former garbage dump, the School Park Garden near DeWitt Clinton Park and the East River proved a highly visible and notably successful project in New York City. *Nature-Study Review* 1 (November 1905): 260. Courtesy of the Wangensteen Historical Library of Biology and Medicine, University of Minnesota.

The curriculum could be time-consuming and difficult, particularly for longtime public school teachers with established lesson plans. Fannie Parsons, assigned to coordinate the DeWitt Clinton school garden program, made the point somewhat harshly when she explained, "We have found the [established] teachers absolutely ignorant of not only the simplest facts in Nature Study but also how to present the question in any shape to the children. They seemed to be seized with terror when the question was broached."[30] Thus, providing specimens and advice to practicing teachers and introducing the subject into normal school training were both critical in the transition to teaching natural science subjects in the public schools. Even with such assistance, nature study encountered the inevitable criticism and resistance that greeted significant shifts in the public school curriculum. It was not an isolated opinion when one school principal complained that "fads" like nature study, music, and drawing distracted students from basic studies and led to failure on standard tests.[31] Despite complaints, a new standard curriculum had been mandated, and a variety of resources were marshaled to assist teachers in its implementation.

Early, informal connections between the schools and urban scientific institutions were made more systematic as nature study was implemented.

At the New York Botanical Garden in the Bronx, former normal school instructor Elizabeth Knight Britton worked through the director, her botanist husband, Nathaniel, to encourage school class tours. As teachers responded with requests for visits and materials, the Brooklyn Botanical Garden hired a nature study instructor, Ellen Eddy Shaw, to coordinate educational programs for visiting pupils. Their activities drew on the model of cooperation between schools and the relatively young American Museum of Natural History under its director Albert Bickmore, yet another student of Louis Agassiz.[32] In a program similar to the Teachers School of Science developed by Lucretia Crocker and Alpheus Hyatt in Boston, the American Museum of Natural History staff supplied specimens, offered lectures for teachers, and instituted displays in the museum specifically designed for children. It did not hurt that the American Museum and similar institutions found they could garner city and state support for educational activities and also attract private funds for such programs (fig. 9).

When state politicians withdrew financial support from the museum's

FIGURE 9. Museums in major cities across the country loaned boxes of rocks, zoological and botanical specimens, and scientific instruments to classroom teachers. The service functioned like a lending library, delivering new boxes on a weekly or monthly basis. In St. Louis, Missouri, a special educational museum was established for this purpose. *Nature-Study Review* 6 (April 1910): 103. Courtesy of the Wangensteen Historical Library of Biology and Medicine, University of Minnesota.

educational programs in 1904, nature study was sufficiently in place that the city picked up the cost of distributing museum collections sent on loan to schools, and the American Museum of Natural History reinstituted free admission for school classes.[33] First-year statistics in 1904 indicated that about 45,000 children heard lectures and spent additional time in the museum, and 235 sets of specimens were loaned and used by nearly 200,000 school children.[34] Escalating demands from teachers in the school system forced museum staff to add additional lectures to accommodate those who arrived without reservations, and to more than double the boxes of specimens circulated among the schools. The commitment seemed firmly in place, and in 1909, the librarian of the museum prepared additional collections of books to go into school libraries, some on birds and native plans and others on ethnology.[35]

Requests for assistance from nature study teachers led the American Museum to hire Agnes L. Roesler as a full-time instructor in 1906. She was assigned to work particularly with the teachers in the Normal College and the New York Training School for Teachers. She also established one of the early museum children's rooms, opening hers in 1909, thus enhancing opportunities to provide museum resources to teachers appropriate to their pupils' interests.[36] Other advocacy groups and educational organizations used museum facilities to present public lectures on nature topics, and commercial venues opened their auditoriums for them as well. Thus, for example, the Sullivant Moss Society heard Elizabeth M. Dunham on "Suggestions for Talks to School Children upon Mosses (illustrated by prepared specimens)" on December 29, 1916, while the Strand Theater hosted a zoological lecture series featuring Raymond Ditmars of the New York Zoological Park that same year.[37]

Nature study might go in and out of favor with administrators, but teachers who found the curriculum effective with their pupils could mobilize public support when a threat appeared. In 1913, the Committee on Studies recommended the elimination of nature study as a separate subject and held a public hearing on the matter. Frederick Holtz, head of the New York Department of Nature Study, wrote to point out the number of classrooms where very successful nature study was underway, and Margaret Knox, representing the Association of New York Principals, argued that nature study engaged children in learning more than most other subjects.[38] A number of prominent New Yorkers wrote letters in nature study's defense, including William Hornaday, director of the New York Zoological Park; C. Stuart Gager, director of the Brooklyn Botanical Garden; T. Gilbert Pearson, secretary of the National Association of Audubon Societies; Charles H. Townsend, director of the New York Aquarium; Maurice Bigelow of Teachers College; and Anna Clark of the New York Training School for Teachers.[39] Even Hornaday, a conservationist sometimes critical of nature study, believed that apparent problems with the curriculum did not justify "the destruction of nature teaching."[40] Gager argued that science was fundamental to modern life and that "to eliminate nature study from our elementary schools would be to turn back the wheels of educational progress twenty-five years." The mobilization of support, meetings with Superintendent Straubenmueller, and stronger coordination with New York City institutions that could supply nature study materials for classes forestalled the challenge.[41] For the educational director of the Brooklyn Botanical Garden, an important measure of success was the enthusiasm of crowds of children who "voluntarily flock to lectures on plant and animal life and on chemical and physical phenomena at our two botanic gardens, at our several museums, and especially at that wonderful institution, the Children's Museum in Brooklyn."[42]

Indeed, across the East River, the expanding Brooklyn Institute of Arts and Sciences had directed its attention toward educational programs. Anna Billings Gallup was certified by Bridgewater State Normal School in 1893 and earned a bachelor's degree in biology from the Massachusetts Institute of Technology in 1901; she joined the Brooklyn Institute in 1902. Gallup had taught at the Hampton Normal and Agricultural Institute in Virginia, from 1893 to 1896, and for one year at the Rhode Island State Normal School. She was thus well aware of the demands made on teachers and the range of abilities among their pupils.[43] Within two years, she was put in charge of the new Children's Museum.[44] With her "please touch" philosophy and her advocacy of creating attractive exhibits with the assistance of the well-known museum taxidermist and curator Frederick W. Lucas, the Brooklyn Children's Museum became an often-cited model for teaching about nature to urban children who had limited countryside to explore.[45] Raising money through the local women's clubs, Gallup hired demonstration staff so that children had somewhere to direct their inquiries and teachers could observe techniques for teaching with natural objects. Gallup would be among the founding members of the Nature-Study Society, and she regularly described her programs in its journal and other widely distributed publications.[46] The New York superintendent of schools noted that among the several museums cooperating with the schools, the Brooklyn Institute of Arts and Sciences stood out because it also "established a School of Pedagogy for our teachers and a Children's Museum that is much frequented by school children of that borough."[47] Such collaborations demonstrated the positive public response to nature study and its integration into multiple educational and cultural settings.

There is little published description of the curriculum in parochial schools, whether in the centralized Catholic systems or the more independently organized Protestant (most often Lutheran) schools, although both accounted for a significant proportion of the school population at the turn of the century. They had less funding and may have found it difficult to enhance the training of their teachers. Nonetheless, by 1911 the New York Catholic School Board had a specified course of study that included nature study through grade five. Hygiene and physiology were mandated and nature study was included for at least an hour each week "with the hope that the seed of knowledge sown by the Christian teacher . . . may bear fruits of kindness and sympathy toward created things and an increase of love for their creator." Recommended projects echoed those of public nature study curricula: aquaria, bird cages, egg shell farms, daily weather maps, and out-of-door activities. The only thing that distinguished this syllabus from others during the same period were the concluding series of Biblical quotations from Genesis and other chapters about God and nature.[48] In the 1920s the Archdiocese in Philadelphia similarly endorsed nature study and provided some rudimentary materials.[49] Particularly enterprising was Sister Mary Azaveda of the Sisters of Notre Dame in Cleveland, Ohio, who compiled a no-nonsense reference textbook for teachers in the Cleveland Catholic schools and encouraged them to take advantage of the Cleveland Museum of Natural History's offer to provide materials and guest lecturers.[50] It would be useful to study as well the curricula of both parochial and private schools, which varied in their emphasis, some still following classic formulations and others specifically responding to new progressive educational ideas. The liberal Ethical Culture School in New York City, for example, was a leader in adopting nature study programs that included field trips to Palisades Park and other regional sites.[51]

FRAMING DISTINCTIVE URBAN NATURE STUDY PROGRAMS THROUGHOUT THE NATION

Cities across the country worked to find the "one best system" of education in standardized programs intended to produce moral citizens able to work in their communities.[52] Nature study could be quickly implemented, often in very distinctive ways, as urban school systems underwent rapid changes brought about by consolidation and centralization, the reduction in size of once powerful and autocratic school boards, and city-wide elections that allowed business leaders and others to choose experts as superintendents of schools.[53] This opened the way for innovative curricula as well, implemented versions of nature study as teachers, administrators, and reformers learned from each other and from an array of publications.

In progressive Cincinnati in 1900, for example, the new administrative leadership resulting from such a political shift emphasized history, civics, nature study, and expression through language and mathematics.[54] The interpretation of nature study might be itself be very broad. In a city with a significant proportion of German immigrants, Cincinnati teachers taught the usual lessons in biological nature study and geography and, perhaps influenced by the ideal of *Heimat*, added human environmental studies that included local and world industries.

Ideas for using nature in education spread like windborne seeds as the program cropped up across the country, suddenly springing up in one community while languishing in another. In the flourishing flour mill city of Minneapolis that served as terminus for the Great Northern railroad, for example, Sarah Louise Arnold, who had studied with Arthur Boyden at Bridgewater Normal School, briefly filled a new position of Supervisor for Primary Work.[55] She introduced nature study in the Minneapolis public schools in the 1890s, even as it was being advocated by Charles Scott across the Mississippi River in the St. Paul public schools.[56] The results, according to a contemporary, were a "larger culture for the teacher, greater intelligence and interest on the part of the children, and a brighter atmosphere over all."[57] For Arnold the initial problem had been, as she put it, that "Many of [the teachers] were sure that no tree except the apple tree ever blossomed, while the noun—bird—stood for their entire knowledge of the feathered creation."[58] Arnold used public lectures and publications to advise primary school teachers in Minneapolis about how to link local specimens to familiar poetry and songs; she published her talks in a series of articles in *Primary School*, a monthly produced by the American Book Company.[59] Seeking advice from botanist Charles Bessey at the University of Nebraska, who had taught briefly at the University of Minnesota and served on science subcommittees of the National Education Association, about just what teachers needed to know, she then initiated a weekly science class to provide teachers with better botanical knowledge.[60] Bessey's advice was ambitious, suggesting that any real work would involve a microscope and that teachers must know Latin names and physiological processes.

In Massachusetts a prestigious board and set of state inspectors that included Alice Freeman Palmer and Kate Gannet Wells wielded influence by offering special teachers training institutes on educational psychology and natural history topics. Training teachers were recognized as the key to change. Arthur C. Boyden was a regular on the lecture circuit, appropriately providing lessons "illustrating the nature of elementary and of scientific knowledge and the relation they bear to each other."[61] By 1892, "nature studies" were part of the state-recommended "Course of Study for the Elementary Grades."[62] Training for teachers was important, the board noted, because "In science, more perhaps than in other departments, teachers are found teaching as they were taught," which might mean very little science or simply formulaic science topics presented in lectures rather than with natural objects in and beyond the classroom. Nonetheless, Massachusetts provides considerable evidence that teachers were asking for and attending nature studies classes voluntarily before the subject became a standard one in the schools.[63]

In the mid-1890s, one of the appointed state "visitors" reflected on the changes that had occurred during his twenty-five years touring the state. Noting that Louis Agassiz had lectured to teachers' institutes and normal school classrooms, he recalled that the best normal schools had already introduced basic mineralogy, botany, and geology during the last half century. But another long-term school visitor, George A. Walton, asserted in 1893 that few teachers had used the "objective method" in presenting anything of the natural sciences they had learned in their normal school training until the advent of the new curriculum of nature study, now quite widely used.[64] This change may have induced Sarah Louise Arnold to return to Massachusetts in 1895 to become supervisor for the Boston primary schools, where she developed a complete course on teaching nature study. But here her teachers faced the particular issues of overcrowded classrooms, special pupil needs, and an already packed curriculum.[65] Her justification for nature study took an increasingly moralistic tone as she confronted the problems in poverty-stricken inner-city Boston schools, where curbing unruly boys and setting higher moral standards seemed to take precedence over issues of content.[66] She encouraged organizations like the Massachusetts Audubon Society and the Massachusetts Society for the Prevention of Cruelty to Animals to have their members volunteer to work with students and to provide materials for classroom study, typically pamphlets oriented toward their particular causes.[67] Although the Boston School Committee reported reassuringly to constituents that its new course of study containing natural science was "really the same old tree, pruned here and there to its shape or grafted a little by way of experiment," historians agree that transformations in child psychology, school management, and curriculum were having significant impact on the main trunk itself.[68]

Teachers in the cities of the upper Midwest faced somewhat different circumstances, perhaps because the divisions between urban and rural life seemed less distinctive and extensive park systems were part of their urban planning. Dietrich Lange, educated in Germany before immigrating to the United States with his family in 1881, had a lifelong interest in natural sciences that he expressed in local newspaper columns, fictional books, and nature hikes with pupils and fellow teachers in St. Paul, Minnesota.[69] Like others in largely rural states, he was active well beyond the Twin Cities. He taught nature study as early as the summer of 1895 in the Minnesota Teachers Training School centrally located in Windom and later helped coordinate a topical program at the headwaters of the Mississippi River at Itasca in 1909.[70] After Charles Scott left St. Paul to go to Oswego Normal School, Lange became the familiar leader of nature study in the city schools of St. Paul from 1897 until 1906, when he took another administrative post. Nature study offered a vehicle for him to pursue his interests in the outdoors, nature, and conservation; he was an avid amateur ornithologist. After his appointment as principal of Humboldt High School, Lange pursued nature interests avocationally, becoming one of the founders of the Ramsey County Boy Scout Council.[71] He remained committed to teaching agricultural skills and spent time developing the school garden movement in St. Paul. Such nature study initiatives, some more long-term than others, sprang up in various parts of the country.

In the industrial town of Worcester, Massachusetts, Clark University had been designed to emphasize research much like Johns Hopkins and Chicago. Its new president, G. Stanley Hall, recruited from Johns Hopkins, expressed deep concerns about how children were being raised in America's increasingly "unnatural" urban society and was attracted to the idea of nature study.[72] Trained in psychological laboratories in Germany, he was particularly interested in child development and was intrigued by the apparent successes in education at Cook County Normal School under Francis Parker. Hall's academic research in the early 1880s involved quantitative assessments of children's knowledge, and his results led some to call him the founder of the child-study movement. Hall also closely followed the career of his former student, John Dewey, as the latter became a leader in educational philosophy. Hall's academic research was only indirectly related to nature study, but he intended Clark University to be in the forefront of educational innovation and child study, and nature study seemed to be ascendant. Hall lectured to local teachers on the value of teaching nature, and at the NEA meeting in 1896 he endorsed David Starr Jordan's argument that there was a correlation between nature study and moral education.[73]

After several leading members of his faculty were lured to the University of Chicago by its new president, William Rainey Harper, Hall invited Clifton Hodge, a recent Ph.D. graduate of Johns Hopkins and former fellow at Clark in biology and psychology, to return as assistant professor in 1892.[74] Hall encouraged Hodge to emphasize science teaching and to contribute to Hall's new journal, *Pedagogical Seminary*, which was published at Clark.[75] Hodge wrote a series of essays entitled "Foundations of Nature Study" and regularly reviewed nature study literature for the educational publication. Hodge was explicitly critical of some textbooks being produced, but he also argued that the topic should be offered because elementary-aged children were at a crucial stage for learning from nature.[76] Putting his research and experience to use, in 1903 he published the textbook *Nature Study and Life*, which would reportedly sell a million copies.[77] The volume had numerous pictures and line drawings and specific ideas about appropriate projects for the classroom, as well as for neighborhoods (fig. 10). Following Hall's lead, Hodge concentrated on the challenges facing teachers in heavily industrial cities like Worcester.[78] He believed that urban industrial society could come to terms with the organic world in which it was actually situated through a carefully conceived approach to nature study. One project, for example, had children become familiar with just one city block in Worcester. Pupils mapped the block, filled in houses and outbuildings, then identified trees and shrubs, and completed the project by noting nests of birds. They also observed and recorded information about which shrubs attracted particular birds by providing shelter and edible fruits. In another class, they built bird houses and feeders and watched the changing density of the bird population over subsequent years. Hodge, who had evident conservationist leanings, pointed out the importance of counting for the bird census and the growth of knowledge about valuable wildlife. He argued that the personal engagement meant the children would "hardly be induced to molest a bird's nest" in the future.[79]

Hodge did not have the advantage of a campus practice school like his colleagues at Chicago and Columbia, but he worked closely with Mary C. Henry, principal of Upsala Street School in Worcester, to test his ideas in her teachers' classrooms.[80] From the outset, his definition of nature study had

FIGURE 10. Worcester schools encouraged gardens with edible and flowering plants, which culminated in a flower show at the end of the growing season. Clifton Hodge, *Nature Study and Life* (Boston: Ginn and Co., 1903), 90.

a frequently quoted theme of human practicality, or, as he put it, "Learning those things in nature that are most worth knowing to the end of doing those things that make life worth living."[81] Undoubtedly it was the engaging yet simple suggestions that made his text popular; it was typically placed high on summer institute reading lists on nature study.[82] Community leaders took notice, too. The president of the Worcester Natural History Society encouraged his members to support the programs in nature study, noting that this new curriculum "created an opportunity to be helpful" to teachers by providing specimens and information.[83]

Hodge, as we shall later see, brought a different and sometimes contentious point of view on nature study and never seemed to enjoy the collegiality of most nature study advocates. Perhaps because he was in a struggling industrial town with a significant immigrant population, he embraced the urban, even industrial possibilities offered by nature study alongside the agricultural. He certainly thought that much teaching would be more effective if the lessons were practical. He anticipated that his "confrères" in the movement might call him "farmer" or "granger'" or "degraded utilitarian" but nonetheless argued that the purpose of nature study was to learn how to control the forces of nature for the highest human happiness and the best human good.[84]

One of his most publicized projects involved Worcester school children working to solve the problem of mosquitoes, recently shown to carry the endemic disease of malaria. Under the guidance of teacher Edna Thayer, elementary and grammar school students reviewed the life history of the mosquito, collected larvae from local stagnant water, and then watched the life cycle unfold in their classroom. They also learned by experiment that oil on the surface of the water, whether in glass containers or in the stagnant pools along nearby Beaver Creek, would destroy the potential pests. As Hodge pointed out, as the children learned, so did their parents, with the result that the community subsequently sponsored a drainage project to eliminate the swamps with their mosquito larvae and thus resolved the disease problem at its source (fig. 11). Hodge argued that the wider public attention was important:

> I am inclined to think that the main reason why for the past year we have not heard one word about nature study being a "fad" or a "waste of time" is to be found in these simple mosquito lessons. And why should not the children of every community grapple with such problems? Why should they not be encouraged to utilize every possible condition which may instill the sense of mutual cooperation and brotherhood and thus grow strong in the principles of intelligent citizenship?[85]

Hodge's rhetorical questions made clear that for many advocates nature study was as much about civic and individual enhancement as it was about the natural sciences; but knowing how to do systematic scientific inquiry was essential.

The mix of motives and the challenges for early twentieth-century educators were particularly evident in the schools of Gary, Indiana, which became an "experiment" in educational possibilities. A virtual company town built at the turn of the century by United States Steel on swampland, sand dunes, and forests of scrub oak along the south shore of Lake Michigan, Gary quickly grew to a city of more than fifty thousand people as the steel industry prospered there.[86] The school board was run by well-established, well-educated, if conservative men self-consciously undertaking an "experiment on a larger scale and with more elaborate facilities" than any previous community.[87] Their goal was to provide a twentieth-century education

FIGURE 11. Clifton Hodge photographed Downing School pupils in the tenement district of Worcester as they worked on a community project to eliminate mosquitoes and thus stem the incidence of malaria. *Nature-Study Review* 3 (February 1907): 34. Courtesy of the Wangensteen Historical Library of Biology and Medicine, University of Minnesota.

that would instill proper values and attitudes toward work and productivity among a population highly diverse in terms of race and ethnicity.[88] Gary's enterprising superintendent of schools was William A. Wirt, a "country boy" at heart and a teacher who had attended summer classes at the University of Chicago. Wirt sought to create a system that was so well equipped that it would attract outstanding teachers from across the country. One primary school remodeled a classroom to accommodate a plant-growing house, an animal house, and a workplace where children could mount specimens and prepare birds' nests and feeders; other primary schools in Gary created "ample and flourishing school gardens."[89] The Gary schools were highly regimented (indeed the program was sometimes called the "platoon system"), with students going to regular classes for half a day and spending the other half day doing athletics, studying art, music, dancing, or dramatics, going on nature field trips, working in school libraries and laboratories, or participating in home economics or workshop projects.[90]

Experimentation in education and the range of school organization in this period make it impossible to fit nature study curriculum into any single

category since it could be and was used in schools that might be characterized as traditional, progressive, or conservative. The Gary schools, which
have been assessed in various ways by historians of education, maintained a
rigid schedule with highly structured programs, while the individual classes
adopted some of the most innovative curricula and methods of group study
and active learning in the country. Progressive educators in private and
public schools were able to find in nature study elements that worked for
them, whether in rural or urban settings or in large or small schools. Nature
study was, for example, part of the well-known and experimental Organic
School of Marietta Johnson in Fairhope, Alabama, founded in 1903.[91] Nature study was also taught by Alice Barrows at the Ethical Culture School
in the heart of New York City, where students were encouraged to explore
independently throughout Central Park and similar green retreats.[92]

EXTENDING NATURE STUDY TRAINING

Faced with the need to supply training and resources for nature study,
school superintendents responded with formal and informal programs. In
the 1890s, participation had been largely voluntary and the efforts typically spearheaded by one or two enthusiastic advocates. Summer school
institutes were the quickest way to disseminate the theory and practice of
the new subject and a cadre of educators willing and able to conduct them
existed, but their own preparation in science and in this new educational
concept varied considerably.[93] Although no state mandate was in place,
courses on nature study began to appear simultaneously in several states
in institutes for teachers taught by normal school faculty, textbook authors,
and administrators. It was through such a program that Dietrich Lange
taught nature study in the summer on the Windom Normal School campus
to regional teachers.[94]

Although some pupils learned about plants and animals in school well
before the 1890s, nature study advocates provided specific programs that allowed such education to be readily incorporated into the classrooms of even
the inner-city schools in New York City and Worcester. Special instructors
in the cities gave teachers techniques to help their pupils investigate nature
closely, recording climate observations and collecting tree leaves, insects,
and other natural objects for detailed study. City schools concentrated on
materials close at hand, but, in more rural areas, the programs sought to
refocus the thinking of rural children about the wildscapes in their vicinity
and the changing and challenging habitats on and around farms.

CHAPTER FOUR

Revitalizing Farm and Country Living

*Where we are thinking particularly of the subject we are studying, and are orga-
nizing our teaching with reference to that subject, we are teaching science. But
when we teach about these things with our thought chiefly upon the child, his
capacities, the nature of his mind, the nature of his interests, then we are teaching nature-study.*

Liberty Hyde Bailey[1]

LIBERTY HYDE BAILEY, a botanist at Cornell University, enjoyed an abundance of the kind of nature he referenced above as he looked out across
Cayuga Lake in upstate New York. But the bucolic scenery did not prevent
this faculty member at Cornell University from recognizing the declining
agricultural economy in his region in the late nineteenth century. While an
increasing stream of summer visitors seeking relief from city temperatures
appreciated the Finger Lake regions' varied landscape and recreational potential, farms were reverting to forest as farm families migrated off the land.
The changing demographics worried faculty in the publicly funded College
of Agriculture affiliated with Cornell, itself a private liberal arts institution,
as well as legislators at the capitol in Albany.

Recent passage of the Hatch Act in 1887 and consequent expansion of the
Agricultural Experiment Station in New York enabled faculty to work with
farmers on specific problems and to improve their methods. Extending this
mission to families and especially to the coming generation of rural children appealed to the progressive thinking of educators, and, in the 1890s,
the methods and rationale of nature study seemed a way to enable teachers

scattered in small one- and two-room schools to educate pupils to understand and appreciate their natural environment.[2] As Bailey later explained: "A name for the movement was necessary. . . . we chose the current and significant phrase 'nature-study' which, while it covers many methods and practices, stands everywhere for the opening of the mind directly to the common phenomena of nature."[3]

While the programs in Chicago, New York City, and other cities were developed primarily for urban pupils who typically lacked much direct access to nature, the rural program at Cornell was designed to give fresh perspective to children already familiar with farm and wild life. The stagnant rural economy pushed the state legislature to provide the faculty with the financial resources sufficient to build and sustain a distinctive program. Within a very few years, Cornell's highly successful rural nature study initiative was imitated in states and provinces across North America. Nature study particularly attracted national attention and local sponsors because its advocates offered a vision of activities that would enhance struggling rural schools and simultaneously improve the quality of life in the schools' communities. That vision appealed not only to educators in upper New England where poverty was driving farmers from the land but also to those in the relatively prosperous small towns and rural areas of the Midwest and the Far West. The Cornell leaders promoted nature study through publications, lecture tours, and intensive courses for teachers in their state and beyond. Some advocates of the new curriculum focused on potential vocational connections because of the extension work already in place through the College of Agriculture. The vision of the Cornell leaders, however, was on out-of-door nature, whether wild or domesticated, and they resisted making nature study work deeply vocational for students in grammar schools, where eight years of education was still a high standard.

THE COMSTOCKS

Among the faculty at the new (founded in 1865) but already well-respected Cornell University in the 1880s was John Henry Comstock, an entomologist with a Cornell Ph.D., who had worked briefly in Washington, DC, for the Bureau of Entomology in the Department of Agriculture. Pursuant to that agency's outreach mandate, the serious young professor remained committed to extension work with farmers on his return, even as he developed Cornell's Department of Entomology into the leading entomology research school in the country.[4] Funds from the New York state legislature and the federal government had, by the late 1880s, made Cornell a leader in extension education, which brought experienced farmers to campus to discuss how to improve the quality of soil, select highly productive seeds, and control pests.[5] Off campus, traveling extension agents offered lectures in county seats and on scattered normal school campuses in what were appropriately termed "short courses." Extension education was also offered by correspondence. College-level courses for future farmers included botany, geology, natural philosophy or physics, chemistry, and zoology, because such sciences were viewed as integral to agriculture in a modern, industrializing society.

Comstock's wife, Anna Botsford Comstock, often traveled with him on lecture tours in the countryside in the late 1880s, speaking to farm women and families.[6] The chestnut-haired former school teacher spoke with polished confidence as she used familiar examples to discuss the importance of both schools and country life. She also worked closely with her husband as a highly skilled illustrator, editor, and coauthor of *A Manual for the Study of Insects*, a book that quickly became the standard reference. Although beyond the capacities of elementary teachers, the college text enhanced the reputation of John Comstock as its senior author.[7] The partners collaborated on any number of projects and Anna typically wrote about "our" laboratory and "our" books (fig. 12). The couple shared an enthusiasm for education, and in 1891 Anna Comstock ran a successful summer school field laboratory where students studied "nature under the most favorable opportunities, when we can see her face to face, not merely books and preserved specimens."[8] She herself had experienced the isolation of teaching in a small rural school, and she was eager to help others who had a more limited education. A contemporary noted that a casual acquaintance with botany, geology, natural philosophy, and chemistry merely allowed teachers to introduce such subjects "incidentally, as a means of culture and for the purpose of keeping up interest, enthusiasm, assisting in governing, and with the hope that some good seeds may be sown which will find proper soil and [stimulate] a future Agassiz or Linnaeus."[9] For Comstock and her colleagues, science was too important to be left to happenstance.

As the depression of the 1890s deepened into the Panic of 1893 in upstate New York and throughout the country, the state legislature appointed a Committee for the Promotion of Agriculture. Anna Comstock was made a member of that committee, which was instructed to address the "abandoned farm" problem as entire families left to look for work in larger cities.[10] Comstock, who had a bachelor's degree in zoology and was familiar

FIGURE 12. Anna Botsford Comstock and John Henry Comstock, a collaborative couple pictured here in the 1910s, wrote books (many were illustrated by Anna Comstock), taught summer field work classes at Cornell, and built a successful publishing company. Courtesy of the Division of Rare and Manuscript Collections, Cornell University Library.

with work at Oswego Normal School, thought the nature study curriculum could be adapted to rural schools. In 1894 the legislature appropriated additional funds for farm extension work at Cornell that specifically mentioned the phrase "nature study," and by 1896 it had allotted $18,000 a year for the work.[11] The Cornell team had a mandate to create a program specifically for the small and often isolated country schools of New York, led by Anna Comstock and her colleague, Liberty Hyde Bailey.

In 1894 Anna Comstock developed a course of nature study for the Westchester County school district. Because, as a woman, she was not at first a regular member of the Cornell faculty, she partnered initially with James Rice and later with Isaac P. Roberts in the School of Agriculture. When Liberty Hyde Bailey became administrator for the nature study program, Anna

Comstock received a full-time appointment as instructor. This was a bitter-sweet event because Comstock was initially appointed as an assistant professor but the Board of Regents overturned her appointment. She remained an instructor until finally promoted to full professor in 1913.[12] Their balance of leadership skills would prove extraordinarily productive, each bringing strengths of oratory, authorship, and administration as they coordinated a rapidly growing curriculum and became visible advocates for nature study throughout the state and beyond.

Anna Comstock, raised an only child in a cabin in Cattaraugus County, readily empathized with the circumstances of rural families. She conversed easily with local teachers and was familiar with the object-lesson approach that had made Oswego Normal School nationally known. A student colleague of John Comstock, Henry Straight, had taught at Oswego briefly before taking his position at Cook County Normal School. Later the Comstocks became acquainted with Charles B. Scott, who in 1895 left the public schools of St. Paul, Minnesota, for Oswego Normal School, where he transformed its object-lesson orientation toward nature study.[13] Scott , who taught nature study to about fifty teachers each year and oversaw the public school program for nearly five hundred pupils in Oswego, became chair of a committee of the state teachers association that sought to introduce "real" nature study and offset the "tendency to be satisfied with mere book work and cramming for examination."[14]

For Anna Comstock, an emphasis on using natural specimens as a basis of study was an appropriate pedagogical strategy for children, and her approach reflected the popular field excursions that John Comstock took with his college and summer school students. Anna's technique was a pragmatic variant on the child-centered philosophy of the progressive educators at the University of Chicago, since both started with the practical intention of using the presumed innate curiosity of children. Nature study in both theory and practice emphasized visual and other sense experiences, and Anna Comstock made those connections quite explicit. For her, the poetic and aesthetic qualities of the environment complemented the practical and familiar outlook toward nature that she and the teachers would promote.[15]

From her base at Cornell, she encouraged teachers and parents using the pamphlet series, *Home Nature-Study Course,* which guided individual correspondence students through various topics by soliciting their outdoor observations. Assignments were intense for both students and teacher, and those who completed ten lessons earned a certificate.[16] Comstock's goal was to train teachers to be observant in a systematic way and then pass their techniques along to their students. The assignments in "Bird Study" not only required physical descriptions of birds but also pointed readers to their behavior, asking readers to note, for example, that the hen and rooster express "at least ten different mental conditions or emotions with perfect distinctiveness" by voice.[17] She argued that the young learned these behaviors from adult fowl, commenting tartly that there should be "a society for the prevention of incubation, or the establishment of asylums for idiot ducklings hatched in them."[18]

Comstock's work had an aesthetic sensibility not much visible in the earlier nature study textbooks by Jackman, Lange, and others. She had studied wood engraving for six months at Cooper Union and became a member of the Society of American Wood Engravers to help illustrate her husband's manual of insects, *Introduction to Entomology*.[19] Anna made many of its over six hundred illustrations. As nature study became a more established curriculum, John published his less-technical *Insect Life*: *An Introduction to Nature Study* in 1897; also illustrated by Anna, it sold well and demonstrated the market for such specialized books.[20] Anna's own insect book, *Ways of the Six-Footed*, did especially well with elementary nature study teachers.[21] With chapters organized by habitat, such as ponds, meadows, and forests, this book also typifies the observant instructor's familiar writing style. Highly productive, she wrote well-illustrated school leaflets, but there she often used photographs rather than the more time-consuming woodcuts her husband preferred for his scientific publications. Her friends in biology also contributed to the pamphlet series, among them Susanna Phelps Gage, wife of another Cornell professor, Simon Gage. Phelps Gage's leaflet, "The True Story of the Little Red-spot," concerned a vermillion-spotted newt.[22] Nature study became quite literally a cottage industry in Ithaca, engaging others with artistic talent, including bird illustrator Louis Agassiz Fuertes. An undergraduate, Effie, recalled that the classes by John and Anna Comstock "changed my life completely." Having studied art for two years, she developed a technique to hand-color lantern slides, an incredibly complex task in terms of precision and projected color. She worked with a watercolor company to find the right translucence, and the product allowed lecturers across the country to show colored slides.[23] Themes of nature's aesthetic, knowledge acquisition, and practical advice were seamlessly interwoven into the leaflets, study guides, and textbooks alongside ideas about democracy and civic life.

Expanding on the instructional materials she had written in the previous decade, in 1911 Anna Comstock produced her *Handbook of Nature Study*.

When the New York State Department of Agriculture declined to publish the eight-hundred-page manuscript, Anna turned to the Comstock Publishing Company, which her husband and his colleague Simon Gage had started in 1893 to control the quality of their own publications and to reissue updated versions regularly. The publishing enterprise grew to fill a neighboring cottage along the Fall Creek Gorge.[24] Although John Comstock was skeptical about the costs involved in producing Anna's large manuscript, the book proved instantly successful, eventually going through more than twenty editions and translation into eight languages; it remains in print today.[25]

Her essays typically started with something familiar and close at hand—a dandelion, a grasshopper, or a smooth stone—and she then encouraged her reader to think about its relationship to others of its own kind and to the complex environment in which it had been found. Her straightforward commentary, which reflected on both the elegant simplicity and yet the diversity of nature, combined with an acute artistic sensibility made her distinctive among the nature study theorists.[26] Accolades for her teaching qualities started early and persisted as students responded to Comstock's humor and humaneness while acknowledging her high intellectual expectations of them. When Anna was recognized as one of twelve outstanding women in the United States in the 1920s, a former student reflected on her teacher's approach:

> It's because she never selected some special line and stuck just with that. She stands back of it all—the study of animal, plant, insect life. She could be a specialist in any one of those branches, but she's taught us and inspired us with the subject as a whole—kept it all related . . . She has made the subject human as no other real scientist has done. I guess it's that and the genius she has for helping others to understand and find out things for themselves.[27]

This exceptional teacher took her growing national reputation in stride and never lost sight of the pupils and teachers who relied on her efforts. Thus she recommended to Bailey that they send postcards to five thousand rural teachers with the hope of getting perhaps five hundred of them into correspondence with her, using the Cornell materials.[28]

Cornell agricultural faculty members concentrated on increasing farm production, but the Cornell nature study program emphasized the wider countryside, including meadows, streams, and forests. Comstock encouraged children to bring small animals, such as turtles from a local creek, to their classes, but urged students to return the specimens to their same habitat rather than letting them "loose to scatter bewildered and helpless over

FIGURE 13. Anna Botsford Comstock (*at left*) and Liberty Hyde Bailey (*at right*), on the proverbial logs, here converse with teachers from the Tompkins County schools while on a field excursion. Courtesy of the Division of Rare and Manuscript Collections, Cornell University Library.

a strong earth." The ecological sensibility in her writing meshed well with the conservation movement gaining strength at the turn of the century. It also attested to her personal environmental perspective, far from what she and others labeled the "specie hunter" outlook.[29]

Having no children (to their apparent disappointment), Anna and John were relatively free to travel, and for several years they taught at Stanford University in Palo Alto, California, during the winter months. Closer to home, she lectured at regional normal schools, Teachers College in New York City, and the summer Chautauqua meetings held in western upstate New York near her family home.[30] Visible, versatile, and engaging, Comstock made contributions to the nature study movement over the course of thirty years, not least by editing its professional journal in her later years.[31] She and her colleagues practiced what they taught, working closely with teachers in Tompkins County, where Cornell was located (fig. 13).

THE NATURE-STUDY IDEA OF LIBERTY
HYDE BAILEY

Liberty Hyde Bailey, who had spend two postgraduate years at Harvard with Asa Gray and then became a horticulturalist at his alma mater, Michigan State University, before going to Cornell in 1888, was an outspoken advocate for country life. Raised in what was at midcentury the frontier of southwestern Michigan, he credited an early teacher in his ungraded country school for telling him, "You are going through a beautiful world with your eyes shut. You see nothing." Urging him to observe carefully on his long walks to and from school, she then led him to learn more about the animals and plants he encountered from a reference book she kept in her classroom.[32] Not surprisingly, Bailey later emphasized the importance of building on children's sensation of discovery. He also carried with him the lingering awareness that enterprising teachers like his own had little formal training in the details of natural history.[33]

Once on the Cornell faculty, the articulate and energetic Bailey gained a public and scientific reputation in horticulture and agricultural science (only partially surpassed in public notoriety by the nonacademic botanist Luther Burbank). He published widely and had frequent speaking engagements on agriculture, horticulture, education, and other aspects of what became labeled as American country life. In 1898, he was appointed director of the Agricultural College and Experiment Station and chief of the Bureau of Nature Study at Cornell, positions that provided a podium to advocate for nature study in and beyond New York State.[34]

Bailey was arguably nature study's most visible public spokesperson, particularly in reference to the rural schools. His relatively slim volume, *The Nature-Study Idea* (fig. 14), was widely cited, frequently reprinted, and, as we shall see, sometimes criticized by his scientific peers.[35] Moreover, his contacts with local teachers and with normal school faculty members like Charles Scott at Oswego, helped him reflect on appropriate methods and content for young pupils.[36] He argued that natural science education could highlight the qualities of rural life for young children, who, like himself, had not appreciated their natural environment.[37] He expressed his educational philosophy to an editor at Macmillan Company with reference to his proposed botanical text *Lessons with Plants*: "I am setting no model ideas or influence or definitions before the student. In every case we begin our lesson with an observation directly from nature (or from a drawing, which is the nearest approach we can make to nature in a book). After the student has studied the object, the inference or definition then comes naturally."[38]

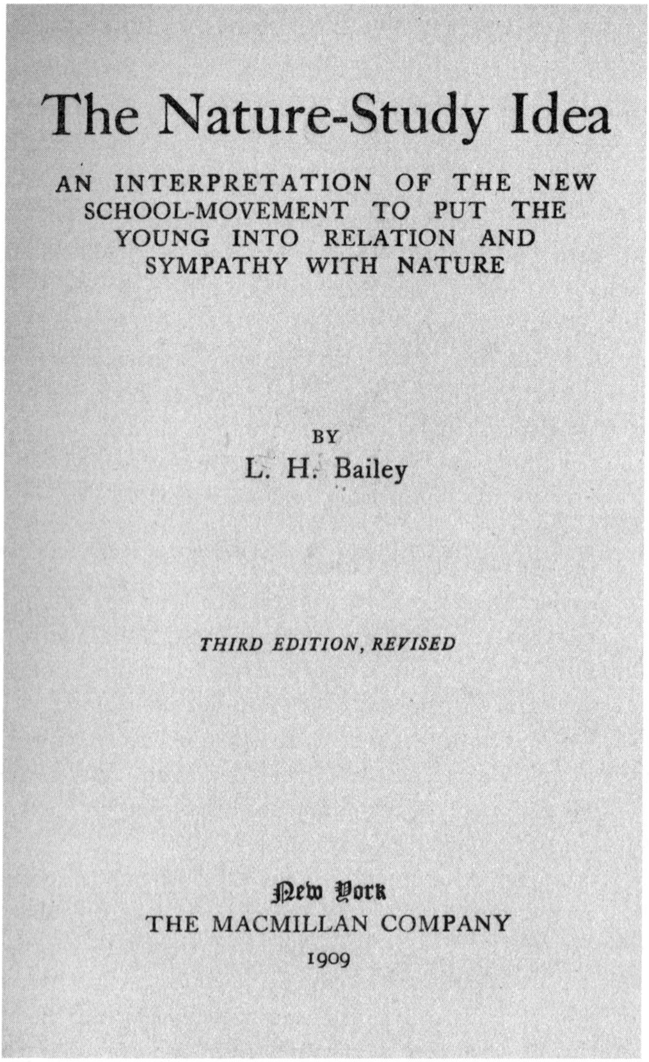

The Nature-Study Idea

AN INTERPRETATION OF THE NEW
SCHOOL-MOVEMENT TO PUT THE
YOUNG INTO RELATION AND
SYMPATHY WITH NATURE

BY

L. H. Bailey

THIRD EDITION, REVISED

New York
THE MACMILLAN COMPANY
1909

FIGURE 14. Liberty Hyde Bailey's *Nature-Study Idea* was widely read by teachers and often assigned in normal school courses, representing the philosophy of the subject rather than practical guidelines for its implementation. Author's copy.

His was not an original approach—indeed it echoed Emerson, Agassiz, and Parker—but Bailey showed its power and provided workable examples through his articles, books, and textbooks, as well as the Cornell School Leaflets. The first leaflet, in December 1896, "How A Squash Plant Gets Out of the Seed," suggested planting seeds in a box and watching them develop, with Bailey noting, "the wonderful difference between the first and later stages of nearly all plants [is similar], and it is only because we know it so well that we do not wonder at it."[39] Such leaflets depended not on new scientific research but on a formula for presentation that took a well-established empirical observation and translated it into a lesson for children and their teachers. Leaflets seemed a particularly good device; being brief and explicit (as well as relatively inexpensive to produce). Bailey argued that they were "guides" rather than texts because they pointed students and teachers toward objects rather than providing explanations to be learned.[40]

Always, for Bailey, specimens were best studied when viewed in their natural setting, so he distinguished his nature study from older "object lessons" introduced at Oswego that typically had used static indoor materials as reference points. Bailey made it clear that the objects presented in the earlier movement, often manufactured or taken from their natural location, provided instruction quite different from genuine nature study that "tries to place the pupil with the objects and phenomena as they occur in nature."[41] Trained as a botanist, he discriminated between nature study and more systematic work in natural history and biology, and this categorization sometimes distinguished him from academic peers elsewhere. Nature study, he argued, would not lead to new truths, but rather to a "sympathetic attitude" toward nature. Bailey polished his rationale and outlook on nature study through his public lectures and published his philosophy in 1903 in *The Nature-Study Idea*. Here he argued expansively for a moral component in nature study in addition to the pragmatic possibilities he had outlined earlier. Thus, he explained, "Nature-Study is not primarily a natural history subject; it is primarily a pedagogical idea. Natural history subjects are the means, not the end. Nature study is not a science. It is not knowledge. It is not facts. It is spirit. It is concerned with the child's outlook on the world."[42] Bailey's interpretation would not be universally shared or approved by all nature study educators or all his fellow scientists, but at the turn of the century it resonated well with his public audience and attracted such prominent supporters as Theodore Roosevelt and later the nature writer Aldo Leopold.[43] Bailey could correctly claim that at the turn of the century, nature study was "the most popular method of teaching natural history subjects."[44]

Bailey presumed that rural school teachers were dedicated to their work but isolated, and he understood that most farm families were literate but had few books or magazines in their homes. His goal was to inspire teachers and parents to encourage nature study and to provide them with the means to do so. That the curriculum might vary from place to place or teacher to teacher did not trouble him: "It isn't necessary at all that all teachers of nature study take the same method in producing their results, provided it's all founded on accurate observation of real things as they are in nature."[45] In fact, he believed that it was teachers' innovative use of familiar and accessible subjects that would arouse pupil enthusiasm.

Cornell's reputation and successes helped persuade the New York State Department of Education to commit to the curricular idea of nature study in its first official report in 1895, and by 1905 it provided a "stated" curriculum for public schools. Such imprimatur was largely rhetorical, since the state could not require a course of study in autonomous school systems; but by that time many city and country schools already offered nature study, so the guide was primarily intended to establish greater uniformity from school to school.[46] What was perhaps more significant and influential was ensuring that the nature study curriculum became standard in state-sponsored normal schools so that their graduates would know how to teach it.[47] Encouraged by public interest and attention from teachers who voluntarily took correspondence classes, Cornell faculty believed that nature study was "taking hold," especially as their invitations increased to give lectures across New York and beyond.[48] Thus, the president of the Chicago Women's Club, impressed by the New York project, invited Bailey to be featured speaker at an event that she hoped would "awaken a fierce desire in our community for a close study of a companionship with nature."[49]

Bailey was an entrepreneur, writing articles for popular periodicals and developing book series that built on public enthusiasm for botany and horticulture. This activity intersected with educational nature study, and he assured his publisher: "We are doing a good deal in this state to encourage nature-study, and I am satisfied that there will be enough demand in this state alone . . . to make the effort [to publish specialized books] worthwhile."[50] Arrangements with the Macmillan publishing house, through which he planned to edit a twelve-volume Rural Science series, a Garden-Craft series, botanical handbooks, and his *Cyclopedia of American Horticulture*, contributed to his private income, even as his connections with nature study teachers and advocates provided him with potential contributors and subscribers.[51]

Under Bailey's articulate leadership and Anna Comstock's management, Cornell's Nature-Study Bureau had produced eleven nature study leaflets within just two years, written by six authors who were not paid but produced them with a kind of missionary zeal to help teachers work more effectively on natural history topics.[52] Publication and distribution of the leaflets, however, were sponsored by special state appropriations; therefore, these leaflets officially circulated only in New York.[53] Bailey's correspondence suggests that he maintained close oversight of the details, reading page proofs and even signing off on the work of his colleagues.[54] As rural free postal delivery advanced in the 1890s, extension leaflets were distributed free to farm families and teachers, bridging the gap between rudimentary elementary schooling and more technical books and periodical literature in science, education, and agriculture. Publications were important, but providing personal examples of how to introduce nature study into classrooms and beyond proved particularly effective as Cornell faculty brought teachers to the campus each summer and used their picturesque countryside for fieldwork.

TEACHING TEACHERS AND REACHING PUPILS

In 1898 and 1900, Cornell offered a nature study summer school to about one hundred practicing teachers, with state appropriations paying for the tuition of those from New York State (fig. 15).[55] Comstock and Bailey, assisted by Isaac Roberts, John Comstock, James G. Needham, A. H. Wright, A. A. Allen, and George Embry, introduced the school teachers to nature study by taking them into the fields and laboratories to work directly with the "wonders" of the natural world.[56] Like the summer program at the Chicago Institute, Cornell's course attracted applications from across the country. Some were from established instructors like Lucia McCulloch of DeFuniak Springs, Florida, who explained, "I have had some training in science but it is as often a hindrance as a help. I am teaching nature study in a normal school and expect to take the same position next year but hope to be better satisfied with my work."[57] Faculty members were recruited from other institutions to assist, including Stanley M. Coulter of Purdue, who one summer taught an innovative course on nature in literature with Anna Comstock.[58] His experience at Cornell led Coulter to establish a short-lived leaflet series for Indiana that was "adapted to the use of teachers of schools in rural districts" in his area.[59] The Cornell faculty did not advertise widely in New York City or in other states after the turn of the century because the

FIGURE 15. Many Cornell faculty members in the College of Agriculture assisted in teaching the large summer nature study courses at the turn of the century. In the above illustration, Liberty Hyde Bailey is the dark haired man in the middle of the second row, and Ann Botsford Comstock is immediately behind him and to the right. Courtesy of the Division of Rare and Manuscript Collections, Cornell University Library.

legislature disapproved of expenditures not directed toward rural life in the state. However, the success of the summer school led the College of Agriculture to introduce a two-year course in nature study education as part of its regular academic program.[60]

In the early twentieth century, Anna Comstock and others spent more of their time visiting state normal schools, where they gave lectures and offered the weekend and summer institutes more easily attended by rural teachers.[61] Reflecting on her activities, she reported to a friend, expressing humor and fatigue, "I am doing extension work in the state teacher's institutes this fall and have already made my plea before six thousand teachers. I'll quit some day and stay at home and wash dishes and make beds and mend Harry's socks."[62] Her threat was an idle one, however, and she remained an active leader of nature study until her death. Bailey had assisted Comstock's professional life by helping her gain a faculty position; by hiring other outstanding women, Bailey also had made his college a leader in the emerging field of home economics. His willingness to bring women into the collegiate teaching program was not unopposed. His strong support may be credited in part to his awareness that most of the attending teachers in the programs would be women, but also to his own early encouragement by women.[63] Undoubtedly, it helped that women faculty at Cornell did not receive salaries as high as their male peers and were willing to do considerable amounts of routine work.[64] Putting nature study into the regular curriculum after 1899 also added significantly to the number of women taking Bailey's regular botany courses at Cornell.[65]

As the activities of the Cornell Nature-Study program continued to grow in response to demand from teachers, Bailey added staff. He originally hired fruit farmer John W. Spencer, a former legislator and Master of the New York State Grange, to develop outreach programs to farmers.[66] Spencer, however, asked to work with young people, arguing that this would give them an early start in building good farming skills. Bailey agreed, and Spencer quickly became "Uncle John" to literally thousands of students as he developed Junior Naturalists Clubs and engaged their members in correspondence.[67] Not unlike the Agassiz Association's network, but much more systematically maintained, the small local groups created membership lists and elected officers; in turn, John Spencer sent them a charter and badges for each member. By 1899 Spencer reported 135 clubs, in cities as well as in the countryside, with forty-five of them outside the state.[68] The former fruit farmer's correspondence with students soon required an assistant, and he turned to standardized letters to maintain contact with the extended network. The reported number of members varied during the first decade, but in 1906 Spencer reported around thirty thousand.[69] Uncle John dispensed down-to-earth advice, reminding students not to memorize or create lists of birds but to concentrate on some specific bird or plant and use it as a way to understand natural life cycles, the relationship between plants and animals, and the impact of humans on the environment.[70] Encouraging one young student correspondent, Jennie L. Olson, to focus her energies, Spencer drew an analogy between someone moving to a new town and an explorer in nature. Rather than try to read the city directory to learn about who lived in town, she should get to know her immediate neighbors, looking into the family affairs of townspeople and "gathering all the gossip you can about them."[71] The parallel was clear: the details of particular lives of creatures residing nearby would be more informative than memorized and dull lists. Spencer's "courage, resourcefulness, and untiring zeal" were "potent factors" in nature study's success in New York, according to Anna Comstock.[72]

Using homey language, Spencer encouraged parents to work closely with younger children. As he told a local Women's Alliance in 1905, "I often compare children to bicycles. Keep them busy and they remain right side up. Check their speed and they wobble. Stop their activities and they fall down."[73] Copies of his talks to teachers and students are full of aphorisms that bespoke country life and a democratic sensibility: "I would prefer to give one thing to a thousand children rather than a thousand things to one child."[74] He was extraordinarily active, writing to local newspapers, advocating for Junior Naturalist Clubs among administrators, as well as teachers, and continuously traveling to visit school children in his clubs and encourage school gardens. One woman from Coopers Plains was inspired by such a visit and followed the program he recommended, even writing to Spencer monthly and keeping her "eyes wide open to see the unusual things he said were free to us all."[75] He was generous in sharing resources, sending letters, leaflets, and specimens to New York City, to schools in Texas, and including the teachers' training schools for African Americans at Tuskegee Institute.[76] Some of the detailed responses of interested teachers and students who sent him letters and pictures are discussed in Ellen E. Doris's dissertation on nature study teachers.[77]

Alice G. McCloskey, a teacher from Saratoga County, was hired to assist Spencer and soon proved helpful on several projects. Under Spencer, she spent much of her time editing *Boys and Girls*, the Junior Naturalist monthly newsletters free to members, and providing other materials for their clubs.[78] She wrote nearly twenty six-to-ten-page leaflets on animals, plants, and rocks indigenous to New York State, using state funds, that went to Junior Naturalists and, later, to its successor Junior Citizenship Clubs. These rural clubs became common across the country, often using variants on the phrase "boys and girls clubs," and, while some were short-lived, those that persisted eventually merged under the 4-H club banner in the 1920s.[79] In the absence of a model school on the Cornell campus, McCloskey also supervised practice work in the schools of Ithaca, where advanced students spent time in the schools as part of their two-year special course on nature study.[80] One of several staff who contributed in a variety of ways, McCloskey worked as an assistant in Nature Study and in the Extension Department from 1899 to 1908, when she completed her A.B. at Cornell and moved to Philadelphia.[81]

The Cornell programs in agriculture were ambitious and overlapping, with staff often working together across divisional lines. Martha Van Rensselaer, a former Commissioner of Schools in Cattaraugus County, came to work directly with farm women ("farmers' wives"), after being recommended by Anna Comstock, her classmate from Chamberlain Institute days.

Van Renssalaer coordinated a Farmers' Wives Reading Course, and her initiatives became the basis for Cornell's pioneering College of Home Economics.[82] Mary Farrand Rogers (later Miller) helped with outreach by distributing resources on nature study through the farmers' institutes and by her lectures to teachers throughout the state.[83] She and another nature study teacher, Julia E. Rogers, worked with Comstock on the labor-intensive nature study correspondence course, subscribed to by over three thousand rural teachers, particularly serving those without the time or financial resources to attend courses at Cornell or the state normal schools.[84]

Although only teachers in New York State were allowed to take the correspondence classes, the staff distributed the leaflets generously while they remained in print. Press coverage of newly issued leaflets resulted in requests from "regions as remote as Minnesota, Kansas, and Kentucky," along with questions about how to produce and disseminate such materials.[85] Staff sent them to teachers from Nova Scotia to California in response to individual requests.[86] By 1904 there were enough materials with various titles — teachers' leaflets, home nature study lessons, and rural nature study leaflets — and calls for reprints of the earliest ones that Cornell reprinted and bound eighty into a durable volume of 606 pages titled *Cornell Nature-Study Leaflets*. The response was enthusiastic, including a letter from Edward F. Bigelow who wrote that the volume "literally FILLS A LONG-FELT WANT"; and the staff was particularly pleased when Melvil Dewey, Director of the State Library in Albany, requested five bound copies for the traveling libraries of the state.[87] Demand was so high that the Commissioner of Agriculture indicated that they had exhausted their supply and requested additional copies.[88]

Cornell staff continually stretched the outreach of nature study into unlikely locations, as when John Spencer visited orphanages throughout New York in an effort to enlist their children to join the Junior Naturalists Clubs, to encourage orphanage teachers to use a nature study curriculum, and to persuade administrators to devote part of what were often extensive residential grounds to school gardens.[89] Like his city counterparts, Bailey thought nature study could help calm troubled youth and therefore urged the founder of George Junior Republic, an innovative facility for juvenile delinquents in Freeville, New York, to incorporate the subject into his rehabilitation program.[90] Cornell arranged with the Chautauqua Institution in western New York State for Alice McCloskey to conduct three-week classes on nature study for a number of years, with her salary paid by Bailey and her expenses and assistant provided by the Chautauqua.[91] Summers proved particularly busy for staff, with county fairs, summer field days, and other exhibitions, in addition to the courses held on and off campus.[92] Instructors in 1898, for example, calculated that they participated in seventy-two institutes on nature study with 14,400 teachers, but concluded that this was more efficient than the intensive classes on campus.[93] Cornell's multilayered program served pupils, teachers, and parents throughout New York State and inspired faculty across the country to emulate its projects.[94]

By 1906 Bailey reported that the "results of our propagandistic work have been good . . . [and] the sentiment is now almost everywhere in favor of nature-study work in the schools."[95] He envisioned nature study as one important component in an array of country life activities where every individual counted, and education on every level could be coordinated with the agricultural college. In fact, Cornell's nature study program was at its high point, with 28,169 children registered in garden programs, 25,111 children enrolled in Junior Naturalist Clubs, and 1,782 teachers engaged in correspondence.[96] Cornell faculty, hoping to institute degrees in education at the university, proposed a model "country school" for the agricultural college, persuaded that the normal school administrators in urban areas did not understand the pervasive way nature study ought to be part of children's daily experience. In this school, children were to have a room adjoining their regular classroom for nature work—a site to bring specimens, build a museum, and hold flower shows, where they would take responsibility for maintaining tools and objects.[97] The model school, however, was never fully implemented and a comprehensive education program not created, but some Cornell faculty members would remain remarkably committed to the vision of educating children through direct contact with nature in and beyond the public schools.

Inevitably, as the program grew, so did the challenges. Bailey's *Idea of Nature-Study*, which was viewed as more "expositional" than scientific and experimental, was criticized by academic colleagues elsewhere.[98] Administrators, concerned with measuring results of expenditures, were not persuaded that a rural curriculum that allowed so much discretion to teachers matched the need for stronger grade standardization across rural and urban systems.[99] The topical leaflets seemed expensive and too idiosyncratic to Albany administrators like Edward S. Goodwin, who discontinued funding from the State Department of Agriculture. He was particularly unhappy that they had proven particularly popular among teachers in New York City, and so resources went to them rather than the rural audience for which they were intended.[100] A frustrated Bailey dropped the Junior Naturalist Club program and in 1907 initiated a somewhat different set of *Rural School*

Leaflets more narrowly aimed at country school teachers.[101] Anna Comstock diverted her energies into the *Home Nature-Study Course* for teachers and parents and the new *Nature-Study Review*.[102]

Despite such problems, Cornell continued its outreach to New York's rural teachers, pupils, and parents. Its program was an enviable model that influenced certain state normal schools, agricultural colleges, and experiment stations across the country.[103] Bailey watched with pleasure as midwestern states like Wisconsin, Nebraska, and Minnesota moved rapidly forward with rural educational programs in which nature study and sometimes agriculture were integrated into well-established normal school systems.[104] F. L. Stevens, a professor of botany and vegetable pathology at the North Carolina College of Agriculture and Mechanic Arts and a biologist at the Experiment Station, invited Bailey to talk to an assembly of teachers in Greensboro with the goal of developing a program. He explained, "We have a Nature Study Society and are attempting to organize in the rural district schools societies similar to those which you have in New York State."[105]

Cornell graduates also were in demand. Not untypical of the high expectations of those hiring instructors was a Virginia normal school president who sought someone to offer courses on nature study, as well as "supervise the work done in the Training School of Eight Grades and . . . prepare pupil teachers to take up the work in the Training School."[106] The principal of the Detroit Normal Training School sought a recommendation specifically for a woman in her late twenties, one with "considerable tact and poise," whose "influence would be good with young lady students" and who would lecture on nature study and geography as well as supervise a class of practicing students in the public schools.[107] A recommendation from Bailey and university certification were important as licensing became more stringent and specialized credentials were required.[108] A Philadelphia supervisor of recreational work for the Board of Education reported his satisfaction with a Cornell graduate, Miss Lewis, who had shown "executive ability and a large portion of common sense" in her school gardening work with children and teachers.[109]

The cadre of graduates who attended Cornell programs, and especially those who received bachelor's or master's degrees from Cornell and participated in the nature study programs, seemed to readily find positions in teachers training institutions, prestigious laboratory and practice schools, and colleges of education across the country where they spread the nature study curriculum. Cornell faculty continued to mentor former students, responding to requests for advice and sharing information, as when

G. A. Crosthwaite took a position at the normal school in Moscow, Idaho, and struggled to establish nature study classes based on quite different ecological conditions.[110] Sometimes former students came back for additional training, as did Joseph W. Hungate. He taught at Washington State Normal School at Cheney, returned to Cornell for an M.S. degree, and then went back to Washington, where he helped develop tracts of land near the school for use as bird refuges and outdoor "laboratories." By 1905 the normal school at Cheney had a full nature study curriculum, having hired John P. Munson to teach nature study and biological sciences.[111] Correspondence files reveal that Bailey helped other former pupils in various ways, lecturing at their institutions, advising about job prospects, and even loaning photographic plates to use in their publications and lectures.[112]

Bailey knew the students and worked closely with his staff, especially "Mrs. Professor Comstock," but his direct involvement declined as he turned his attention to other administrative tasks and to the national Country Life movement.[113] He was aware that university administrators were ambivalent about what role the education of teachers ought to play at Cornell and reluctant to respond to the increasing expectations about collegiate and advanced degrees in that field. Although nature study was struggling, the program initiated by Martha Van Rensselaer in a basement room of Morrill Hall was becoming a leader in home economics. Its success was aided by the fact that graduates did not need education degrees, their work could be directly connected to issues of poverty and rural life, and campus research was eligible for funds from the U.S. Department of Agriculture.[114]

When Theodore Roosevelt established a Country Life Commission, Liberty Hyde Bailey, by then the Dean of Agriculture at Cornell, was his obvious choice to chair it. The commission concentrated on the economic and social issues confronting farm families who were living in what was now an identifiable, if vaguely demarcated, space: the country.[115] The issues, however, were elusive and complex, and education was proving a slow solution that required a generation or more to bring about change.[116] Just what to do about country schools, identified as one of the problems but also offering potential solutions, was not clear. A special issue published by the American Academy of Political and Social Science on "Country Life" included an article that endorsed nature study, which provided "a substantial background for the whole scheme of farm education.[117] Country life advocates found it difficult to measure the impact of nature study after a decade of effort but found no better alternative.[118] In fact, during the first decade of the twentieth century, nature study had become a highly visible movement that spread through and beyond the United States and had attracted a wide range of advocates in rural areas from Maine to California.

DISSEMINATING RURAL NATURE STUDY

Lines of influence reached out from the Cornell program, particularly to state universities and normal schools, where the most dynamic nature study participants became regional catalysts, notably in the Midwest and West Coast. Ambitious faculty at land grant schools with mandates for agricultural education and extension work had watched the New York situation carefully. When Oswego's Charles Scott went on a lecture tour to normal schools beyond the Mississippi River in 1898, he reported to Bailey: "Western schools are way ahead of ours in nature study. But the work of your Bureau and of other agencies is telling. The interest is increasing rapidly."[119]

In fact, this direction had been anticipated by a short teacher's guide for nature study in Lincoln produced in 1893. Charles Bessey, like Bailey a graduate of Michigan State University, was building a botany program at the University of Nebraska, where his research addressed both fundamental questions and practical issues facing farmers on the prairie.[120] He readily accepted invitations to teach summer institute courses on nature study for teachers not only in his home state but also in neighboring ones.[121] Bessey appreciated that working with regional teachers was important and made one of his own goals that of preparing future college students to engage in "the best [scientific] instruction in Agriculture."[122] Responding to Bessey's commitment, the University of Nebraska introduced a nature study program at the end of the century. Bessey was identified as "one of the pioneers in the nature-study movement" based on his work with teachers, but he found it hard to match Cornell's resources.[123] While seeking funds for a permanent head of the Nebraska nature study program, Bessey exclaimed: "similar work under the auspices of Cornell University has yielded great results and . . . institutions in other states are moving towards the same end."[124] Bessey's colleague, professor of agriculture Thomas L. Lyon, invited Bailey to come and do "missionary work" with the Nebraska state legislature in hopes that he would use "some of that hypnotic power that drew $250,000 out of the New York assembly" for agriculture, including nature study.[125] The outreach work was intensive as state elementary school administrators asked Bessey to visit their schools and encourage their elementary teachers in nature study work; his correspondence reveals that he was generous in responding to queries about botany from individual teachers.[126] Bessey's enthusiasm for science and education were contagious, and one colleague pointed out that he had "worked up the Nebraska teachers from a botanical standpoint."[127] Nonetheless, Bessey grew somewhat skeptical, perhaps envious, of the popularity of the nature study movement and claimed, "I began such work in the public schools of Michigan long before the term 'Nature Study' was invented. . . . I find that what I did then was quite like the work which we have been trying to introduce into the public schools within the last few years."[128] His claim was legitimate because indeed there had been numerous predecessors to the more formally identified nature study curriculum.

Bessey was fortunate in recruiting Ruth Marshall as a graduate student in biology. After earning her Ph.D. in 1907, she stayed briefly to coordinate the nature study program at Nebraska. Her training insured genuine scientific expertise in the educational curriculum, a factor important to Bessey.[129] Although the university tended to teach high school teachers and leave elementary school training to normal schools, nature study was mentioned in its course offerings, often in conjunction with another subject like ornithology or agriculture, from 1901 to 1910.[130] Bessey encouraged this program because he was concerned about scientific rigor in the programs for elementary teachers being trained in normal schools and some of his graduates were likely to teach in those institutions. Over time Bessey took a less sanguine view of nature study teaching, offering a tart criticism of Bailey's popular book, *The Nature-Study Idea*.[131] His early encouragement, however, had led Nebraska to be among the early states to implement such study.

In California, personal connections and experience at Cornell positioned the faculty at the young Leland Stanford, Jr., University in Palo Alto to encourage nature study. Underpinning the nature study initiative was a long-standing relationship between David Starr Jordan, Stanford's president, and Cornell faculty, especially the Comstocks. Anna and John Comstock had entertained Jane and Leland Stanford in Ithaca. The Comstocks endorsed their friend and colleague for the presidential position.[132] They also encouraged entomologist Vernon Kellogg, who had studied and collaborated with John Comstock, to take a permanent position at the new institution.[133] When the Comstocks spent winter 1899 at Stanford, Anna joined Kellogg and Oliver P. Jenkins, a physiologist, in lecturing to nature study teachers in central California.[134] Well connected in Washington, Jordan and the Comstocks brought national leaders like Alfred True from the U.S. Department of Agriculture to "preach" nature study to students at Stanford and teachers in the Palo Alto area.[135]

Oliver P. Jenkins became a key collaborator of Kellogg in courses on nature study between 1899 and 1906. Having taught in a country school in Indiana, Jenkins found it easy to work with the California Teachers Association when it established a nature study section. His graduate work at Northwestern University perhaps also acquainted him with the educational initiatives at the University of Chicago. He worked in various ways with over a hundred Oakland area primary and grammar school teachers from the mid-1890s in a recommended but voluntary program for the teachers.[136] At Stanford he gave at least thirty-two lectures on "Nature Study in the Elementary Schools" during the 1898–99 academic year. Jenkins also invited teachers to participate in summer programs at the Hopkins Seaside Laboratory at Pacific Grove, then affiliated with Stanford.[137] Because Jenkins was also the director of the Hopkins Seaside Laboratory, he encouraged nature study teachers to participate alongside regular graduate and undergraduate students in its early years, an echo of Agassiz's school at Penikese. Bertha Chapman registered in 1897, and half the attendees were women during the first decade. Joseph Grinnell, later director of the Museum of Vertebrate Zoology at Berkeley, offered a class on general ornithology in summer 1903 that was "a practical course of especial value to teachers in Nature Study." After 1906, however, the renamed Marine Biological Laboratory was expressly directed at Stanford undergraduates and visiting researchers, echoing the exclusion (or segregation) teachers experienced at Woods Hole. Collaboration with Isabel McCracken of the East Oakland schools led to a large exhibit in Sacramento in 1899, and Jenkins frequently spoke at other association meetings about the purposes and techniques for nature study.[138] In subsequent years Jenkins and Kellogg wrote a regular column, "Department of Nature Study," in the *Western Journal of Education*, which provided practical tips for teachers. Some of these were combined into their book, *Lessons in Nature Study*, which identified natural history detail intended specifically for California teachers.[139] Several of the early women graduates of Stanford became nature study leaders, including Effie Belle McFadden, Helen Swett, and Bertha Chapman, as well as Charles Lincoln Edwards, who directed nature study in Los Angeles for over two decades (see appendix B).

California's enthusiasm for nature study was influenced by educational administrators who followed closely the reform activities elsewhere, noting the effective rural program at Cornell and the pedagogical methods at Chicago, to insure that the increasingly populated state kept up.[140] In the

1890s a special California Council on Education produced a series of studies to engage teachers in discussions about curriculum. The final report gave nature study a positive review but did not provide a specific program, leaving the matter to the normal schools and local school districts and opening the way for collegiate faculty members like Jenkins and Kellogg.[141] This undoubtedly encouraged the Los Angeles Normal School to introduce nature study methods in its training school in 1899, and in 1904 the school hired Loye H. Miller to teach. He offered a term of nature study that included "presentation of the pedagogical, or child-study basis for the subject, review of the great facts of animal and plant life which must be kept in mind in teaching, and a discussion of the course in actual operation in the Training School."[142] The dimensions of nature study in relationship to the urban schools of Southern California will be discussed later. The Cornell influence, however, was also directly visible across the northern border of the United States and spread to other Commonwealth countries in the early twentieth century.

INTEREST IN CANADA AND ABROAD

Canadian educators in Ontario, which bordered New York State across Lake Erie and the St. Lawrence Seaway, adopted aspects of nature study shortly after Cornell began its multifaceted program. Thanks to the generous philanthropy of Sir William Macdonald, the Ontario Agricultural College at Guelph established a "model" program in nature study.[143] After making a fortune in Canadian tobacco during the American Civil War, Macdonald turned to philanthropy and became a close friend of educational reformer James W. Robertson, who shared his understanding of the value of rural life and his zeal for making it more efficient. Hired to direct special projects, including a seed competition, Robertson drew explicitly on the Cornell model as a way to influence both experienced and aspiring teachers not only in Ontario but also further across Canada.

Eventually the practices were formalized at the Macdonald Institute at Guelph and later at the Macdonald College for teachers at McGill University in Quebec. Classroom horticulture, entomology, forestry, physics, and other subjects were balanced by hands-on experiences managing school gardens, identifying local woodland plants, and participating in open-air field trips each weekend.[144] The program, like the ones at Chicago and Cornell, were not without precedent and reflected earlier initiatives, including efforts in Nova Scotia to teach more natural science in the 1880s; and Macdonald was familiar with them.[145] But nature study gave a name and momentum to the diverse nature-oriented activities incorporated into teacher training programs in the five eastern provinces, with the formulation of theory and pedagogy centered at Guelph. Hoping to synthesize the best practices from the United States, Robertson selected representative teachers in each of the provinces and sent them on a study tour in 1902, starting with a course on nature study at the University of Chicago, then on to Cornell for topical courses, and finally to Teachers College at Columbia University for special training in the "new education."[146]

Canadian authors took some cues from their neighbors but produced a number of significant manuals and textbooks that emphasized regional insects, birds, and other topics particular to Canada's more northern climate. In 1902, for example, Mattie Rose Crawford, who taught in the Provincial Model School in Toronto, wrote one of the earliest and most thorough Canadian texts, *Guide to Nature-Study for the Use of Teachers*, in 1902.[147] A comprehensive introductory text, it started with the pedagogics of nature study, then gave short life histories of animals and provided lessons on familiar topics in nature study, all drawn from Canadian natural history. Her interests clearly echoed those of the Cornell program with its emphasis on creating an affinity for nature through careful observation.

The model educational museum at Guelph demonstrated to prospective teachers the value of teaching from objects, although much about the Ontario program took on an agricultural, even vocational, orientation.[148] That farm sensibility helped the nature study movement take root in the ambitious communities that populated western provinces, bringing farm families and with them teachers from eastern Canada.[149] Once normal schools were established in Saskatchewan, at Regina and Saskatoon, the provincial director of agriculture appointed men to teach agriculture, nature study, school gardening, and household science. The presumption was that most teacher training school graduates would at least start their careers in rural schools, and the normal school faculty would provide already practicing teachers with special institute courses on these topics.[150] A report in 1917 indicated that school gardens had become common and that most schools received shrubs and trees to beautify their school grounds. A report the following year found that 65 percent of the 1,600 schools in the province had school gardens and nearly 16 percent had active programs for home gardens. Many of the rural schools in Saskatchewan were open through the summer rather than during the rough prairie winter, and that schedule enabled active school gardening.[151]

The North American initiatives captured attention even abroad after the turn of the century, particularly in the English-speaking parts of the world. One route for the spread of nature study ideas was via school displays that appeared regularly at the international world's fairs leading up to the 1904 St. Louis Exhibition, which focused on education.[152] In Britain, two urban planners, Patrick Geddes and J. Arthur Thompson, late nineteenth-century leaders in regional development, became fervent advocates.[153] Geddes, for example, published six lectures as an *Introductory Course of Nature Study* and promoted nature study as part of urban development.[154] Thompson, on the faculty at Aberdeen University and a particularly outspoken proponent, declared that nature study revitalized the "old broad wholesome Natural History."[155] As a regional planner he was especially attracted by nature study's attention to location, and he cited American theorists in arguing that such study necessarily had to "vary its garb" in different places. Following Bailey, he maintained that it "matters not what the predominant subject is if it is congruent with the locality, if it not be too narrow, and if it not be dwelt on so persistently that the youthful mind becomes bored."[156] Geddes's and Thompson's support seems to have been particularly influential in Scotland, where nature study was a recommended curriculum in 1901.

Further south in Britain, a special London Nature Study Exposition in 1902 was the first nationwide conference on nature study.[157] Shortly thereafter British teachers of nature study organized a School Nature Study Union as a professional base for British teachers engaged in nature study and a place for discussion of this curriculum.[158] The Moseley Educational Commission sent Henry E. Armstrong to observe the movement in North America in 1904–5. *Nature-Study Review* summarized the comments he published and acknowledged his specific criticisms, even as English schools experimented with their own versions of nature study.[159] Support came from adult groups, like British field clubs, who were interested in the outdoor themes of such study; and British public (i.e., tuition-based) schools seemed particularly responsive to the movement. Macmillan in London published Ernest Stenhouse's *An Introduction to Nature Study*, and other British publications followed. Efforts were also made to launch a journal teachers engaged in nature study.[160]

The Irish Free State required nature study classes into the late 1920s, and its descriptive handbook noted that, whether called "Nature Study, Elementary Science and Nature Study, School Gardening, [or] Rural Science and School Gardening," the objective of all such programs was "the general development of the intellect and the character of the pupils [in order to] bring their education into close relationship with their lives and surroundings."[161] As enthusiasm spread throughout the current and former British colonies, a request for issues of the American *Nature-Study Review* arrived from as far away as Pretoria in South Africa from a woman who had studied in Leeds.[162] H. J. Bhabha, an educator from Mysore, India, came to Cornell to learn the methods of nature study and school gardens.[163] The nature study idea also took hold in Australia and New Zealand where it persisted until well after World War II—and long after the term nature study had fallen into disuse in most North American schools.[164] These two countries originally pursued the outdoor schemes developed at Cornell and its Guelph counterpart, making short-term forays into physiology and hygiene, and then settled into their own distinctive forms of the educational program.

NATURE STUDY AND AGRICULTURAL EDUCATION

From the outset, nature study education had an ambiguous relationship to the emerging vocational education movement, which translated into agricultural education in rural areas. Anna Botsford Comstock and her colleagues understood that "the Cornell University nature-study movement is primarily an agricultural movement" sponsored by the State Department of Agriculture and directed at children on farms and in small towns. But the Cornell staff simultaneously thought that elementary and grammar school pupils should be taught to observe and think independently as they were introduced to the natural world. Comstock resisted the idea of didactic instruction in practical agriculture, arguing that "nature-study is the alphabet[,] and words of one syllable" and must first be acquired before applying any basic knowledge.[165] Rural children already knew firsthand about farm work, whether on farms or in the small towns that served farmers, so she wanted nature study to provide a genuinely new perspective on environment, and she persistently resisted narrow vocational training.

Other leaders, like St. Paul naturalist and public school teacher Dietrich Lange, were less cautious and encouraged some relationship between nature study and practical applications for pupils who might not otherwise learn them. He wrote in the preface to his textbook, "The study of nature with a view to understand the relations of plant and animal life to the welfare and happiness of man, needs no justification in this age of scientific agriculture and applied sciences. All our most progressive teachers agree that Elementary Science or Nature Study should have a place on the programme of every graded and ungraded school in the land."[166] Lange then proposed that nature study was and should be explicitly linked to agriculture as well as technology, and his textbook listed sources for nature study materials, including the federal Department of Agriculture, Cornell's nature study program, and state agricultural experiment stations. His forceful observation was a theme that others developed more fully during a period when vocational education gained considerable public attention.

Political concern about rapid demographic changes and the steady drumbeat of agrarian agitation in the Grange and the Populist movements had spurred the Hatch Act of 1887. The act not only established state agricultural experiment stations but also mandated diffusion of research results and other useful information to the public. Schools and colleges were an evident mechanism, and, as its bureaucratic base grew in Washington, the Department of Agriculture began to provide instructional material and in-service training for teachers of agriculture.[167] The second Morrill Act in 1890 further focused agricultural colleges, sometimes to the detriment of their classical courses, and spurred state legislators with large numbers of agricultural constituents to grant additional funds.[168] In the 1890s Secretary of Agriculture Alfred C. True, a friend and former colleague of John Comstock, recognized the possible implications of nature study for rural schools. He published courses of study and bibliographies of nature study and also furnished teachers with seeds, plants, and even insects (those both helpful and injurious to plants) on request.[169] In 1897 True addressed the Grange with "A Plea for High School Courses in Agriculture," in which he suggested that elementary school children should appreciate and observe plant and animal life as preparation for more advanced study. Specifically, he argued that nature study gave "direction to the movement for the introduction of agriculture into secondary schools" by fostering basic knowledge and observational skills.[170]

Liberty Hyde Bailey, like Comstock, resisted the conflation of nature study and agriculture education, even as they both recognized that the "economic" or practical application of nature study could stem any criticism of nature study by rural parents and politicians.[171] Nature study had a distinct observational approach that drew on the specific aptitudes of younger pupils, and even at more advanced levels Bailey wanted "to teach the subjects and the objects themselves," using the field and the laboratory.[172] Thus, he argued for "agricultural nature study" rather than "nature study agriculture."[173]

But the vocational trend was powerful and attractive to colleagues in other agricultural colleges. One correspondent in North Carolina acknowledged Bailey's distinction but reminded him that indeed nature study "opens the way to agriculture in the schools, by awakening interest and quickening observation, and creating a love for all out-doors, but it is not agriculture."[174] Another argued that nature study "does for agriculture what Manual Training does for technical and industrial education. It furnishes a wide basis of general intelligence and ability from which to specialize towards particular occupations."[175] A few faculty members were skeptical about these claims for nature study's relevance for agriculture. A textbook by North Dakotan Charles A. Schmidt derided a "sentimental and essentially urban style of nature study," advocating instead that a very practical nature study be presented to his presumed audience of farm children who had "actual contact" with nature.[176]

In 1905 Alice McCloskey felt she needed her supervisor, Liberty Hyde Bailey, to be clear that "our future Nature-Study will be strongly Agricultural Nature-Study."[177] The shift from Cornell's *Teachers' Leaflets* to *Rural School Leaflets* had been compelled by political circumstances and marked a deliberate decision to address rural farm needs.[178] As dean of the Agricultural College, Bailey was well aware that in Europe, especially Belgium and France, as well as in western and southern states and Canada, school courses in agriculture were successful and expanding.[179] Moreover, he was ready to establish constructive links with the Bureau of Education in Washington, which was growing and would eventually be better equipped to address some issues in early education than the Department of Agriculture.[180] As a result, the compromise for Bailey and his nature study colleagues was to try to keep agriculture distinct from grammar school nature study while acknowledging that advanced agricultural instruction had a definite place in high schools. Undergirding the latter movement was direct federal support through the Smith-Lever Act of 1914 and the Smith-Hughes Act of 1917, which provided subsidies for agricultural education and for extension work.[181]

The apprehension that nature study might be subordinated to vocational work was vindicated by events in schools established in the South to teach agricultural and industrial arts to African Americans. Hampton Normal and Agricultural Institute in Virginia and Tuskegee Institute in Alabama, both of which had support from northern philanthropists, included teacher training programs and practice schools.[182] In general, southern public schools lagged well behind those in the rest of the country, and the segregated schools that existed for African American pupils were even worse, a situation exacerbated by the lack of teacher training.[183] The exceptions were

FIGURE 16. African American pupils were introduced to school gardening at a very young age at the Whittier School of the Hampton Institute in Virginia in a system that emphasized practical training over the aesthetics pursued in many nature study programs. Courtesy of Hampton University Archives.

Tuskegee and Hampton, and a program at the latter was expanded to help offset a shortage of teachers on American Indian reservations. In 1894 the administrators reported that Hampton had 294 students taking classes in the normal school, 89 pupils in its practice school, and 170 pupils in its Whittier Training School for African Americans (fig. 16).[184]

Major northern philanthropists, including John D. Rockefeller and George Peabody, were particularly interested in teacher training programs and sent experts like James Earl Russell of Teachers College, Columbia, to work with the staff of Hampton and Tuskegee.[185] Both schools introduced nature study, and Hampton began publishing its own monthly series of nature study pamphlets designed for teachers in southern states; it maintained its program for several years. The early pamphlets were part of a "Nature Study Library" with topics like seed planting, the life history of a butterfly, and beautifying schoolhouses and yards; and a later series paid much more attention to health and nutrition.[186] George Washington Carver at Tuskegee also initiated a pamphlet series with his "Suggestions for Progressive and Correlative Nature Study."[187] John Spencer sent copies of Cornell's leaflets to Tuskegee and sponsored correspondence exchange between pupils in its training school and those in schools in New York State.[188] Both Hampton and Tuskegee sent students north to learn nature study techniques and agricultural education, and two Cornell graduates were involved in establishing regional teachers groups interested in nature study.[189]

Given the serious financial problems of rural and segregated schools in the South, however, it appears that nature study education was not widely or consistently implemented for white or black pupils, although the pamphlets may have been welcomed by teachers who often had no textbooks for their classrooms.[190] In 1915 the U.S. Department of Agriculture produced a sixty-four-page pamphlet, *Exercises with Plants and Animals for Southern Rural Schools*, which read like many of the commercial nature study textbooks and presented topics by the month. It was available free to those who requested it, but there is no evidence of how widely it was advertised or distributed.[191]

Initially established to educate African Americans, Hampton and Tuskegee had emphasized industrial and agricultural classes under the direction of highly directive philanthropists. In the early twentieth century, educational administrators continued to concentrate on "vocational training, guidance, and measurement . . . [rather] than social amelioration," and they expanded training for teachers by establishing practice schools.[192] Moreover, these schools tended to have pupils doing farm and domestic work that contributed to the maintenance of the school.[193] The decision was conscious. An instructor in agriculture at Hampton observed that educators friendly to nature study tended to "oppose instruction in agriculture, claiming that the teaching of 'practical things' bores the pupils and defeats the objects of nature teaching." Nonetheless, Charles L. Goodrich took up the subject that "is widely engrossing the attention of the teaching worlds" and sought to integrate it into his agricultural instruction.[194] This emphasis on agricultural productivity left little time for the kinds of outdoor and exploratory projects enjoyed by primarily white students in the schools of upstate New York and elsewhere. In fact, Hampton faculty discouraged George Peabody from supporting biological nature study at Spellman College, suggesting agriculture would be better.[195]

The situation was similar at schools dedicated to education for American

Indians. Brenda Childs pointed out that such education was not sufficiently vocational to prepare graduates for the better paying jobs which many parents hoped their children would obtain after attending boarding and summer outing schools.[196] Over the first two decades of the twentieth century, the Department of Agriculture and the Bureau of Education, still subsumed under the Department of the Interior, vied with each other over authority for agricultural education and nature study, both gathering data and publishing nationwide data on pupils, schools, and teachers.[197] The Bureau of Education was assigned responsibility for the residential and day schools for American Indians, especially the highly visible programs at Hampton and at Carlisle in Pennsylvania, as well as other schools in the prairie states and Southwest. Here, too, the practical aspects of nature study dominated, with little of the aesthetic sensibility found in schools in middle-class neighborhoods or even progressive inner-city ones.[198]

Estelle Reel, Superintendent of Indian Schools, provided a detailed set of mandates with a centralized organizational structure that reached into the boarding schools in Hampton; Pine Ridge, South Dakota; Tomah, Wisconsin; and Newport, Oregon, as well as the day schools on and near reservations. Her office produced an official "Course of Study for Indian Schools" that specified nature study but actually offered only modest agricultural exercises, with none of the outlook encouraged by Bailey and those who sought to put children "in sympathy with nature."[199] She reminded agents, superintendents, and teachers in a circular in spring 1902: "The work in nature study as laid down in the course of study should be prosecuted vigorously as this is the beginning of the more advantageous season of the work."[200] Nothing in the circulars suggested study of the plants themselves. American Indian students' gardens were to provide food for the school and establish character: "Garden work, properly directed, promotes industry, attention, judgment, skill and self reliance."[201] These boarding and reservation schools held up a distorted mirror of the more standard nature study curriculum as they minimized children's self-expression and enlarged the didactic and patronizing outlook found throughout their curriculum.[202] There were exceptions where education in natural sciences was comparable to that in neighboring white schools, and one of these seems to have been the Santee Normal Training School in Nebraska, administered by the American Missionary Association and perhaps influenced by the active nature study program in that state. Here the teachers were introduced to Bailey's nature study ideas and encouraged to study natural objects growing without cultivation in their area.[203]

RURAL NATURE STUDY

Where rural and small town nature study was implemented and sustained, it drew its support from politicians, parents, local voluntary societies, and teachers. Many nature study founders had grown up in farm families and been attentive to the plants and animals in their environment. Most shared the assumption of Francis Parker that "every child is born a naturalist."[204] Cornell University, with considerable state support, pioneered in developing new techniques for reaching its broad rural public. In building its program between 1896 and 1905, Cornell faculty had depended heavily on material and promotional support from the state legislature and the Department of Agriculture and on expanding federal initiatives directed at rural life. Rural organizations like the Grange and Farm Bureau had largely concentrated on economic issues, but both had advocated better training and salaries for teachers to insure that their children received up-to-date schooling.[205] The result was that, as Wayne Fuller put it, educational administrators "grafted nature study and agriculture onto the courses of study used in rural schools throughout the Middle West with such fervor that many of them must truly have believed that nature study would keep country children on the farm."[206] There was ambivalence about whether nature study should be tightly linked to agriculture, and the intensity of the connections varied widely even as rural high schools adopted such vocational work.[207]

Teachers were, of course, essential in establishing nature study in practice. Many, typically from region where they taught, seemed to welcome the opportunity to teach subject matter, as one put it, "very near to my heart."[208] They responded to the technique of dealing with familiar objects in a new way and found it fruitful. One of John Spencer's correspondents pointed out that an effective nature study teacher might be "overjoyed to see her pupils devoted to such a lowly plant as an onion. . . . She would begin at once a study of the lily family and make them familiar with more aristocratic members, lily of the valley, hyacinth, day lily, tulip."[209] Various indicators suggested that the Cornell faculty members were effective in their educational program, including the fully enrolled classes, the extensive correspondence that followed John Spencer's institute lectures, and the distributions of leaflets during nature study's first decade.

Bailey's *Nature-Study Idea* was a visionary statement, not a handbook or course outline. But by the time his book was widely distributed, a set of textbooks, teaching manuals, curriculum materials, and programs were in place to implement his child-centered outlook and to guide teachers into working with their students to explore the intricacies of plant and animal life. In Santa Ana, California, for example, the *Manual for Public Schools* stated reassuringly, "Teachers seem to believe a scientific education necessary to teach Nature Study, while it is a fact that any intelligent teacher can easily prepare an abundance of material and present it effectively. For the purpose is not simply to teach the pupil facts but to teach him to observe the wonders of the great world in which his eyes open."[210] The economic circumstances facing the country and the evident problems of rural populations might have dictated a highly practical program, but many of the leaders of nature study recognized the importance of raising spirits as well as increasing talents. Bailey, Comstock, and their colleagues posed a wide-eyed approach to nature study that was distinctive, influential, and positioned for further consideration among other approaches to the popular but sometimes controversial curriculum.

CHAPTER FIVE

Deliberating Theory, Texts, and Topics

Nature-study as a thing in itself has had its ups and downs. . . . Today it spreads its tents and unfurls its banners; tomorrow nothing remains but the litter and muss of encampment. One superintendent drags it in and another drags it out. . . . Some teachers love it, some hate it, and others preserve an armed neutrality. . . . Neither the successes [n]or the failure of the great nature-study movement depends upon it having a special and permanent place upon the daily programme. . . . Books and apparatus, sand tables and museums, aquariums and window boxes, excursions and experimental farming, this is the path and course of events along which and through which the nature-study spirit has led. . . . It has converted field, forest and stream into a laboratory, a demonstration room. It has made the school a part of the life that is.

Francis G. Blair[1]

AS FRANCIS BLAIR'S candid assessment in 1908 acknowledged, by measures of stability, definition, and consistency, nature study was struggling. His conclusion, however, embraced the diversity and ambiguity of nature study programs by suggesting their sometimes unstable dynamics were part of the inevitable reality of "the life that is." Although Wilbur Jackman had confidently outlined a structured and detailed program a decade earlier, as his core idea was rapidly adopted, his insistence on local resources meant that nature study required innovative implementation in diverse environments and communities across North America. A basic premise of nature study was that its methods were fundamentally different from the older reliance on rote memorization, but the prescription for what should be substituted was less clear. As one school superintendent fretted, "If I should insist upon it, schedule it, it wouldn't be 'Nature Study,'—you know what I mean, the informal heartfelt sort,—it would be a perfunctory, cut-and-dried task or a bit of elementary or dilute science. If I don't insist upon it, don't allot a certain time for it, I don't get anything."[2] His frustration expressed a common concern about how to insure that this new curriculum was implemented regularly and with spontaneity.

Froebel's injunction "Lead the child out into the world of nature—it is his native air" proved deceptively simple. In effect, Froebel invited teachers to bring their personal educational experiences and theoretical training to the study of the natural environment in and around their schools. The challenge was to frame the curriculum so that its learning outcome added both skills and information relevant to their pupils' lives. In his preface to a widely used manual for teachers, Francis Parker identified the paradox of equipping teachers with the requisite knowledge about the natural world while asking them to avoid didactic and rote presentations of what they already knew. Teachers should, he insisted, use a textbook only to go well beyond its content.[3] Confident that teachers were capable and well prepared by normal school training, Parker encouraged them to exercise considerable discretion to explore the multiple possibilities offered by field trips, correlation of school subjects, classroom projects, and nature literature.[4]

Educational reformers generally shared that optimism about the potential of teachers, and they sought to equip them with educational philosophy and child psychology on which to build their classroom practice. Nature study advocates adopted four significant, sometimes overlapping educational approaches, typically articulated in textbooks written by faculty who trained teachers to present nature study in public classrooms. The initial foundations for nature study under Jackman and his Chicago colleagues had built on their own scientific interests with somewhat casual connection to the well-established and child-centered pedagogy associated with European educational philosophers Pestalozzi and Froebel. These ideas proved compatible with a second, distinct theoretical line of argument advanced by professors at Illinois Normal in the 1890s, particularly the McMurry brothers, Charles and Frank, who had studied the educational philosophy of Johann Friedrich Herbert at the Pedagogical Seminar of William Rein at Jena. A third approach, represented by Clifton Hodge, was more empirical and more pragmatic. Hodge had been influenced by the research of G. Stanley Hall on age-related learning, and he actively collaborated with public school teachers in overcrowded urban schools in the industrial slums of

Worcester to provide experiences appropriate to their school setting. The educational challenges of industrial cities offered a significant contrast to rural and small town communities where teachers taught in the isolated one- and two-room schools. The fourth outlook, articulated by Liberty Hyde Bailey and Anna Botsford Comstock, assumed pupils in the country were in some sense adapted to their outdoor world. But their familiarity with the particulars of growing crops had not opened them to aesthetic or scientific ways of understanding the domesticated and wilder landscapes in which they lived; they needed nature study that was simultaneously creative and systematic. Exponents of these philosophical approaches found in the new curriculum, often for quite different reasons, rich potential for teaching children through and about nature. In turn, nature study became a vehicle to demonstrate and extend their particular pedagogical outlooks. Educational leaders presented their theories and prescribed applications through widely distributed published materials that could be taught in normal schools and used by practicing teachers.

NATURE STUDY AND BOOK CULTURE

Progressive-era reformers typically rejected teaching by rote memorization and recitation, which they considered to be ineffective. McGuffey readers, which had sold fifty million copies between 1870 and 1890 and dominated the school book trade, were being displaced, but not by a single, comparable basic reader.[5] Instead, a proliferation of textbooks not only for reading and arithmetic but also for writing, geography, spelling, nature study, and other topics vied for school board authorization and thus sales.[6] With an emphasis on "nature, not books," advocates of nature study were concerned that "ready-made lessons" and even detailed examples might have "superfluous or mischievous" results.[7] Nonetheless, it was important to give sufficient advice to elementary teachers to enable them to introduce this new subject; and manuals, readers, and other publications flooded the educational market in the latter decades of the century.

The entering wedges for introducing the natural sciences into public schools, Kim Tolley has argued, were geography textbooks.[8] The subject had been growing in popularity under the skillful hand of authors like Jane Andrews, who described the lives and comparative natural environments of "seven little sisters" from around the globe.[9] Although her books were meant for a young readership, nature study educators saw a connection of their subject to this "mother of all sciences" that was then being introduced into secondary schools.[10] Geography was an integral part of the original course Wilbur Jackman outlined, connected specifically to geology and local sites as part of the nature study curriculum.[11] Coming from a somewhat different perspective, Charles Scott suggested that the German model of geography, with its links to demography, urbanization, and economics, offered a particularly effective introduction to nature study for young people.[12] Frank Owen Payne, who had introduced a program much like Jackman's into the Massachusetts schools, shared a similar outlook, introducing physical geography and then occupations, transportation, and ethnic diversity to "stimulate a desire for more knowledge and broader views of the world about us."[13] Academic geography was going through a period of reconfiguration, reflected in professional realignments and in secondary school curriculum. The nature study advocates who sought to include geography often found it difficult to determine what aspects of the redefining field could be best integrated into the grammar schools.[14]

The nature study curriculum spread as commercial publishers fed the broadening market for textbooks with a proliferation of materials. Historians agree that textbook publishing became wildly competitive in the late nineteenth century, with large presses absorbing smaller ones in an attempt to capitalize on the demand for public school materials. To try to control the competitive field of textbook publishing, in 1890 a number of leading publishers joined in a trust, the American Book Company (ABC).[15] For a brief time ABC had 90 percent of the pre–high school book trade, until one of the leading participants, Appleton Press, withdrew in 1895 and thus undermined the monopoly.[16] The ABC published for the nature study market, but no single publisher ever dominated it. Payne's book, *Geographical Nature Studies*, was reasonably popular, but few teachers adopted Horace H. Cummings's unimaginative set of outlines for primary grades (1–3), elementary grades (4–5), and grammar grades (6–8).[17] Another ABC text by Marion H. Carter, *Nature Study with Common Things: An Elementary Laboratory Manual*, was intended to train the "power of observation."[18] Carter's common things were indeed common, foods found in home gardens and local shops such as cabbages and berries. Each item was presented with line drawings, a series of questions, and information that students could consult as they actually worked with an object for the day.[19] These somewhat uninspired productions meant that ABC never enjoyed the sustained success of older established publishers and a few niche publishers who had identified authors who wrote well and effectively for the educational audience.

The turbulent marketplace was disconcerting to authors and perhaps confusing to consumers. John M. Coulter wrote to his colleague David Starr Jordan that he was concerned about his copyrights because colleagues had lost their contracts when the ABC had absorbed books managed by Harper.[20] Appleton Press struggled when it first became independent again, but in 1903 it got the contract to produce the required *Course of Study for the New York City Schools*.[21] This competitive but open market may, in fact, have made it easier for nature study authors to publish, as competing companies scrambled to stay current with the most recent trends in the highly lucrative text and education market.[22] The larger companies used itinerant book salesmen and other means to promote their textbooks, readers, specialized notebooks, and reference manuals to school supervisors and teachers.[23] They were eager to capitalize on new subjects and encouraged books by university faculty members, normal school instructors, and anyone whose topic might have a readership. Few publishers made fine distinctions among elementary science, nature study, and popular nature writing for children and adults as they sent out book agents. The deluge of books led John M. Coulter to grumble, even as he published yet another textbook himself, that "books about nature study have become numerous, perhaps more numerous than good teachers of it."[24]

Normal school nature study courses and summer institutes aroused considerable interest in books appropriate for the new school subject. Liberty Hyde Bailey worked primarily with Macmillan Company, which also took over the publication of Jackman's *Nature Study and the Child*. Macmillan produced several widely used texts, including Dietrich Lange's *Handbook of Nature Study* (1898), Lucy Langdon Williams Wilson's *Nature Study in Elementary Schools: A Manual for Teachers* (1897), Lida Brown McMurry's *Nature Study Lessons for Primary Grades* (1905), and Bailey's *Outlook to Nature* (1905). Lange, perhaps because he was an active teacher in the St. Paul schools, produced a particularly popular early text that evinced an eye for local detail. More typically, Macmillan's authors were normal school faculty or supervisory teachers in practice schools, where their courses addressed more abstractly the values of nature study and wrestled with questions about the best pedagogical methods.

The enthusiasm for nature study also stimulated secondary markets. Publishers experimented with leaflets, pamphlets, illustrations, and hanging wall charts, while leaving it to the teachers (or special nature study supervisors) to find natural objects. Bardeen, an education publisher in Syracuse, New York, produced *Flora*, a botanical game for matching flowers that shared common family names.[25] The apparently short-lived Nature-Study Company in Chicago offered a line of animal illustrations for posting in classrooms.[26] Various entrepreneurs produced what have become rare ephemera, including sketch pads, guides with questions to be answered while on nature walks, and field handbooks to provide descriptions and identifications.[27] As noted earlier, some states distributed free leaflets, but such materials typically had short-term sponsorships; even Cornell's original leaflet series lasted less than a decade. Demand for free materials was high. Bertha Chapman Cady, teaching at Chico Normal School in 1919, had received over eight thousand orders from across California for her nature study bulletin before the school's president canceled the program.[28] Publishers found a waiting readership for useful course materials.

University scientists added to their annual salaries by publishing books that could find a reasonable balance between technical and familiar information. They contributed to the steady stream of books and pamphlets that provided up-to-date references on plants, animals, minerals, geography, and other subjects relevant for teaching. Like publications for adult amateurs, many resorted to an encyclopedic approach for identifying specimens and tried to encompass broad geographical areas. John M. Coulter produced several books on botany, David Starr Jordan wrote on fishes and other animals, and Anna (sometimes with John) Comstock wrote books on insects and butterflies, working to find just the right level of detail useful to teachers with limited biological background.[29] Academic authors hoped that, particularly at more advanced levels, "biological nature study" would be based on observations that could be incorporated into field and laboratory analysis done in high school and college.[30] A few of these faculty members wrote personal reminiscences of their interest in nature, perhaps inspired by the success of Burroughs and others, but these retrospective accounts reveal surprisingly little about their often substantial contributions made to primary and secondary education.[31]

Many teachers who taught nature study relied on systemwide "courses of study" that were developed specifically to use their locale and to keep graded classes moving along parallel lines. These became common after the turn of the century, particularly in larger urban systems, but some states also developed them for rural schools. Those who were conscious of the problems with previous educational practices were careful to note that this material "does not aim to be a cut-and-dried syllabus but rather a suggestive outline from which the teacher may select what is best suited to the conditions to be met in a particular school."[32] Only the most ambitious of the relatively poorly paid teachers subscribed to the various journals on primary education, but those who did found that there were nearly three times as many articles on nature study as any other topic in the last half of the 1890s.[33] Textbook and professional literature followed along distinctive lines, enlarging on or sometimes challenging the perspectives being advanced by leading educational theorists, who hoped to demonstrate the intellectual grounding of nature study.

CENTERING THE CHILD

Reforming educators emphasized the importance of activating each child's intrinsic powers of observation and reason, capacities that they felt were stifled in a traditional classroom that required memorization of knowledge provided by others. Parker wrote "If a pupil be permitted to carefully examine an object or a set of conditions, and then be required to interpret what he sees, he is from that moment ever after stronger than before. By that act, no matter how trivial, he begins the great work of self-emancipation from the rule of chance in so far as his interpretation has taught him how the forces around him may be resisted, guided, and controlled."[34] Trained naturalists like Henry Straight and Wilbur Jackman responded to Parker's call to the study of nature as a way of advancing children's capacities and endorsed the discipline as a way to encourage more pupils to pursue science well beyond the elementary grades. Jackman remarked in the preface to his *Nature Study for the Common Schools* that science teaching had been "gradually working itself downward from the college and high schools to the common grades" but without any real guidance about how to teach "beginners."[35] Nature study offered the final, albeit elementary, level that would allow for a comprehensive incorporation of natural sciences into education.

Jackman envisioned nature study as a site where all the sciences could be linked to and through life experiences. He was particularly interested in a child-centered approach that stressed the importance of self-expression, using writing, drawing, modeling, and oral presentations; for him, objects were only a starting point. His first text was organized by month, starting with the school year in September, and suggested specific topics and materials in zoology, botany, meteorology, astronomy, geography, geology, and mineralogy.[36] Lessons were to begin with an observation about a specimen from nature objects directly available to the pupils. The physical science lessons might require materials supplied by the teacher, such as common salts that crystallize, but these objects were also to be handled by the pupils.

Jackman produced a steady stream of articles and books full of suggestions about how to enliven classes, giving examples of hands-on activities that the newly minted teachers might adapt once they entered practice and laboratory schools. Like anthropologists and others at the end of the century, Jackman identified childhood as a distinctive period in which individuals recapitulated earlier epochs of civilization. Children, like primitive peoples, learned primarily by experience and through objects in their environment. Nature study, he argued, was particularly suited for providing familiar objects about which pupils could develop self-awareness and socially integrative skills. By using their imagination and testing their assumptions, students would "repeat, in a measure, the experience of the race."[37] Jackman realized that while most pupils were not future scientists, they would benefit in other ways from nature study methods. His text elaborated on techniques to help children make and record observations, noting wryly that their responses might be expected to differ since "uniformity could not be expected from a trained body of scientific observers either."[38]

Jackman extended his rationale for nature study as he encountered other educational theories. Through his colleagues at the Illinois Normal (located in Normal, Illinois), he became familiar with current German ideas on child psychology that reinforced his integrative approach. His *Nature Study and Related Subjects for the Common Schools* (1896) incorporated discussion of correlated studies, where two or more subjects were taught in an integrated way, and more commentary on educational philosophy. Having taught in front of a classroom, he recognized that too often teachers needed to repeat directions or have pupils help each other, so he published written materials to be used by pupils as well.[39] The Chicago area enjoyed dramatic seasonal change, and Jackman organized his texts to take advantage of materials locally available from September through June. He recommended additional books, ranging from specialized texts to more imaginative works like Arabella Buckley's *Fairyland of Science*. His frequently reprinted manuals served as initial reference points, but other authors freely adapted the malleable concept of nature study in their publications. Indeed, twenty years after he had introduced nature study, a commentary on nature study books concluded that Jackman's texts were "out of line with the present interpretation of nature study."[40]

Charles Scott, who had written up nature study for the report of the national Committee of Ten, produced *One Hundred Lessons in Nature Study Around My School* (1895) and *Nature Study and the Child* (1900).[41] His textbooks were deliberately distinctive from those of Jackman, with a topical organization that relied on a relatively limited set of familiar objects. *Nature Study and the Child* opened with an extended discussion of the dandelion. Each pupil was to pick a dandelion to study carefully. This common plant would allow children, as they examined its structure and compared it to other flowering plants, to learn the rudimentary parts, including root, stem, leaf shape, as well as the stamen and pistil of the flower. But that discussion was only the beginning of thinking about the way that the often dismissed plant actually invaded yards and self-seeded in unwanted places, tracing the geography of plant distribution. Teachers were to ask about usage, encouraging pupils to describe how leaves could provide a healthy, inexpensive vegetable to meals and mentioning that the plant had medicinal properties. If that range of related topics was not enough, Scott pointed out that not all people viewed the dandelion negatively, and he provided stanzas from poets like James Russell Lowell, who described the yellow flower's bright, engaging properties. Thus, pupils were to look closely at this denigrated weed in terms of its structure, relationship with other plants, value to humans, and even its aesthetic qualities. The goal was to stimulate the imagination and encourage the open-ended and investigatory *spirit* of the nature study movement.[42] Drawing on educational theories that emphasized sense perception, Scott encouraged modeling with clay and sketching with various media so that that pupil's "expressive" qualities were drawn out as they manually articulated form, size, and descriptive characteristics like color.[43] Ironically, Scott's manuals and texts had few or no illustrations. For him as for Jackman, the generalized theories of Pestalozzi and Froebel, which advocated a focus on the capacities of children for learning rather than on content, were fundamental.

Physical sciences were in the purview of nature study advocates, evident with Jackman's early textbook, but most publications and essays tended to emphasize the natural sciences, especially botany, zoology, and geology. The *Nature-Study Review* would include physical science topics only occasionally.[44] There are multiple explanations. Kim Tolley notes that at the turn of the century women had gained a stronger foothold in the biological sciences. Thus, women educators were most comfortable teaching about botany and zoology and were, in limited ways, more accepted in those fields.[45] Orra Underhill comments that in the 1890s committees reviewing curriculum in the schools believed that young children responded better to zoology and botany.[46] Some contemporaries suggested that advocates for science education in the natural sciences had been "more zealous" than those in physics and chemistry in encouraging and supporting teachers in the lower grades.[47] Certainly most nature study textbooks catered to teachers who were interested in botanical, zoological, and meteorological topics rather than those wrestling with less familiar concepts in physics, chemistry, or even geology. Since living nature was both available and engaging, it fit a child-centered pedagogy that focused on using objects both familiar to and of interest to children.

CORRELATING SUBJECTS IN THE CLASSROOM

Recognizing that the curriculum could not simply be driven by pupils' established interests, educators were drawn to the ideas of Friedrich Herbart, which advocated active correlation of subjects within an integrated curriculum. Thus children interested in nature, for example, could be encouraged to connect it to literature, art, geography, and other subjects, and vice versa. A number of American scholars, several of whom had taken advanced degrees in Germany, helped introduce Herbart's philosophy into normal school training. Particularly prominent was Illinois Normal faculty member Charles A. McMurry, whose research was strongly endorsed by his school president, John W. Cook. Proud of his European-trained faculty, Cook enthused, "this work is comparatively new in the west and has already enlisted the warm interests of many prominent men."[48] McMurry, with his background in educational psychology and philosophy, understood that nature study provided a particularly appropriate mechanism for working out Herbartian principles in classrooms. McMurry spent two years in Germany, earned a Ph.D. at Halle, and subsequently joined the famous Pedagogical Seminar of William Rein at Jena, where "thirty students from a dozen countries discussed actual classroom practice."[49] On his return to Illinois he joined with his brother Frank and several prominent educators such as John Dewey at Chicago to form a National Herbart Society, which would later become the National Society for the Scientific Study of Education.[50] Charles McMurry served as secretary of the society, actively engaging in theoretical discussions, writing textbooks, and training prospective teachers at Illinois Normal in ways that helped make it a leader among Midwestern normal schools.[51]

Charles McMurry expressed Herbartian educational goals clearly in his introduction to a nature study text by his sister-in-law Lida Brown McMurry. In a nature study lesson, he wrote, "The teacher's questions and suggestions are designed to throw the children constantly upon their own power to observe, to experiment, to find out." The goal was to stimulate self-activity and thoughtfulness that would lay the "foundation for an alert and self-reliant mind." The teacher's role was to push pupils to observe more closely, inquire more definitely, and be inquisitive about things that they had not yet considered. Lida McMurry's textbook offered a concrete example with evocative questions on specific plants and animals:

> The lesson on the apple, for example, has students observing the fruit on a tree, picking it and cutting it in half, observing the parts of the apple, and questioning how the color attracts birds and humans, how the tastes of different apples appeal to different consumers, how the seeds are in a protective chamber, and how they are distributed. The teacher then probes: "Did you ever eat a core? How do you like it? What trouble does eating the core sometimes give you? (The scales of the cells stick in the throat.) What is the use of these scales? Let us examine these apple cores. (The scales make room for the seeds.) How many scales does it take to make one room? How are the scales fitted together? How many seeds are inside a room? (The children will find that the number differs. . . .).[52]

Open-ended questions might anticipate answers, but they were not narrow or fixed.

Charles McMurry also engaged in debate with the more idealistic Hegelian W. T. Harris at the National Educational Association and other professional meetings in the 1890s. Charles was part of the group who, according to yet another Herbartian, emphasized that the "center and starting-point of early education is not the grown man, but the immature child—his apprehension, his actual and potential interests, the limited objectives of his most rapid mental development, the correlation of his studies so that each reinforces and expands every other, and at the same time makes its contribution to intellectual, moral, and aesthetic growth."[53] A fellow Herbartian, Charles DeGarmo, who was a one-time colleague of McMurry's at Northern Illinois, led in articulating the moral values of natural science and advocating ethical training in the public schools.[54] Herbartians understood intellectual growth to entail the integration of social, moral, ethical, and substantive thinking and behavior, and they believed this was best accomplished when teachers consciously connected various subjects in ways that intersected and reinforced the individual threads of knowledge and experience.

Although McMurry was not fully persuaded by Parker's argument that nature study should be at the center of classroom curriculum—McMurry favored history and geography—he was nonetheless enthusiastic about using the natural world for pedagogical purposes. He insisted on occasionally teaching in the practice school at Illinois Normal and creatively used such natural objects as specimens at the museum of the Illinois Geological Survey (now the Illinois State Museum), corn in nearby fields, and an unanticipated beehive in the school attic to demonstrate and test his pedagogical ideas.[55] Influenced by the ecologist and nature study enthusiast Stephen Forbes, he involved his normal school students in a "type study" of the riverscape and landscape along the nearby Des Plaines River to arouse their interest in biology and geography. Such excursions also introduced them to Forbes's prescient ecological approach, reflected in his influential and pioneering report that had described the holistic way a lake was itself a "microcosm."[56] Jackman, too, was familiar with the emerging field of ecology through Chicago faculty members like Henry Cowles and John M. Coulter, as well as his downstate friend Stephen Forbes, and introduced the fundamentals of plant succession and habitat into his own texts.[57]

Charles McMurry translated German philosophy texts and articles to give them a wider audience and to advance the National Herbart Society, which flourished for about a decade after 1892.[58] Members of the association understood that their "debt to German thinkers for an organization of fundamental ideas" was substantial but further argued that they were promoting ideas and practices that sprang "out of American conditions." Thus, McMurry's course of study for the first eight grades claimed that "while strongly influenced by Herbart's principles, ... it is the outgrowth of several years' continuous work with the classes of children in all the grades of the common school."[59] His normal school colleagues agreed that "[t]he child is the center of the teacher's endeavor. He should come to know the world of man and nature."[60] Frank McMurry, who also taught for a time at Illinois Normal, explained, "Children should not become interested merely to acquire knowledge better; on the contrary, they should acquire knowledge in order to become interested. Ideas are merely the means by which lively interest may be aroused."[61] Both brothers combined theory and practice as they became outspoken advocates of the study of children and children's learning. When Frank joined the faculty at Teachers College in New York, he temporarily reunited with his old Illinois colleague Charles DeGarmo, and they introduced Herbartian theory to that faculty.[62]

The third in the family-connected trio of Illinois Normal Herbartians was Lida Brown McMurry, who had married a younger McMurry brother, William (who was not in education). She was by all accounts an outstanding critic teacher, who supervised and evaluated student teachers in the practice schools at Normal and then at Northern Illinois as part of its founding faculty. She actively worked to integrate the curriculum and was particularly interested in literature.[63] Among her textbooks was *Nature Study Lessons for Primary Grades* (1905), a textbook book for which Charles wrote the introduction.[64] She also integrated imaginative materials, introducing metaphors that elided the differences between humans and the nature they observed. She unabashedly suggested that trees in the fall "wore" yellow dresses and their leaves "forgot" to hang on.[65] Her brother-in-law asserted that the immature minds of little children were "easily and properly absorbed in the objects and their surroundings" by such language and that "there will be plenty of time later to hunt out the deeper truths."[66] Her example encouraged other textbook authors to use metaphor and simile, finding value in the integration of literature and nature study.

Lucy Langdon William Wilson, supervisor of nature work at the Philadelphia Normal School for Girls, similarly found nature study a popular topic with pupils, and her textbooks on nature study often included literary materials.[67] She quoted John Milton in the frontispiece to *Nature Study in the Elementary Schools*: "Till by experience taught the mind shall learn / That, not to know at large of things remote / From use, obscure and subtle, but to know / That which before lies in daily life / Is the prime wisdom."[68] Her teachers' manual for nature study, written in 1897, followed the September-to-June format that had become known as Jackman's "rolling year."[69] It emphasized "equip[ping] the student for practical, every-day work with little children" in ways distinctive from particular training in science. Noting that sense training was the "shibboleth" of the day in education, Wilson argued that observation should not be an end in itself but rather the means to judgment and imagination.[70]

Wilson's advice was specific yet open-ended as she guided teachers into using nature out-of-doors whenever possible. She described taking many excursions — eighty in one year — with a class of thirty-five children so that they could "see for themselves." She reported from her experience that reading descriptions of nature subjects was "stultifying" to younger children, whereas poems or prose, "real literature" related to the subject, proved more profitable, as did oral discussion.[71] Her textbook's bibliography incorporated topical books and poetry that would be readily accessible to teachers.

Wilson provided specific examples that teachers might adapt, reminding them, for example, that "The stories of Mercury, child of Jupiter and of Maia, in whose footsteps grew beautiful flowers, make a most happy introduction to the study of winds and clouds." She recounted a dramatic story of theft and reconciliation, in which Mercury became herdsman for his half-brother Apollo and enjoyed driving Apollo's cattle "in the great blue meadow which surrounds the whole earth." Having invested the story with "all the possible charms of literature and art," the teacher could, Wilson suggested, choose an excursion day when the sky was full of billowy cumulus clouds to observe their speed and direction of movement, adding other details once the children have had their attention drawn to the phenomena.[72] She integrated her own experiences, suggesting that color work was particularly good with younger children fascinated by swatches of bright color, while older children found "color work with plants and animals a great waste of time," because they saw more interesting detail when they attempted an accurate drawing. Lively and practical, Wilson's manual encouraged the correlation of studies and was one of those most referenced after the turn of the century.[73]

EDUCATION FOR THE NEIGHBORHOOD

Although the goal of correlating different parts of the curriculum remained strong, active discussion of Herbert's philosophy faded after the turn of the century. Other psychological approaches, like that of G. Stanley Hall, whose German Ph.D. made him a well-respected authority, put increasing emphasis on discovering the specific physical and cognitive capacities of children. A growing child study movement sought to gauge what children knew, and could learn, at specific ages.[74] After working with the psychologist Wilhelm Wundt, Hall also spent time at Jena in an effort to link psychology with education and returned to the United States determined to study child development.[75] He conducted pioneering studies of young children in the 1880s before taking a position in psychology at the new Johns Hopkins University.[76] He had become familiar with Francis Parker's educational results in Quincy and visited Cook County Normal School to keep up with evolving practices there. But Hall also sought more systematic analyses and explanations for the success of Parker's approach.[77] It seems that like Charles McMurry he admired Parker but found him a "rough iconoclast" and "not a close logical thinker."[78] Hall was also concerned that education should enable pupils to function well in the industrial world that lay ahead in the twentieth century.

After he was appointed president at Clark University, Hall founded a new journal, *Pedagogical Seminary*, and encouraged potential authors to submit empirical studies. His goal was to evaluate Herbartian concepts like apperception, testing to see whether it was true that children's minds learned best in relationship to things learned before. Hall encouraged one of his faculty members, Clifton Hodge, to work on the nature study curriculum in conjunction with the teachers at the Upsala Street primary school in Worcester. Clifton Hodge shared with Hall a concern about the impact of industrializing society on children and both rejected "sterile" object lessons. Hodge, speaking out against "dust covered mementoes," insisted that personal experience with "out-of-door life" was fundamental for pupils; and he designed nature study projects to address their immediate circumstances as well as their capacity to learn.[79] While child study would be challenged on various grounds by John Dewey, William James, and Edward Thorndike, Hall and Hodge gave it high visibility among nature study educators. They argued for its basic message that course materials should be age specific and sensitive to children's interests and capacities even as they prepared pupils for later studies and for life.[80] For Hodge nature study meant something more as well. Nature was not simply out there beyond the city or even sequestered in parks, but was an intimate part of part of city life rooted in everyday encounters and commodities. He felt that no pristine distinction existed between nature and culture or between the country and the city, since in Massachusetts, at least, the entire countryside showed the imprint of human activity.

Clifton Hodge produced a detailed teachers' manual, *Nature Study and Life*, for which Hall wrote an effusive introduction. Hall quite rightly pointed out that the text "differs not only in all respects from some" other nature study texts, but "in some respects from all."[81] Perhaps because it was so distinctive and clearly written, with a thoughtful awareness of its teaching readership, the textbook was widely reviewed and often chosen for teachers' reading circles. Hodge focused on the study of living things, suggesting that animate forms interest children before inanimate ones. He avoided simply following the seasons in some routine way and seemed to delight in being a contrarian and happy to be called a "degraded utilitarian."[82] His often-quoted view was that "nature study is learning those things in nature that are best worth knowing, to the end of doing those things that make life most worth living."[83]

Hodge's outlook constituted a highly individual mix of spirituality and industrial sensibility.[84] Like other nature study leaders, he knew nature firsthand as a child. His book's dedication was to his father, "who gave me my first animal and pets, my first garden plot and little farm, who left the big oak uncut for its beauty and the wild prairie unplowed for its wild flowers, who set the elm tree by the porch and the red moss rose in the old home garden."

Nonetheless, there is little sentimentality in this book, which looked directly at contemporary issues, especially economic ones, and suggested ways that nature study could address them. Hodge acknowledged Herbart's doctrine of apperception and Froebel's ideas about self-generated activities of children as important because they shared the common assumption that an active method of instruction was essential to children's learning. Hodge began his text by urging the teacher to ask about pets, a topic of immediate importance to children. While he suggested points to address with pupils, he reminded teachers to ask first what children thought were important features to know about such living creatures.[85] From the friendly pets, Hodge moved on to less welcome insects in the household, thus emphasizing how much could be learned by reexamining what seemed familiar. Hodge presented nothing by chance and little by formula but followed children's likely experiences in specific environments of the classroom and neighborhood.

Generously filled with sketches and charts for the teacher to emulate, as well as pictures and photographs, the five-hundred-page book seems somehow shorter than that. Hodge's final topic was flowerless plants and his concluding example, bacteria. After discussing those powerful single-celled creatures, Hodge provided a graph that showed the recurrence patterns of contagion for measles, diphtheria, and scarlet fever over a five-year period. Identifying the significance of "germs," he then described an experiment at the Upsala Street school. All feather dusters were collected, burned, and replaced by dusting cloths given to two girls in each classroom. The designated cleaners were given instructions on dusting: "come fifteen minutes early each morning, take their cloths to the sink, moisten them, wipe the desks and furniture of their school room, rinse their cloths and hang them up to dry." Once each week the cloths were laundered. The result of this "health brigade" was that during a year-long experiment, Hodge wrote, not one case of contagion was reported in a school of 435 pupils.[86] Such practical classroom projects of progressive educators seemed to reflect gender stereotypes more often than the out-of-door activities.

His chapters were arranged by topic, with advice for assignments, readings, and experiential activities. He recommended myths, legends, stories, and poems as especially appropriate for the third through fifth grades, levels at which children could read material for themselves and understand imaginative literature as such. Hodge stressed projects that had civic implications: Were once-common passenger pigeons truly extinct? Could mosquitoes be eradicated as pests? What were the sources of industrial pollution?[87] *Nature Study and Life* was widely and positively reviewed and reprinted several times, and its dynamic style drew teachers from across the country to Hodge's summer program at Clark University.[88] Hodge emphasized that "The nature-study point of view is to teach the child. The scientific point of view is to teach the subject"—and he believed that only the former would work with younger pupils.[89]

In an effort to help readers of *Nature-Study Review* sort through the extensive literature, Maurice Bigelow of Columbia invited them to send in their list of the ten best books on nature study. Hodge and Liberty Hyde Bailey were the top-ranked authors. Anna Comstock's *Handbook of Nature Study* was not yet published but would eventually become the best seller in the field.[90] Bigelow expressed surprise at how many different books were identified and especially by the number of specialized books mentioned. In the category of supplementary books, Neltje Blanchan (discussed later) on birds was at the top, joined by coauthors David Starr Jordan's and Vernon Kellogg's *Animal Life*, John M. Coulter's books on plant relations, Frank M. Chapman's books on birds, and John and Anna Comstocks' books on insects. Bigelow admitted that the lack of clear consensus on just one leading textbook could be interpreted as confusion, but, he argued, it also indicated the importance of local and regional materials, despite a curriculum that had become truly national.

PROVIDING A VISION

Liberty Hyde Bailey was perhaps the most publicly recognized name in nature study, but he never wrote a teaching manual. Instead, his book *The Natures-Study Idea*, published in 1903, served as a kind of inspirational text that he wrote to explain the movement and to inspire and empower teachers. *The Outlook to Nature* (1905) provided an even more personal commentary and reflection on the potential for such study in the lives of pupils. Bailey also explained that, in his view, there was no conflict between evolution and his religion, but he and others did not bring up the topic directly in textbooks, perhaps because they thought the idea was too complex for younger children.[91] For Bailey, it was the "spirit" of nature study that mattered, and he meant the term to evoke its aesthetic dimensions—artistic representations, photographs, and poetry—as well as the text in the practical little leaflets that he and his colleagues at Cornell produced. With John Burroughs, he resisted approaching nature in an "exact, calculating, tabulating, mercantile spirit"; he preferred to see in its practice "love and friendship."[92] Bailey wrote poems published in *Nature-Study Review* that stressed either "nature-sympathy" or portrayed the persisting struggle of farmers who met nature with both appreciation and courage.[93] The last stanza of his poem "The Farmer's Challenge" concludes with a haunting comment on persistence: "For these are my lands / And these are my hands / And I am bone of the folk that resistlessly stands."[94] He also admired Walt Whitman and often quoted him, particularly this passage:

> "When I heard the learn'd astronomer,
> When the proofs, the figures, were ranged in columns before me,
> . . . . I became tired and sick.
> . . . . Til rising and gliding out I wander'd off by myself
> In the mystical and moist night-air, and from time to time,
> Look'd up in perfect silence at the stars."[95]

Bailey, dean of the Cornell College of Agriculture, was deeply committed to systematic scientific inquiry, but he insisted that children, indeed all human beings, also needed aesthetic and spiritual experiences. Others picked up this theme, including Edward Bigelow, the stalwart coordinator of the later Agassiz Clubs, whose marginalia in his copy of Bailey's *Nature-Study Idea* makes clear that he endorsed the emotive claims for nature study.[96]

Not surprisingly, Cornell nature study leaflets often expressed an aesthetic as well as systematic approach to specific topics. By the time the several series of leaflets were collected in two volumes in 1904, Bailey's introduction emphasized the outlook as well as the steps involved in genuine nature study: "It is a fundamental educational process, because it begins with the concrete and simple, develops the power of observation, relates the child to its environment, [and] develops sympathy for the common and near-at-hand."[97] Like these leaflets, later course guides for school systems, whether only a few pages or fifty or more, balanced visual elements, humane commentary, and practical advice as they guided teachers to plants, birds, small mammals, amphibians, and insects likely to be available in their neighborhoods.[98]

Most striking was that the leaflets, unlike earlier teachers' manuals by Jackman, Lange, and others, were generously illustrated. The manuals might have occasional black-and-white sketches to indicate, for example, the parts of plants, but Cornell's leaflets contained woodcuts, etchings, and photography. These may have inspired imitation in the early twentieth century, as more textbooks included visual materials, some even having colored plates. Anna Comstock reused illustrations from her insect books, but she added less expensive photographs when she adapted the material for her large *Handbook of Nature Study*, published in 1911, which soon rivaled Hodge's text in sales. Giving children such illustrations seemed to her an alternative way to "capture" nature. Frederick Holtz argued in his *Nature-Study: A Manual for Teachers and for Students* that the subtlety of color and detail "clarifies and makes more definite the mental image the child may have gotten from the object in the nature lesson."[99] His emphasis on visualization was evident in the more than 170 illustrations in his text, and he encouraged pupils to express themselves in ink-wash drawing, crayon or pencil work, and clay modeling. Many illustrations were provided by women, often wives, following a long-standing tradition of women artists collaborating in the scientific work of their husbands. Katharine Elizabeth Schmucker, for example, produced most of the illustrations for Samuel C. Schmucker's *Study of Nature*.[100] As faster film and lenses improved photography, photolithographs became more common, and Perl Stella Gehrs provided those for her husband's *Agricultural Nature Study*.[101]

Besides communicating information, sketches and drawings could be presented as intrinsically valuable for stimulating children's curiosity, although the issue of direct contact with nature remained an issue.[102] Comstock, for example, produced a charming five-by-nine-inch portfolio-styled lesson book, *My Own Book of Three Flowers Which Bloom in April*, with a short description of each flower, a series of questions ("What insects visit your flowers?"), a page for an original story by the pupil, a page for a child's own watercolor, and an embroidery design that could be transferred to a pillowcase or tablecloth.[103] Observation was thus expanded to lead to activities and composition wherein pupils might find their voices on subjects that interested them. In fact, as one textbook suggested, "the correlation of nature study with language lessons [was] almost inevitable."[104] Helping students learn self-expression required arousing their imagination, and child-centered pedagogy pushed teachers to find creative ways to stimulate children to self-discovery. Textbook authors were quick to point out that children must realize that what they were discovering had undoubtedly also been seen by others, but argued that their arriving at this common understanding could reinforce a sense of mastery. The nature study promoted by Bailey and Comstock, as M. H. Carter noted, required only the "simplest apparatus" and "cheapest materials" to stimulate wonderful visual, written, and oral responses.[105] Moreover, this strand of nature study thinking, as promoted by Liberty Hyde Bailey and the Cornell educators, resonated with teachers because it allowed them to reinforce the child-centered approach, encourage correlation of studies, and rely on the immediate environment of home and school, albeit one more rural than that encountered by Hodge.

Other crosscutting themes defy easy categorization. References to God were often woven into texts and couched in spiritual, humanitarian, or even philosophical vocabulary.[106] These echoed generalized statements in public classrooms, courtrooms, or ceremonial events at the turn of the century.[107] Leaders of nature study seldom articulated any conflict between scientific understanding of nature as an evolutionary process and a religious and moral outlook; their language casually assumed that God was in nature. Wilbur Jackman, for example, advocated teaching nature study in Sunday schools in his essay on "Nature Study and Religious Training," and he often suggested that children would be led to God by seeing the wonders of creation.[108] Attention to nature in context meant referring to an intelligent designer who had so marvelously integrated species and their environment, and textbook authors presumed that both aspects of nature changed over time. In its more abstract version, this outlook assumed that nature elicited a deep personal spirituality whose claims were broad and nonsectarian. When Charles Scott suggested that nature was a "manuscript" of God, he returned to eighteenth-century themes about the author of design in nature.[109]

Specific biblical references were usually to the Old Testament, with attention to the moral responsibility for creation given to Adam and Eve as they were assigned to name and oversee all other creatures. The often iconoclastic Clifton Hodge, for example, opened his *Nature-Study and Life* with a quotation from Genesis to emphasize the human moral responsibility assigned in the creation story.[110] In a period when invocations were common in college ceremonies, Liberty Hyde Bailey quoted words of Jesus from the New Testament: "Raise the stone and there men shall find me; cleave the wood and there am I."[111] In rare instances, religion was central, as in the school text designed for rural districts in Utah, *First Book of Nature*, by James Talmage. He opened with a religious segment from a poem by Samuel Taylor Coleridge and used religious language throughout to discuss "created beings" and referring to "animal, vegetable and mineral kingdoms" that were organized under the "heavens."[112] Religious and secular outlooks seemed to reside easily side by side in early nature study.

Nature study was also mobilized to educate students with special needs as they gained increasing attention in the public school system. The American Foundation for the Blind, for example, reported on a project at the Perkins Institute in *The Blind Child in the World of Nature*.[113] Again, a patron of the American Museum of Natural History in New York, Henry Phipps, created a fund to provide transportation for children with disabilities to visit the museum where they enjoyed jungle stories told by the prominent zoologist

Carl E. Akeley. Other times they were treated to special lectures with lantern slides and, by the 1920s, to motion pictures that were intended to bring the out-of-doors vividly to life.[114] Nature study was evidently viewed as one subject that had educational potential for all pupils.

Nature study became well-integrated into normal schools' curriculum in part because it was malleable and worked with disparate ideas about child psychology and educational philosophy. Prominent educational reformers promoted it among teachers in training, at professional meetings of educators, and through textbooks designed for teachers or their pupils. Advocates and teachers agreed that nature study was most effectively taught by those who had some background in science, but they also sought to use techniques that were not didactic and that encouraged pupils to explore their world directly. Establishing agreed-upon parameters of such open-ended nature study proved increasingly difficult. One issue related to the use of imaginative literature. Other concerns were raised by the efforts of outside organizations, such as those concerned with conservation and hygiene, eager to get their message into classrooms using nature study as the vehicle.

ANIMAL STORIES

Stories in which animals were a central feature antedated the nature study movement as an established genre in both children's and adult literature, but an explosion of nature literature mirrored the textbook frenzy at the end of the century.[115] The long end of the nineteenth century has been described as a golden age for children's literature, a period when the childhood experience was being reframed as important in its own right.[116] Some of this juvenile literature, often well illustrated, was designed to complement or extend direct observation of the living and natural environment, much as did nature study. Elementary teachers, following the recommendation of their instructors and textbook authors like Lida McMurry, Lucy Wilson, and Clifton Hodge read poetry, essays, and stories to pupils and often found that these provoked better responses than descriptive texts with too much detail for younger pupils to absorb.

Some preferred the classics, agreeing with Charles B. Scott, that "the best literature is not too good for children,"[117] whereas others turned to contemporary nature writing for adults or to fictional and nonfiction literature on animals being written specifically for children. The range of what teachers considered classic varied from Greek mythology to nineteenth-century authors. The literature was not intended primarily to teach details about the subjects but, as Lucy Wilson suggested, to stimulate imagination. Regular textbooks might include excerpts from such well-established British poets as William Wordsworth and Robert Burns, as well as such familiar American authors as James Russell Lowell and Henry Wadsworth Longfellow. Some textbook authors liked to quote Henry David Thoreau and Walt Whitman. They particularly valued vivid and clear expressions of personal responses to nature.[118]

Teachers were also drawn to contemporary nature writers whose books were best sellers and whose names were familiar. As Henry Holt recalled, there was a lively "nature book fashion" at the turn of the century.[119] This fiction presented wild animals as not only exotic but also independent and resourceful. The genre of stories relating to wild animals promised, as Ralph Lutts has argued, to reveal nature "as experienced by animals for their own goals."[120] Animals that lived independent of humans were portrayed with their own desires, fears, and thoughts, a technique that minimized the distinction between animals and humans. Nearly all authors of literature for children took seriously the mandate to start with actual animal behavior, and much of the literature concentrated on the birds, frogs, turtles, and other small wildlife likely to be available to suburban and rural children. Even these might seem exotic to children in inner cities.

Imported books were also popular, with narrative and animated animal stories from Britain that became new classics.[121] Arabella B. Buckley knew her nature sources intimately and wrote in a way that encouraged children "to observe, to bring in specimens, and to ask questions."[122] Her style was familiar, empathetic, and matter-of-fact as she described a young gnat on the water surface: "[H]e stood tiptoe on the empty skin, which floated like a boat on the water. He spread his wings, and then he was safe and flew away. Sometimes the wind blows him over before he can get out, and then he is drowned."[123] The life she portrayed had cycles, opportunities, and failures. Even more popular in the United States was Beatrix Potter, with her *Tale of Peter Rabbit* (1903). Her stories are best remembered for their wisdom about human nature and Potter's attention to accuracy about animal behavior.[124]

Literary scholars have analyzed the genre of American nature writing, but few have noted its relationship to nature study or investigated the parallel phenomena in children's literature at the turn of the century.[125] John Muir's popular essays in the 1870s, dealing primarily with California and the Far West, had helped develop public interest in nature books. The highly popular genre of adult books on outdoor life ranged from best sellers like Jack London's fictional *Call of the Wild* to John Burroughs's *Little*

Nature Studies.[126] Burroughs, who wrote about wildlife in New England and upstate New York, attracted a particularly strong following. His philosophy coincided with the goals of nature study advocates: "To interpret nature is not to improve upon her; it is to draw her out." As an example, Burroughs suggested that if he were to describe all the birds on his walk, readers would become bored, but if he related one bird "in some way to human life, to my own life,—show what it is to me and what it is in the landscape and season,—then do I give my reader a live bird and not a labeled specimen."[127] Teacher Mary E. Burt found Burroughs's essays highly effective with her pupils and, with his permission, edited and reproduced them in readers for elementary children.[128]

Burroughs, the so-called sage of Slabsides, had caught the wave of nature study with his descriptions, which were vivid while eschewing the kinds of drama that made some nature writers controversial. His book of essays, evocatively entitled *Birds and Bees* and published as part of the Riverside Press Literature Series, sold well and steadily, with over 207,000 copies purchased between 1896 and 1907. By contrast, his adult essays in *A Year in the Fields* sold only 9,149 copies in the same period.[129] Later, when the Ohio Reading Circle of teachers selected his *In American Fields and Forests* as one of their books for the year, he sold more than 6,000 copies just to its members in contrast to the 1,108 copies sold through trade booksellers around the country.[130] The audience of teachers working to prepare for nature study classes provided a significant consumer base.

The single most frequent topic in specialty books on nature study for children was birds. This may have been in part because of the popularity of books by Neltje Blanchan, wife of the publisher Frank Nelson Doubleday. Blanchan's titles included *Birds that Every Child Should Know* (1907).[131] She illustrated her books generously with captivating photographs of birds, many of them nesting, and her stories and detailed descriptions are rich with practical, factual observations. She carefully crafted language accessible to children and used familiar images, as in her discussion of bluebirds:

> The bluebirds hunt for a cavity in a fence rail or a hole in some old tree, preferably in the orchard, shortly after their arrival, and proceed to line it with grass. From three to six pale blue eggs are laid. At first the babies are blind, helpless, and almost naked. Then they grow a suit of dark feathers with speckled, thrush-like vests similar to their cousins, "the baby robins"; and it is not until they are able to fly that the lovely deep blue shades gradually appear on their grayish upper parts. Then their throat, breast, and sides turn rusty red. While creatures are helpless, a prey for any enemy to pounce upon, Nature does not

dress them conspicuously, you may be sure. Adult birds, that are able to look
out for themselves, may be very gaily dressed, but their children must wear
sombre clothes until they grow strong and wise.[132]

The images were familiar to children and parents, with deliberate nest building and suits of feathers, and she avoided simplistic moralizing even as her
text put Nature in charge. However, just such metaphorical language made
some critics question the value of literary nature study.[133] Nature making
deliberate choices and small birds growing wiser with age seemed romantic
and vague to natural and social scientists as they moved toward more dispassionate, objective language and quantified analysis. Nonetheless, most
nature study textbooks recommended literature as supplementary to other
classroom materials and readily used literary devices.

Authors also cultivated adults among the education-minded audience, as
did Olive Thorne Miller, an early member of the Audubon Society, who lectured to teachers and wrote books she thought they could use.[134] Teachers
found publication a convenient way to design materials for their own teaching and to supplement still quite meager salaries. Flora Juliette Cooke's *Nature Myths and Stories* sold nearly four thousand copies and paid royalties
of two cents per book. For teachers like her who earned about $1,000 a
year, this additional income was nearly equivalent to a month's salary.[135]
Others trained in nature study found they could earn a living by providing
materials for others. Julia E. Rogers, who had taught with Comstock and
Bailey in correspondence and summer school courses, completed her M.A.
in 1902 and moved to New Jersey, where she concentrated on writing specialized books on trees, shells, and useful plants.[136] She also coordinated a
series of fifteen books entitled the *Nature Library* that involved scientifically
respected authors, including David Starr Jordan.[137] After several years at
home, Rogers sought to return to teaching, considering normal school positions because she thought she "could do more good in institute and other
extension work for teachers than anywhere else."[138] Writing, teaching, and
participating in a variety of other related activities made nature study an attractive place for women whose lives and family responsibilities often made
any "standard" track complicated.

Academic faculty men also produced books that would sell to nature
study teachers and their pupils. Addison Verrill, who taught zoology at
Brown University, deliberately titled his illustrated book *True Nature Stories*
to distinguish it from those of the "nature fakers" and directed it at boys
interested in natural history.[139] Vernon L. Kellogg's *Insect Stories*, however,

contains tales of drama about contests between the "narrow-waisted mother wasp" and a spider, the industry of ants, and the competitive role of beetles on orange trees told from the point of view of the author and his observant young friend Mary.[140] Kellogg, a Stanford faculty member who later coordinated relief work in Belgium under Herbert Hoover, a Stanford graduate, during World War I and subsequently headed the National Research Council in the 1920s, had studied with the Comstocks and participated in the nature study movement in California in the early part of the century. His interest in writing animal stories for children came after the birth of his own daughter.[141] Kellogg's attitude toward nature study had been somewhat ambivalent. In his systematic text on *American Insects* he noted that, while the volume might require its readers to do some serious work, they would learn "more satisfactorily than could be done with that utter freedom from effort with which some [clearly not his own] Nature-study books try to disseminate knowledge."[142] Other zoologists also addressed the nature study market. His colleague in nature study from his Stanford days, Oliver P. Jenkins, wrote a book entitled *Interesting Neighbors*, nominally for his children.[143] William Hornaday, a leading museum taxidermist who subsequently headed the New York Zoological Garden, produced an encyclopedic book of animals specifically addressed to young adults, normal school students, and teachers. His *American Natural History*, however, offered an arid alternative to more narrative and imaginative styles.[144]

Such scientific authors, women as well as men, were often aware of the competition for children's attention from "blood curdling novels." Many created drama in animal stories based on observed behavior, using language and metaphor to heighten the reader's interest.[145] Among the most prolific and popular scientific authors for children was Maine's state entomologist, Edith M. Patch, who wrote lively stories that were simultaneously highly scientifically detailed, including "The Strange House of Cecid Cido Domy" about the cecidomyiid that left behind the familiar pine-cone gall.[146] Kellogg's former student Bertha Chapman Cady wrote *Tami, The Story of a Chipmunk*, a cautionary tale about careless destruction of habitat in the wild, as well as the capacity of children to care for wild things (fig. 17).[147] Many of the school readers designed to be correlated with nature study sought to create an empathetic bond between nature and the audience of children. Mother Nature was a familiar figure, dispensing explanations as well as overseeing processes: "Mother Nature thought he [a caterpillar] needed a new suit of clothes, and so he crawled out of his baby dress, and then, when he had rested, what do you think he did? Why, he ate up his old clothes—yes, every

TAMI

The Story of a Chipmunk

BY

BERTHA CHAPMAN CADY

with foreword by
Anna Botsford Comstock

Frontispiece by
Robert Bruce Horsfall

Marginal Sketches by
The Author and her daughter Carol

THE COMSTOCK PUBLISHING CO.

Ithaca   New York

FIGURE 17. Bertha Cady Chapman was interested in the humane treatment of animals and in conservation, an outlook also reflected in her story of the chipmunk, Tami, a pet acquired when two children killed her mother through carelessness. Chapman, *Tami, The Story of a Chipmunk* (Ithaca: Comstock Publishing Co., 1927), title page.

bit."[148] Sometimes children were put into the picture. In Marietta Warren's text, children who study crayfish handle them carefully and reassure the tiny creatures that they will be returned "back to the water in which you live." In a less happy tale, a narrator dreams that he is a frog and recounts a nightmarish experience being chased by boys with stones.[149]

In its early days, the nature study movement and the enthusiasm for nature literature were mutually reinforcing projects. Anthropomorphism and animating nature came increasingly under criticism by scientists, and by the 1920s, as Ralph Lutts has demonstrated, the market for even realistic wild animal stories aimed at children as well as adults declined.[150] Attention shifted to other, often activist topics concerning larger issues of conservation and preservation in which narratives about individual animals were replaced by discussions of species and their habitats.

NATURE STUDY AND CIVIC REFORM

The versatility of nature study and its content attracted activists for a number of public reform issues, ranging from conservation to sexual hygiene. Encouraged by the emphasis on botany and zoology in the nature study curriculum, advocates for park preserves in the West or for humane treatment of the animals in their midst sought to have those topics explicitly introduced into elementary classrooms. Many nature study leaders agreed and, whether subtly or directly, reinforced a humane outlook toward living creatures. Attempts by progressive educational reformers to link nature study's emphasis on living things to human public health and hygiene met with less success. For those most focused on nature study as a child-centered pedagogy, the turn to civic issues seemed awkward and even inappropriate, but several in the movement not only endorsed addressing such topics in nature study but also helped lead the efforts to adopt them.

Nature study's goal of stimulating each child's imagination and "bring[ing] him into an intelligent sympathy with his environment, thereby broadening and enriching his life," resonated with many reforming groups.[151] Nature study texts and teachers frequently advocated humane treatment of animals, those being studied directly in classrooms as well as those in the wild endangered by human behavior. Even those to be used for food were to be treated respectfully while alive, according to most agricultural texts. Dietrich Lange thought that a central tenet of nature study should be to teach kindness to animals, arguing that "the boy who has learned that animals feel pain, that many of them think, work, and play . . . will no longer crush and club everything that creeps or walks."[152] Anna Comstock stressed that pupils trained in nature study would "enjoy nature through seeing how creatures live rather than by watching them die."[153] Even more explicit were pamphlets from such private organizations as the Society for the Prevention of Cruelty to Animals (SPCA), some of which were produced specifically for nature study classes.[154] By the 1910s, animal welfare advocates had successfully lobbied for city and state laws that mandated "instruction in the humane treatment of birds and animals." Active organizations published literature on such subjects as the usefulness of insect-eating birds, the benefit of bees for food and pollination, and the correct treatment of pets like guinea pigs in classrooms.[155] The pamphlets produced by the well-funded SPCA reflected factual and fictional literature written by or for children, and SPCA volunteers lectured in classrooms.[156] The best-selling author and naturalist Gene Stratton Porter published her first story with Boston's humane society while still a teenager.[157]

Pamphlets and bulletins about caring for animals in domestic and wild settings were produced and distributed by several federal agencies, including the Department of Agriculture, the Forest Service, the Weather Bureau, the Bureau of Biological Survey, and the Bureau of Education. These were supplemented by such private associations as the American Game Protection Association, the National Association of Audubon Societies, and the National Wildflower Preservation Society. Even large corporations like International Harvester contributed to the advocacy.[158] The National Geographic Society added global issues of endangered species and habitat conservation to its agenda, producing charts, graphs, and pictorial material that teachers could use in their classrooms.[159] Nature study teachers could initiate or be drawn into the public programs generated by national organizations, which found in schools an opportunity to advance their agenda by educating future generations. Arbor Day, established shortly after the Civil War, provided elaborate rituals to plant trees on school grounds and in local parks. Its success led to Bird Day in the 1890s, with the encouragement of the Department of Agriculture and the Audubon Society. Organizers specifically distanced themselves from older natural history that involved collecting and thus destroying habitats, specimens, or eggs. Rather, public advocates encouraged a moral outlook that stressed the preservation and the renewal of nature as they maneuvered their programs into the public schools.[160]

Nature study textbook writers also addressed contemporary problems raised by conservationists and preservationists. Clifton Hodge wrote on species extinction, advocating for protection of birds in particular. He sought to use school children as observers to see if any passenger pigeons could be found alive, and he wrote a lead article for *Nature-Study Review* in a special issue on the species.[161] Other educators sought to mobilize children in the campaign against the frivolous killing of birds for women's hats.[162] William Hornaday, a sometime critic of nature study, recognized that its curriculum easily supported conservation ideas, including his own mission to preserve disappearing buffalo.[163]

Bailey, Comstock, Hodge, and other highly visible nature study advocates wanted teachers to encourage pupils to think constructively about their role in nature. This might involve quiet observation in a wooded or scenic setting.[164] John M. Coulter, for example, emphasized field trips that eschewed specimen hunting of birds or wild flowers in favor of a distinctive and nonintrusive mode of analysis of species in context.[165] Respectful interest in living landscapes coincided with preservation ideals, and advocacy could be explicit. Schools across the country were mobilized to participate in Arbor Day. The Forest Service provided materials to explain its role on public lands,[166] and a conference in North Carolina highlighted the relationship between forestry and nature study.[167] As forested land was taken over by suburban homes, Ruth Marshall became involved in a local movement to preserve one border landscape near Chicago for study and recreational use.[168] The advocacy for conservation, as for the humane treatment of animals, was grounded in the argument for biblical patrimony as much as in discussions of resource management.[169] Moral and civic virtues were made coincident.

Nature study faculty and teachers were more ambivalent about introducing health and hygiene in their curriculum, although the goals of ameliorating health problems among poor children, eliminating contagious diseases, and improving public sanitation were hardly contested.[170] Resistance was largely pragmatic because of the limited time allocated for outdoor study, which required close attention to the description, function, and aesthetics of the nature found there. Nonetheless, a few of the second generation of nature study advocates believed that human health and hygiene were topics well suited to their work on all living things.

Teaching physiology and hygiene in public schools antedated nature study. Normal schools began to teach personal hygiene in conjunction with physiology in the 1880s and 1890s, largely in response to city or state requirements. Using pamphlets and textbooks written for teachers and secondary students, these classes presented "the science and art of the preservation of health" both "individually and in relation to the community," along with a basic introduction to human physiology.[171] As states extended their requirements for teacher certification, some, like the Pennsylvania State Department of Education, responded to scientific ideas about germs and disease and developed a standardized curriculum that included hygiene and public sanitation.[172] The interweaving of social, political, and moral ideas was evident in the description of hygiene in a contemporary encyclopedia of education: "Hygiene in a broad sense includes also the study of conditions that favor the healthful development of the human species (eugenics); the conditions affecting the health, preservation, and development of special races (racial hygiene); and the conditions that favor the healthful development of human society."[173] Increasing attention was also paid to the health qualities of school buildings.[174] Normal school classes relating to physiology and health were expected to trickle down to school classrooms, but advocates of particular reforms wanted more direct influence.

Various groups, most visibly the Department of Scientific Instruction of the national Women's Christian Temperance Union (WCTU), formulated the argument that alcoholism was a disease, one that could be avoided by informed individuals who chose not to drink. Scholars who debate whether the WCTU was progressive or repressive nonetheless agree that its state and local branches were powerfully influential, not only in changing public policy but also in positioning issues of abstinence and social hygiene in the school curriculum.[175] Their materials and those related to issues of alcohol and tobacco emphasized self-control of that "wonderful and complex machine, the human body" and often were presented alongside other health literature.[176]

With physiology and hygiene came an extensive self-help literature on body functions, exercise, cleanliness, and sexual hygiene, building on but also distinct from professionalization in those fields.[177] Hygiene's incorporation into nature study was initially evident in major cities with large immigrant populations. However, by the mid-1910s Mankato State Normal School, serving rural Minnesota, added hygiene into its campus elementary school's nature study curriculum, arguing that "nature-study has many opportunities to teach children how they may best perform their social duties and cooperate with others to promote the best welfare of all."[178]

In the post-Pasteur era, scientific experts wanted to introduce cleanliness and sanitation topics early in life and identified nature study as an appropriate setting. Winfield S. Hall, a professor of physiology at Northwestern University's medical school and affiliated with the Society of Social Prophylaxis, published "Social Hygiene: Its Pedagogic Aspects and Its Relation to General Hygiene and Health" in *Nature-Study Review* as part of an effort to encourage such work by teachers.[179] W. J. McNeal of the University of Illinois offered very direct advice about cleanliness and other matters in his essay "What Teachers May Do to Promote Personal Hygiene and Public Health." Stories about the epidemic outbreak of tuberculosis, the presence of bacteria in school settings, and the potential danger of disease from pets were all part of the discussion in "Hygiene as Nature Study," and a few nature study leaders committed themselves wholeheartedly.[180]

Nature study leaders in progressive institutions that emphasized social betterment extended health hygiene to sex hygiene, which proved more controversial. Maurice A. Bigelow introduced hygiene at Teachers College and, like Bertha Chapman Cady, became particularly prominent in sex education, producing a number of textbooks and pamphlets throughout the 1910s and 1920s.[181] Bertha and her husband Vernon Cady lectured on sexual hygiene and wrote a textbook, *The Way Life Begins: An Introduction to Sex Education*, which went through multiple editions.[182] Bigelow had begun work with well-known spokesman Prince Morrow on sex hygiene for a Congress on Hygiene in 1912, and during that decade he spent much of his time advancing work on applied biology, eugenics, health education, and social hygiene, even as he retained his active membership in the Nature-Study Society.[183] The link to eugenics themes in sex hygiene manuals could be explicit, as when H. Royston asked rhetorically after a discussion of animal breeding, "Are we not bound at least to eliminate from the rights of procreation those of our own kind who possess undesirable characteristics which experience shows will be handed on to their unfortunate children?"[184] Such explicit eugenic statements were, however, relatively rare in nature study more generally.

For Bigelow and Cady, acquainted through the *Nature-Study Review*, interlocking nature study with sex hygiene education initially seemed self-evident. Cady moved to New York City with her husband and two small daughters in 1914, taking a position with the American Federation for Sexual Hygiene, which later joined with other groups to become the American Social Hygiene Association. In 1915 she organized a display on social hygiene for the Panama-Pacific International Exhibition in San Francisco. Her evocative and aesthetic nature study display of plants and animals drew large crowds, to whom she then lectured on social hygiene. That summer she taught a related course for teachers with William F. Snow in nearby Berkeley. Written with her husband, *The Way Life Begins* argued that sex education offered "the deeper meaning of nature study," drawing on the reproduction in plants and animals in anticipation of thinking about human experience.[185] The book, described by reviewers as simple, clear, straightforward, and uninhibited, sold well at $1.25 a copy and was reprinted a number of times.[186] Leaders were aware, however, that sexual hygiene was also a highly controversial topic, especially as providing sex education impinged on parental prerogatives to introduce this topic, which may also have dampened the enthusiasm of some nature study teachers.[187]

In fact, the sexual hygiene movement made it easier for nature study teachers to discuss topics that related to physiology and sexual reproduction, but it also had the unintended consequence of opening the door for other kinds of authors to write on sexual themes.[188] In his textbook *Sex Education* (1918), Bigelow observed in frustration the discomforting reality that much of the interest in sexual issues seemed to focus on the abnormal, perversity, and even bestial vulgarity, rather than on the "natural" aspects of the topic.[189] By the late 1920s, the sometimes awkward link was gone. Hygiene had its own established place in the curriculum and was decidedly distinct from nature study, which, where it persisted, had retained or returned to early themes of experiences with plants and animals and the outdoor world.

AN ELASTIC CURRICULUM

While nature study textbooks drew on what one historian of education termed "the fine pedagogical sense" of midcentury writers like Almira Hart Lincoln Phelps and Eliza Youmans, they were also considerably different from the early volumes on botany, chemistry, and other scientific subjects.[190] Textbook authors on nature study were typically self-conscious about theory, undoubtedly more than were the teachers and administrators who implemented it. Indeed, practitioners found that "teaching with nature" could be readily adapted to changing pedagogical philosophies and practices.

Among the initial generation of advocates, typically those educated before 1890, nature study echoed themes of nineteenth-century natural history and concentrated on observation rather than application. But over time what the optimistic Bailey had called the "missionary and altruistic spirit" of nature study shifted toward social goals.[191] A younger cohort, often with advanced degrees and seeking to use education to accomplish social change, defined programs and prescribed outcomes somewhat differently. The McMurrys in Illinois, influenced by German philosophy, retained a firm commitment to thinking in terms of child development and promoted an integrative curriculum that emphasized building on apperception and correlation, an approach that seemed to some skeptics to make nature study simply a convenient adjunct to other subjects rather than valued for its own content. Clifton Hodge, whose advanced degree was in biology and who was working with psychologist G. Stanley Hall, found himself drawn to think about nature study quite differently, owing to the circumstances in the public schools neighboring Clark University. By the 1910s, Hodge had largely left nature study behind as he moved to the University of Oregon to teach "social biology," and he subsequently wrote a high school textbook with Jean Dawson entitled *Civic Biology*.[192]

These multiple strands meant that nature study method and content were never encapsulated and standardized. A single, prescriptive curriculum would, in fact, have contradicted the assumption that nature study should be based on local materials and allow a creative teacher to use the theories and practices that worked best in her classroom. Many nature study advocates agreed with Frederick Holtz, who was content to have nature study "pass on the relative values and educational usefulness of present practice, guiding the subject along" on the assumption that best tendencies might ultimately lead to greater agreement.[193] Definitions of nature, too, were remarkably fluid among these educators, but most came back to the experience of children in relationship to the life found in their immediate environment.

Where nature study flourished, educators seemed willing to tolerate its multiple meaning and varied programs. Those who developed the New Orleans Course of Study, for example, were satisfied with broad goals when they stated:

> The object of Nature Study is not to make scientists, but to develop powers of quick and accurate observation and exact description; to impart a knowledge and love of nature; to lead children to become keen, accurate, and sympathetic observers of the world around them. It should be the aim of this study to call forth the self-activity of the child through interest, and to arouse in him a sympathy for living things through the care of plants and animals.[194]

Their description anticipated multiple outcomes based on the self-initiated activity of children, including the acquisition of new knowledge and a humane and moral outlook on the natural world. Nature study was not intended to build specialized expertise in young pupils but to direct them toward a civic understanding of their natural environment based on familiarity with its particularities. While this outlook sufficed for practicing teachers, administrators, and school boards, university educators and scientists were becoming more demanding and insisted on creating standards and measuring learning outcomes. In order to establish a stronger and more consistent voice nature study leaders formulated *Nature-Study Review* and the Nature-Study Society, establishing some pattern in the "zigzag progress" of a field with multiple advocates, instructors, and outlooks.[195]

CHAPTER SIX

Establishing Professional Identities

Another trend that is on the increase in large cities and in platoon schools is the employment of special teachers for nature work. Desirable as this may be we cannot hope to have it become universal. The extra expense makes it prohibitory in a large percent of our town and village schools and always in the rural schools. After all, if we can see to it that the regular room teachers are trained to do effective work the results so far as the children are concerned may be quite as much worthwhile.

Alice Jean Patterson[1]

INTRODUCING NATURE study into the curriculum at the turn of the century was simultaneously challenging and exhilarating as normal schools, school systems, and individual teachers experimented with ways to define and shape the subject. Although normal schools and college departments of education differed in their admission standards, length of training, course requirements, and placement patterns, both played a central role in establishing the progressive agenda in public schools. Their faculties produced much of the core material for their own classrooms and campus practice schools, even as they educated a cohort of the best-trained teachers who, in turn, became informal leaders as they entered into public schools and became administrators. As Alice Patterson notes above, a few would even become specialized nature study supervisors, well equipped by both education and experience in classrooms to integrate the philosophical roots of nature study with the practical issues that they encountered in urban schools and some rural school systems. Because a majority of normal school graduates werc women, the opportunity to coordinate and administer nature study programs opened a significant career niche for women in science.[2] Nonetheless, gendered hierarchies and assumptions were operative in this field as in others during the early twentieth century.

Positions for women had opened up in private schools for girls earlier in the nineteenth century. During and especially after the Civil War, women moved on in ever-increasing numbers into urban, rural, and suburban public schools.[3] Advanced educational opportunities expanded in women's academies and colleges, in the network of public and private normal schools in more populated states, and at the coeducational state universities founded under the Morrill Act of 1862 where there was low or free tuition.[4] Young women of the aspiring lower and middle classes were drawn by the prospect of respected, paid employment. Young men, too, were attracted to teaching, but they had a wider array of options in their course of study and the positions they might attain.[5] For women, regardless of their particular interests and skills, teaching ranked high on a list of limited choices. Both men and women benefited from the public enthusiasm for education, new pedagogies that offered an increasingly sophisticated set of tools for those trained to teach. Published accounts and private letters often expressed the pride and excitement teachers felt about their preparation and practice.

Historians have noted that a substantial proportion of young white women spent some part of their lives as teachers in the late nineteenth century, an occupation that constituted the second highest concentration of employment for women after domestic service.[6] College science graduates typically sought jobs in the areas of their expertise. Margaret Rossiter has documented the employment patterns of women who had studied science as they sought to enter standard career paths in colleges, government agencies, museums, and industry. These women found employment as illustrators, editors, writers, librarians, researchers, taxidermists, and normal school instructors, among other roles.[7] In the emerging field of home economics, women trained in chemistry established a number of subspecialties that emphasized health, hygiene, and nutrition as primarily women's domains.[8] The nature study movement provided a similar niche for women, one where they could not only teach scientific subjects in public and private schools but also write textbooks, join the faculties at normal schools and women's colleges, and become local administrators and national leaders.

TEACHERS IN TRAINING

Preparation for teaching was not standardized in the last half of the nineteenth century and varied considerably in different regions and between urban and rural settings, although steadily increasing attention was being given to pedagogy as well as to curriculum content as the basis of sound training.[9] Moreover, no national or state agency was able to mandate content or method, so that the style and quality of elementary schooling was dictated by local school boards that selected their own teachers, textbooks, and curriculum. During the last third of the nineteenth century many rural pupils were taught in one- or two-room schoolhouses that remained ungraded, although urban systems with larger student concentrations developed eight-year, graded elementary schools intended to insure common preparation for those going on to public high schools.[10] But as counties and cities gradually consolidated into systems that could oversee school attendance and burgeoning enrollments required more teachers, and as elected boards responding to demand sought normal school graduates, state legislatures followed Massachusetts's example and established additional normal schools in the late 1880s and 1890s. There post–high school students encountered botany, astronomy, and other sciences, subjects in which some found unanticipated talents that they could pursue in teaching, in extracurricular activities, or even in advanced study. Well-trained teachers brought new ideas with them into those systems.[11]

Normal school students, however, remained a privileged minority. Well into the first quarter of the twentieth century, the majority of teachers, especially in country schools where the population ran thin, still gained positions by completing grammar or secondary school and attending short-term institutes or affiliating intermittently with local colleges, religious seminaries, or training schools to gain appropriate credentials. Teaching positions attracted primarily single young women who sought to earn income by teaching for a few years before marrying, but a significant number made teaching their career.[12] Such teachers initially recapitulated their own educational experiences and used whatever textbooks were at hand to teach basic literacy and numeracy, adding other subjects at their own discretion or when encouraged by a supervising school board or principal.[13] As counties and then states required that prospective teachers pass examinations, particular subjects became standard and licenses covering a specified period were issued.[14] Typically the certification tests emphasized knowledge content rather than teaching skills, and practicing teachers were encouraged to supplement their training at institutes provided in local county seats or at normal schools and state colleges.[15] As institutes reached their height of influence between about 1890 and 1910, programs might extend for a week or more and incorporate subjects for which teachers had not previously been trained, such as physical education and hygiene, music, drawing, history, and geography, as well as various kinds of elementary science and nature study.[16] Summer and weekend institutes were sometimes required and not always appreciated, with one educator recalling those in the 1880s as "awful things . . . Even to this day I think of a country institute in August as about the worst possible punishment connected with teaching."[17] Experiences varied, but most teachers reported positively in their diaries and letters about the social life at longer institutes on campuses and the significant friendships that sustained them when they returned to their isolated rural schools.[18] Sarah Jane Price reflected in her diary with some nostalgia, "I feel a slight reaction since the institute closed and a longing for the companionship of those with whom I became acquainted there."[19] Teachers who could not afford to matriculate at a college or normal school seemed particularly to appreciate longer institutes held in the summer at normal schools and colleges where they could use laboratories, libraries, and dormitories.[20] In some cases, voluntary attendance at institutes led ambitious teachers to return subsequently to normal schools that offered six-month, one-year, or two-year teacher training courses.[21]

Those who attended normal school could be transformed by their experience, although not everyone enjoyed academic life. A young and somewhat insecure Lottie Howard of Minnesota reported to her mother in evident surprise that she "liked reading" after all, once she had discovered the writing of George Eliot.[22] Others identified the subjects that they liked less well, as when Louise Bailey recorded in her diary, "Our Botany class rambled about the grounds this morning under the espionage of Miss Rogers, it killed the recitation period and that was the point desired I think. I'm quite disgusted with it."[23]

Often normal school students were somewhat older, following the pattern of the earliest normal schools in Massachusetts; they arrived for training after several years of teaching or administrative experience. Lida Brown McMurry recalled, "My brother, Isaac Eddy Brown, and I both had the State Normal University in mind sometime before we were able to attend it. We both taught with that aim in view.—At last it seemed possible to attend for one year at least. We entered in the fall of '71."[24] After acquiring new skills, better educated teachers might enjoy greater autonomy in what they taught

FIGURE 18. The grounds around and behind Teachers College of Columbia University, as at many normal schools, were used for an extensive garden, where prospective teachers could learn basic horticulture and work with living plants in the greenhouse during winter months. *Nature-Study Review* 4 (March 1908): 82. Courtesy of the Wangensteen Historical Library of Biology and Medicine, University of Minnesota.

in rural schools, move to more attractive or better paying jobs, or even equip themselves for further study in a college or university.

Despite many common elements in normal schools, which spread westward across the country, there were certain regional tendencies in the teacher training schools. In the East, normal schools prepared primarily elementary teachers, while in the Midwest, normal school faculty sought to train not only elementary and secondary teachers but also principals and superintendents.[25] The expanding network of state normal schools, often impressively housed in substantial brick or stone buildings, typically hired faculty members from other well-established institutions or those with university training. In an informal way, this hiring practice spread the theories and practices of new pedagogy and spurred competitive innovation (fig. 18).

NORMAL SCHOOL PRACTICES

One of the most striking initiatives was that of Northern State Normal School in DeKalb, Illinois. In 1899 John Cook was offered the position as the first president, but he accepted only after Charles McMurry and others from Illinois Normal had agreed to join him in its highly experimental program. Both men were persuaded that supervised teaching—McMurry disliked the term "practice teaching" because it implied "the clumsy work of a beginner" rather than the requirement to "get mastery of a subject, organize it into a lesson with illustrative material, and then put the lesson into practice"—was the key to preparing teachers.[26] They formulated a "DeKalb Plan" that integrated the new normal school with the city public school system, making the head of the practice school also the superintendent of the city's schools. Charles McMurry accepted that dual appointment. The plan was ambitious and ultimately proved unworkable, but it was a way for Cook to demonstrate his basic idea that a normal school should be built around the observation and experience of actual teaching in a standard classroom. [27]

To emphasize further the importance of a training school that operated in conjunction with the public schools, Cook hired Lida Brown McMurry to be the lead critic at the on-campus program, with the unusually high salary of $2,500.[28] She was well respected as a supervising teacher and was herself "especially adept at working with small children, for she knew how to select and use materials to develop a child's interest."[29] Her own specialty was nature study, but always with the Herbartian outlook that integrated reading, singing, acting, and playing out-of-doors. She suggested, for example, having children go outside, observe, and then imitate little birds: "Fly to the fields. Pick up seeds. Take a drink. Bathe in the creek. Preen your feathers. Fly home. Perch on a twig. Sing: We are little birdies/ Happy we, happy we. / We are little birdies/ singing in a tree."[30] Presumably there were limits to this imitation, with caution about high tree limbs and flying.

Northern Illinois offered multiple options to its trainees, providing two-, three-, and four-year courses of study, plus a one-year teaching practicum for graduates of four-year colleges. By 1904 nature study was firmly embedded in the curriculum and reflected explicitly the influence of the Herbartians, who had carried their philosophy with them to Northern:

> In the lower grades, the studies are of an informal nature, aiming to acquaint the pupil with the more familiar features of his environment and to foster in

him a sympathetic interest and spirit of inquiry. In the intermediate classes there is an increasing opportunity to elucidate and enrich the other lines of study, and to exhibit the practical bearings of scientific knowledge on the conditions of living and on commercial and industrial processes. . . . In each grade some form of individual nature-study notebook is kept, serving as a record of the work done; nature notes and calendars are preserved from year to year.[31]

Fred L. Charles took responsibility for the teaching of biology and nature study at Northern. He created a Nature Study Club to encourage the teachers in training to take to the fields in and around the campus, and he prepared a series of leaflets teaching nature study for his students and local teachers.[32] In 1909 faculty members, responding to an "insistent demand today from the public schools for specific and concrete direction rather than inspiration or theorizing," produced a bulletin completely devoted to nature study, one based on ten years of experience in DeKalb.[33]

Back at Illinois Normal, the faculty member responsible for teaching the natural sciences was Buel P. Colton, who had studied biology with H. Newell Martin at Johns Hopkins and often spoke of the influence of Louis Agassiz on his thinking.[34] Among his colleagues was Alice Jean Patterson, who graduated from Normal in 1890 and then, while a teacher and elementary school principal, intermittently completed a B.S. degree in biology at the University of Chicago.[35] In 1897 she returned to teach at Normal High School, joining the normal school faculty when Colton died in 1906. Patterson continually redesigned her curriculum for Illinois teachers while keeping in mind the range of resources they could find locally. Normal school graduates who planned to teach in rural schools needed, she believed, a curriculum that would let them closely follow the seasons and explain plant development in terms of biology and the environment around them. Those who were going to teach in towns and cities would need to rely on window boxes and field experiences that took advantage of parks and museums.

As larger school systems began to hire special nature study supervisors, Patterson made special arrangements to train those who might be appointed to assist classroom teachers, organize systemwide curricula, and supervise the implementation of a nature study curriculum. She had a greenhouse built and established a school garden on campus so that students could learn how to coordinate larger projects. Like many of the nature study faculty, Patterson believed that community engagement would reinforce classroom lessons. She organized a local children's garden club and became active in the national school garden movement.[36] John Merle Coulter and his son, John G. Coulter, who received his Ph.D. at the University of Chicago, collaborated with Patterson in writing a nature study textbook.[37] Alice Patterson was unusually motivated and successful, and her career pattern of continuing education and job mobility reflected the possibilities open to the most ambitious nature study leaders at the turn of the century.

Nature study moved quickly across the country, and Californians were particularly responsive to its outdoor and object-centered approach and enthusiastic about their distinctive biotic landscape. The California Normal School at Los Angeles introduced nature study in the 1890s. Edward Hyatt, superintendent of schools at Riverside from 1895 to 1907, was eager to hire its graduates, as were schools in the rapidly growing network of cities in the Los Angeles area. Hyatt had a science degree from Ohio State University and was keenly interested in nature study. He lectured frequently to teachers on the "benefits of nature study" and other nature topics, and his public enthusiasm for innovation in education helped him become the state superintendent of schools in 1907.[38] A decade later, Charles Lincoln Edwards was the established coordinator of nature study for the Los Angeles schools, through which he organized a nature study exhibition for the dozens of schools in his system and carried out daylong seaside excursions involving literally hundreds of children (fig. 19). These mass outings could draw enthusiastic pupils and dozens of teachers and nature study supervisors, and Edwards conducted specialized tours, one with eight supervisors and several classrooms of children to Mt. Wilson to study plants of chaparral and canyons. In 1914 the Los Angeles system had 127 schools with nature study clubs.[39] In northern California the superintendent of schools in Stockton, Edward R. Hughes, reminded teachers, "A common mistake by teachers and supervisors is the attempt to make *science* of the subject. The child sees one thing at a time well. . . . Our knowledge comes to us as isolated facts and experiences. The classification into a body of knowledge is the work of maturer years."[40] He, too, emphasized outdoor activities for teachers to make them comfortable in teaching about nature, taking advantage of the local climate for field trips. As the nature study curriculum leapfrogged across the country to the West Coast, the same issues regarding preparation of teachers, classroom methods and content, and pedagogical underpinnings reoccurred as educators sought to keep their schools current with "the new education."

The situation in the South was different, shaped by rural poverty, segregation, and skepticism about innovations from elsewhere, although certainly reforming educators in some cities and in university towns made an

FIGURE 19. Charles L. Edwards, supervisor of nature study for the Los Angeles public schools in the 1910s, engaged literally thousands of city school children in excursions to the tide pools at Point Firmin, California. *Nature-Study Review* 12 (February 1916): frontispiece. Courtesy of the Wangensteen Historical Library of Biology and Medicine, University of Minnesota.

effort to keep up with national trends.[41] The United States Commissioner of Education reported in 1888 that school terms ran 50–60 days in the rural South, compared to 100–150 in the North and West; and many children, black and even some white, had no access to the kinds of neighborhood schools that dotted the other sections of the country.

Normal schools in southern states were typically segregated by race and sometimes by sex. The most visible and well-supported school for training white teachers was the Peabody Normal College (now George Peabody College of Vanderbilt University) in Nashville, officially founded in 1875 by the George Peabody Educational Fund. Its mandate was to draw students from all the former Confederate and some border states.[42] After an initial effort to establish public schools in population centers in the South, the trustees of the fund had turned to teacher training as a way to have a greater impact. Their principal project became a regional "demonstration" normal college in Nashville that provided scholarships to outstanding students from the area. Students supported by the fund made a pledge to teach for two years, and many of them became leaders in their respective states.[43] In 1891 the normal school, with support from the Peabody Fund, established the Winthrop Model School, named for the president of the fund, Robert C. Winthrop. It was not intended for practice teaching but was rather a literal demonstration model of a well-run school for prospective teachers to observe, suggesting that, in the 1890s at least, this teachers college was following an older model rather than introducing the practice or experimental schools being implemented by leading northern normal schools. Other southern normal schools, like Winthrop Normal College in Columbia, South Carolina, used a similar representative system, admitting sixty-six students, two each from the state's thirty-three counties during its early years.[44]

Peabody Normal School, especially during the 1890s and very early 1900s, sponsored summer institutes for teachers, some in Nashville but most throughout the southern states. These were typically segregated, with reportedly identical programs being offered in adjoining buildings for African American teachers and white teachers. Some states reported that the teachers attended at considerable personal sacrifice, indicating that while the course was free, participants needed to pay personally for their transportation and subsistence. Many southern states lagged behind in terms of the length of time spent in preparation for teaching. Others worked to be progressive. Arkansas continued to have a three-, five-, and finally in 1893, nine-month program; it also reported that "the color line was not drawn" in some institutions and there was no friction.[45]

Biological sciences were not added to the Peabody curriculum until after the turn of the century, when new faculty member, Thomas W. Galloway, introduced a course on "Nature Study: Its Theory and Practice" to prospective teachers in 1903.[46] The following summer Mary Phillips Jones, who had graduated from Cook County Normal under Parker and had assisted in the Horace Mann School at Teachers College, Columbia, taught an eight-week course that included nature study for primary school teachers.[47] Offered intermittently over the next decade, nature study was revived during the significant reorganization of Peabody College in the 1910s, which brought such prominent faculty members as Yetta S. Schoninger, who had a B.S. from Columbia's Teachers College and had attended summer school at the University of Chicago, and Charles McMurry from Northern Illinois.[48] Peabody College, housed directly across the street from Vanderbilt University, remained exceptional, especially in the South.[49]

There were other philanthropic efforts that encouraged nature study, including textbooks distributed through the Carnegie Free Traveling School

Library, which reportedly introduced nature study books into eight hundred rural schools. The impact of such efforts is difficult to assess.[50] Normal schools provided instruction, and a list of nature study instructors in 1916 identified several teaching in North and South Carolina, Virginia, West Virginia, Arkansas, and Kentucky; Julia Mae Williams, for example, taught nature study at the Colored Normal, Industrial, Agricultural and Mechanical College of South Carolina.[51] Additional research might demonstrate the extent to which southern teachers introduced nature study into their classrooms, but most teachers were preoccupied with basic literacy and numeracy, especially in schools where children were seldom in age-specific grades, resources were severely limited, and attendance was sporadic.[52]

In the upper Midwest and Northwest, most normal schools taught nature study to teachers alongside future administrators and increasingly hired staff with advanced degrees. William H. Sherzer, for example, had a Ph.D. in geology, worked for the Michigan State Geological Survey, and then took a position at the Eastern Michigan Normal School where he became involved in organizing its Nature Study Club as part of his responsibility for nature study education. He later introduced scouting to the Ypsilanti community, convinced that its extracurricular emphasis on the out-of-doors would supplement nature study work in the schools.[53] John P. Munson, a Cornell graduate and faculty member at Washington State Normal School, emphasized the biological components of nature study in his textbook for teachers in training in 1903 and suggested the rehabilitative value of studying biology.[54] Some of the women attending graduate school for scientific study also translated their expertise into broader nature study work. Ruth Marshall found that the nature study movement had opened up higher educational opportunities, but she apparently was not encouraged to pursue options beyond teaching at a women's college. Born in Wisconsin, she attended and then taught in rural schools, gradually working toward her B.S. in 1892 at the University of Wisconsin.[55] She took a high school position and spent one summer at the oceanographic research center at Woods Hole, where she began an investigation of water mites, a subject on which she would publish more than thirty papers. Attracted to the University of Nebraska by the work of Charles Bessey, Marshall took a doctorate there at the turn of the century, and Bessey encouraged her to help develop nature study curricula throughout the state, following the model of Cornell.[56]

Balancing the research training in science and the child-centered approach of nature study was not easy. Marshall's colleague Robert Wolcott taught zoology at Nebraska and agreed with Bailey's admonition that nature study be nontechnical. He followed Bailey's suggestion that books should primarily be used to "corroborate and check results" after the child has made independent observations.[57] Marshall, however, articulated a stricter scientific line when she argued,

> Forget, if we can, the pedagogics, or keep it well to the background, and lay aside the trite sayings that nature-study trains the process of observation, fosters love of nature, trains the logical power, and many of the other things that cumber the subject, the so-called principles which so many school people seem to think essential to place conspicuously at the head of the course as a kind of thesis to be defended. The thing we want to get is a knowledge of nature.[58]

Her outlook coincided well with that of Bessey and of the American Academy of Medicine, which encouraged nature study but insisted on expertise: "The best work by general teachers was found in those cities where a biologist supervised their work. In these places, the teachers relied upon the supervisor's visits and stimulus."[59]

Marshall accepted a science teaching position at Rockford College. It was a woman's college known for educating teachers, and some graduates, like Jane Addams, pursued informal teaching in settlement houses.[60] At Rockford a colleague, Norma Nelson, recalled Marshall as a demanding and inspiring teacher whose "love of nature was deep-seated"; Nelson also noted her leadership in the Rockford Nature Study Society, which lobbied for a forest preserve near the city.[61] Marshall was scarcely a sentimental nature lover, but her empirical commitment integrated systematic inquiry with an intimacy with natural things. Under her direction, the local Nature-Study Society identified several examples of each species of tree with zinc labels provided by the Rockford Park Board, and marked geological details in Black Hawk Park. Activism in and for parks and preserves was common among the supervisors, with a bird sanctuary in Glencoe, Illinois, named for Clara Dietz and Atlanta Memorial Park in Georgia commemorating Hattie Rainwater for her landscaping efforts that had been completed with the help of school children.[62]

The expanding opportunities in nature study at the turn into the twentieth century's first two decades enabled a number of men and women to follow career trajectories that moved across boundaries among normal schools, colleges and universities, and public institutions. Ellen Eddy Shaw, for example, was teaching nature study at the normal school in New Paltz, New York, in 1908 before she became educational director for the Brooklyn Botanical Garden, a position she held until she retired in the 1940s. Others may have found administrative work frustrating and, like Alice Rich

Northrop, spent only a few years doing it.[63] Some turned to writing nature columns for local newspapers or producing textbooks.[64] Others pursued advanced degrees and careers in science, such as Alice Catherine Evans, who was so exhilarated by science studies at the tuition-free nature study course for rural teachers at Cornell that she embarked on a career in science that included pioneering work in bacteriology. She eventually was elected the first woman president of the Society of American Bacteriologists.[65] Most of those who learned their nature study science in normal schools or institutes found it fully satisfying to implement it in classrooms across the country and some shared these experiences in *Nature-Study Review*.

NATURE STUDY TEACHERS

Scientists, educational philosophers, administrators, and school reformers advocated for nature study, and elementary teachers responded. Perhaps Wilbur Jackman—and certainly some of his natural science colleagues—envisioned a model in which they would suggest content and experts in education and psychology would determine pedagogical methods, leaving normal school and supervising teachers to oversee a translation for teachers who implemented their ideas. The result of this outlook, as educational philosopher John Dewey noted critically, was that natural scientists and other experts often viewed teachers as "channels of reception and transmission." He called for a more reciprocal collaboration that took into account the practical experiences of elementary and secondary educators.[66] In fact, during this sometimes-contentious period in the history of education, teachers actively asserted their right to both help design curriculum and shape the practice of teaching.[67] Ultimately, nature study was introduced widely because many teachers enjoyed teaching the subject and many children enjoyed learning it.

The "New Education," characterized in the press as promoting "unity, continuity, self activities and freedom," drew many teachers into designing courses for their schools and providing explicit advice about such topics as nature study, school gardening, geography, and hygiene.[68] Evidence of the day-to-day work of teachers is rare, typically caught as glimpses of particularly effective classroom outcomes rather than as accounts of the weeks and months of routine work.[69] Occasionally, personal correspondence and private journals reveal motives for teaching. Educational careers provided income and a certain status in the community, for example, as well as the flexibility to leave and reenter classroom teaching according to rhythms of personal lives, especially as districts abandoned laws prohibiting married women from teaching.[70] Teachers also expressed deep commitments to their pupils and to the community that supported their work, writing about nurturing students and inculcating social and civic values. Nature study seemed particularly open to such possibilities.

Cynthia Richardson's account of Cordelia Stanwood provides insight into the career of one New England nature study teacher. Coming from Maine to be educated in Providence, Rhode Island, Stanwood taught nature and art in several schools for nearly two decades, including some time spent with the nature study supervisor Fannie Stebbins in Springfield, Massachusetts. The work was sometimes stable and satisfying, but conducting classes was challenging. Eventually Stanwood returned to Maine to spend full time as an innovative photographer and writer. Details of her classroom activities remain obscure, but private correspondence and indirect evidence suggest that she found nature study teaching to be demanding and rewarding, but ultimately distracting from the work she herself wanted to pursue in nature.[71]

Incorporating nature study into an already full curriculum was a challenge and the results were not always satisfying. Flora J. Cooke, who had taught at Cook County Normal School and became principal of the independent school in Chicago named for Colonel Parker in 1901, undertook a tour to see how nature study was taught in the eastern and western states. Her conclusion was sobering: "Even judged by a most friendly eye" she found the teaching was "uniformly poor," with a few outstanding exceptions like the Ethical Culture School in New York City and the projects initiated by Clifton Hodge in Worcester.[72] In fact, she argued, in large cities where the public schools provided a supervisor to help provide materials and advice, the result was often that the teacher presented isolated objects from nature and that the correlated studies in language or art actually obscured learning more about nature itself. Cooke highlighted the tension teachers felt in the early twentieth century as they simultaneously sought to retain autonomy in their classrooms and asked for more support for special projects within a densely packed curriculum.[73] Nature study was presented in the schools, but its implementation did not necessarily meet the expectations of the early advocates or even of the teachers themselves.

One of the techniques that enhanced the capacity of practicing teachers was the local reading circle, sometimes promoted by administrators and book publishers and often initiated by teachers themselves. Teachers' reading circles may well have been an outgrowth of the self-help outlook of Chautauqua Literary and Scientific Circles founded in the late 1870s, but they became a phenomenon of their own in subsequent decades. The Ohio Teachers Association, an example carefully analyzed by Eric Lupfer, inaugurated a reading program with the goal of helping young teachers through a "course of reading, partly professional, partly literary."[74] Many Ohio teachers responded to the informal continuing education. Founders reassured newcomers that they need not be embarrassed by what they did not know in the circle or in their classrooms; indeed, they should encourage their own pupils to see them as "older and more experienced" students who could help find the answers to unexpected but interesting questions.[75] Reading circles for teachers were established throughout the country at the turn of the century, and their reading lists help identify topics considered to be current and significant in educational circles, with nature study figuring prominently in the first decade of the twentieth century.[76]

There was an edge to this extraction of extra effort from teachers, a reminder of their inadequacies, often articulated by the advertisers who sold them journal subscriptions and textbooks and by administrators who kept expanding their responsibilities.[77] Yet teachers gamely participated in institutes and reading groups, a signal of genuine commitment and of professional camaraderie. In Michigan, for example, a group of teachers produced a compendium of examples to share with each other, detailing the equipment and books needed for a nature study project and then describing how to conduct it. June Lewis of Kalamazoo described her efforts in detail in "Our Nature Room and Its Equipment." Rather than creating a complete reference collection, she and her colleagues allowed the designated room to develop slowly as children brought in objects and decided how to organize them. The result, Lewis wrote, was that "[i]n three years time, our room has changed into a veritable Nature Curio Shop or a small Natural Museum. Here the children are likely to wander around and become acquainted with the things that have lived about them but that they have never really seen before."[78] The nature room had specimens of all types, including an aquarium with polliwogs and a bird-feeding station on the fire escape just outside the window. Sponsors of the volume noted that the teachers' essays captured on paper "the joy of their children."

Opportunities to teach and to specialize in nature supervision were publicized through informal placement practices that relied primarily on interpersonal networks of normal school and university faculty. Mary L. Cheney, Appointment Secretary at the University of California at Berkeley, also served as a placement officer and relied on the faculty to tell her about promising candidates. Joseph Grinnell, founding director of Berkeley's Museum of Vertebrate Zoology, worked with students like Lulu Marie Newton, who was experienced on birding trips and had "a charming personality, and enthusiasm."[79] He facilitated the placement of nature study teachers and leaders and remained in contact with some graduates who went into education. Violet D. Holgersen reported to her mentor, for example, that her first six weeks on the job in Pasadena had been "more than a trial—it was a nightmare." She faced five classes of tenth graders and a total of 115 students. Some boys tried the "usual pranks," but when she adeptly handled a snake "like a snake charmer," the problems began to resolve. She missed the Bay Area and wanted to move back, even though she felt that nature work was much better supported in Pasadena than in Oakland and Berkeley. Her correspondence also provides a favorable glimpse of the effective supervision of a Miss Pierson over nature study in Pasadena's system.[80] Another Berkeley student taught in a county school in Los Angeles and found that her upper-grade pupils were each assigned a plot ten feet wide and thirty feet long. She reported to Grinnell, "See what I got myself into! Fifty-six children with that many gardens! Well, I didn't know much about agriculture, nor much about the Japanese method of ditching, but I learned it."[81]

In the late 1910s, however, as teachers took three- and four-year normal school degrees and nature study was more widely offered, their involvement in summer institutes, reading circles, and other supplementary programs slackened. By 1919, when Bigelow tried to convince his former teacher Charles Davenport that he should offer summer courses in natural history for nature study and elementary science at Cold Spring Harbor, the latter replied that he had offered such a course for two or three years and concluded there was "no real market."[82] Some of the special summer courses and programs for teachers were successful and persistent, however, especially at strategically located schools, like Massachusetts Normal at Hyannis or California State Normal School at Chico; they could readily attract teachers from well beyond their local constituencies to summer programs at the seaside or in the countryside near Mt. Shasta in northern California.[83] Ongoing summer programs in more ordinary locations incorporated new topics like hygiene or school gardens as these "came into vogue." Teachers who had been in the classroom for a number of years needed to learn the new, sometimes required subjects.[84] As schools consolidated and systems became larger, however, administrators hired experts who could more directly assist and advise classroom teachers during the actual school year.

SUPERVISORS OF NATURE STUDY

By the turn of the century, a significant number of schools had established a distinctive position of "supervisor" for nature study. The appointee might be an experienced teacher or well-trained biologist who visited schools to advise on curriculum, train teachers through special classes, or provide local materials.[85] In what Patricia Graham has called a "golden age for women school administrators" because of the number who reached prominent positions, several women (and some men) found significant careers in the nature study movement, thus creating yet another niche of "women's work in science."[86] (See appendix B for a partial list.) The school consolidation movement created more administrative tiers, and thus openings for ambitious teachers who wanted to pursue special educational interests beyond the classroom. Those engaged in nature study varied as to their status, salary, and responsibilities, with some supervisors in central administrative offices while others were specialists who worked directly with children in classrooms.[87] Whether encouraging hesitant teachers or marshaling nature study specimens, the supervisors initiated projects and facilitated classroom work.

Massachusetts, where Lucretia Crocker had initiated a museum-based program for Boston-area teachers, had one of the earliest models for encouraging teachers and providing them with materials. In 1890 Sarah E. Brassil was appointed to oversee nature work in nearby Quincy. Neighboring systems soon followed in Brookline, Newton, and other growing streetcar suburbs surrounding Boston. Sarah Louise Arnold held the post in Boston in the late 1890s, having established a reputation for such work in Minneapolis.[88] Most of the early supervisors had normal school certification, but college graduates also pursued this career. Mary Dartt, with degrees in botany and zoology from Wellesley College and with course work in psychology as well, wrote of her aspirations to become a supervisor of nature work in the grammar grades to John Spencer at Cornell. Dartt felt that nature study was still "something of a fad except in the more progressive communities" but expressed optimism that the movement was growing.[89]

On the West Coast, as local school principals found that the new nature study curriculum required additional resources and leadership, they hired graduates from Stanford and from the University of California at Berkeley. A number of Stanford graduates became supervisors in the central California region, including Effie Belle McFadden, who organized nature study courses for the Oakland schools from 1897 to 1900, and her successor, Bertha Chapman (later Cady).[90] A number of the Stanford botany graduates went directly into high school teaching, including Elizabeth Merrill Bab-cock who taught at Fresno High School and Maud Whitcomb Mercy who taught at John Marshall High School.

Effie McFadden's responsibilities included visiting the rooms of partici-pating elementary teachers, holding meetings with them every two weeks to discuss details of classroom work, helping them gather materials for pupils to use, and solving problems they encountered. She also took some school classes on short field trips as a demonstration for their teachers.[91] The Oak-land program became a regional prototype for other schools adopting na-ture study, including the city of Stockton, whose published materials gained national attention at the St. Louis World's Fair.[92] When McFadden joined the faculty at San Francisco Normal School, she brought her enthusiasm for nature study and, through Jordan at Stanford, arranged for Anna Comstock to lecture to a standing-room-only audience of San Francisco teachers.[93] One of McFadden's former nature study teaching colleagues found the San Francisco training school housed in a dingy old building but reported favor-ably on the aquaria, "froggeries," and insect breeding cages in every room, noting "much has been done by way of pictures and other decorations to make the rooms attractive" and thus encourage teachers to visit.[94]

McFadden's successor at Oakland was Bertha Chapman, who graduated from Stanford in 1895 with a degree in English, then studied entomology under Vernon Kellogg, completing her masters degree in biology in 1902. She also studied economic entomology at the University of California at Berkeley and spent a summer in the marine laboratory in Pacific Grove before taking a teaching post at Paso Robles.[95] While teaching in the high school, she supervised elementary nature study work, experience that led to her full-time appointment as supervisor of nature study in the Oakland schools in 1901. In the meantime, Kellogg introduced Chapman to Anna Comstock and the two remained lifelong friends and colleagues. Taking over for McFadden, Chapman adapted the "Course of Study" for the Oak-land teachers and began her own local Junior Naturalist leaflet series for school clubs.[96] Within a few years she "provided a splendid object lesson in the use of school gardens as a working center around which to group most of the nature study activities of the schools."[97] She coordinated efforts to help "refugees" from the San Francisco earthquake of 1906 develop school gardens, projects intended to occupy the displaced children and produce food for their families.[98] Chapman apparently considered attending Cor-nell for a graduate degree but ultimately chose to go to Chicago in 1907 to work under Otis W. Caldwell. In 1908 she married Vernon Mosher Cady and moved temporarily to Kansas City, Missouri.[99]

Bertha Chapman Cady's career, like that of many women, was not linear, as she followed her personal interests, tested potential alternatives, coordinated her career with that of her husband, and responded to the demands of raising children. She accepted positions that took her from California to New York and even spent some time lecturing on social hygiene throughout the South. The illness of her parents brought her back to California, where she began taking classes at Berkeley. When Cyril Stebbins, professor at the Chico State Normal School, left for Washington to work with the National School Garden Committee during World War I (fig. 20), Cady took his position temporarily and taught nature study along with health and hygiene.[100] The return to the West was bittersweet, for she reported that she was having a "happy time" at her old occupation of nature study but felt "rather far away up here in a new part of California among strangers," characteristically softening her comment by adding "but they are lovely to me." By 1921 she was determined to complete her Ph.D. in education and did so at Stanford in 1923. She returned to New York City, lecturing to parents and teachers on social hygiene. This turn in her career led her to a final professional position, offered by her former Stanford classmate, Lou Henry Hoover, who had become national president of the Girl Scouts. Cady was appointed Girl Scout Naturalist and directed its programs relating to nature until her retirement in 1935. With a grant from the Laura Spellman Rockefeller Fund, she helped establish the Coordinating Center on Nature Activities, which supported camping and the national parks. She also picked up the mantle of Anna Comstock by serving as Nature-Study Society president.[101]

Another enterprising Stanford graduate, Helen Swett, became supervisor in the nearby Alameda City School District from 1900 to 1902 and enthusiastically promoted the curriculum beyond her district and at meetings of the California State Teachers Association.[102] Swett's father had been superintendent of the California school system, but she independently found a position working on nature study with elementary teachers while she taught botany and zoology in the high school.

Responsibilities could be daunting as she prepared special classes for teachers, addressed state teachers' association meetings, and conducted classes in various short-term institutes.[103] She believed she was underpaid because she was expected to teach a half day at the high school and use the rest of her time for nature study supervision while earning less for both tasks than she would have been paid for a full-time position.[104] She carried on with monthly earnings of $50 for the high school biology and $30 for the nature study work because the latter "means more to me than the high school experience as far as nature study publishing is concerned." Her early

FIGURE 20. The school garden movement was well underway in nature study programs across the country and thus provided a significant base for the School Garden Army during World War I, as shown here in a propaganda poster. World War Poster Collection, Manuscripts Division, University of Minnesota Libraries.

career is recorded in a wonderfully descriptive series of letters written to her fiancé while she finished her education and began to teach school. She drew on her acquaintance with more experienced teachers like Chapman and McFadden and visited the Stockton schools, which already had a well-established nature study curriculum.[105]

During and after school hours Swett gave lessons to nature study teachers, went on expeditions to acquire "critters" for her own classes and those of her teachers, and provided demonstration sessions. One Saturday, for example, she records:

> I started off on my wheel toward Hayward, alone, in search of sunshine, fresh air, and a creek. I found them all without having to go much more than about three miles out of town. I had my lunch and lots of bottles, etc. along so I ate the one and collected critters in the others to my heart's content. My captures were as follows (most of them were brought home alive): four stickle-backs, two small toads, one water snake; two small lizards; one pair of handsome orb-weaving spiders, three other spiders, one back-swimmer; three water boatmen; four grasshoppers of a species I have never seen before; watercress seed; a kind of green algae new to me; a new water grass; a dragon-fly, caught in a spider's web; and several small water forms which I have not yet had a chance to examine under the microscope. How's that for one day's haul?[106]

She also used the local library to get information and to answer questions about specimens. Once she became so absorbed in her studies that she was accidentally locked in and had to escape through a window after the library closed.[107]

Taking care of her specimens was often challenging. Swett's silkworm eggs began to hatch almost before the mulberry and Osage orange leaves needed for their food had begun to bud, and more than once she had to hurry down to San Francisco Bay because she needed additional saltwater for her "sea-critters."[108] She kept Vernon Kellogg apprised of her efforts, and she arranged to get copies of Cornell leaflets to use with her teachers. She usually bound up the individual leaflets she had received because "if they have a distinctive cover on them, they are much more likely to get returned to me than if I loaned them in their present form."[109] As she was becoming an active leader in the movement in California, presenting at the Nature Study section of the state teachers association meeting, for example, her fiancé died very unexpectedly. Devastated, she soon left the school system for charity work and apparently never returned to her nature study teaching.[110]

FIGURE 21. Alice Rich Northrop, a graduate of Hunter College, had a long-standing interest in nature study and coordinated numerous field trips for New York City school teachers, like those pictured above waiting at the West Park Station. Schlesinger Library, Radcliffe Institute, Harvard University.

In New York City, Alice Rich Northrop coordinated a group of Hunter College alumnae and worked with Henry Fairfield Osborn at the American Museum of Natural History. Osborn had been her zoological husband's supervisor at Columbia University when he was killed by a laboratory explosion, and Northrop and Osborn produced a book on her husband's fieldwork in the Bahamas. While she had no official appointment, she worked persistently with nature study teachers, coordinating weekend field trips for the inner city teachers, helping organize school exhibitions, and encouraging coordination between schools and other New York institutions, including the museum and botanical gardens (fig. 21).[111]

Not all nature study supervisors were women. Both Dr. Gustave Straubenmueller and Dietrich Lange were active naturalists who carved out significant careers coordinating aspects of nature study. Straubenmueller worked on various projects related to science teaching for fifty years in New York City.[112] There he held a number of administrative positions in the city school system, including supervisor of nature study, and served as vice president of the School Nature League. The League furnished "nature rooms" in the city schools, held biannual flower shows, and created permanent "School Service" space at the American Museum of Natural History in the 1920s. Straubenmueller helped nature study teachers connect with city programs and institutions, both publicly supported and philanthropic, including vacation schools and school gardens. Dietrich Lange started with a normal school degree and then earned a bachelor's degree from the University of Minnesota while teaching in the St. Paul schools. Involved in nature work through excursions for teachers and participation in summer school programs at regional normal schools, he was officially Supervisor of Nature Study from 1904 to 1907, when he became the principal of the Humboldt School.[113]

In the 1920s, some of these supervising women and men organized a National Council of Supervisors of Nature Study and Gardening to provide a forum for discussing their professional roles as administrators.[114] The original group formed as the National Council of Garden Teachers but changed its name in 1926. The society's history is elusive, but it may have been formed in reaction to Van Evrie Kilpatrick's leadership of the School Garden Association. A continuing antagonism apparently kept the two groups from working collaboratively, at least that was the opinion of the National Council's president in 1926, E. Laurence Palmer of Cornell.[115] However, the supervisors maintained a cordial relationship with the Nature-Study Society and met with the National Education Association in conjunction with its Department of Superintendence; it joined with other societies in 1929 to publish *Science Education*.[116] Its presidents were typically normal school faculty and, besides Palmer, included John L. Randall of Fitchberg, Clarence M. Weed of Lowell, Lester S. Ivens of Kent, Alice Jean Patterson of Illinois Normal, John A. Hollinger of Pittsburgh, and Theodosia Hadley of Kalamazoo in the 1920s.[117]

Aside from formal supervision, nature study educators provided other resources and instruction that served elementary teachers and their pupils. The well-respected horticulturalist Kate Sessions, for example, supervised school gardens and taught private classes on botany for teachers in San Diego.[118] She also developed educational programs for the San Diego Zoo in Balboa Park, whose school bus brought classes to visit one of the largest and best-stocked zoos in the country.[119] Ellor E. Carlisle may not have had an official supervisory appointment, but he coordinated supplies for the teachers in New Haven, Connecticut.[120] In a period that exhibited considerable civic interest in nature studies, there was considerable permeability and exchange among institutions that provided material and intellectual resources for teachers as well as for pupils, including natural history museums, botanical gardens, public parks, zoos, and even private nature reserves.

Something of the public interest and visibility of nature study education is captured in one of the best-selling fictional books for young adults in the early twentieth century, *A Girl of the Limberlost* (1909). The tales of the Limberlost swamp area in north-central Indiana sold in the tens of thousands to readers also caught in the borderland transition between rural and urban life.[121] The author, Gene Stratton Porter, admired Anna Comstock for her nature sensibility, and the protagonist, aptly named Elnora Comstock, is an adventurous young woman who becomes intimately familiar with her local endangered landscape and makes a modest income by her ability to identify and prepare natural history specimens for adult collectors. Fortuitously, this leads to her appointment as advisor on natural history for the city teachers of Onabasha. Her responsibility is to "spend two hours a week in each of the grade schools exhibiting and explaining specimens of the most prominent local objects in nature: animals, birds, insects, flowers, vines, shrubs, bushes, and trees." These specimens and lectures are carefully "appropriate to the seasons and the comprehension of the grades."[122] Elnora Comstock rises to this and other challenges, a modest but strong heroine who makes good use of her particular talents. Eventually this young woman of the Limberlost becomes a teacher and acquaints her own pupils with the details and natural attractions of their locale. The narrative paralleled aspects of Porter's own life and interests, although she taught primarily through her books.

WOMEN, GENDER, AND NATURE STUDY

Threading through this chapter's account of teachers and supervisors is an account of women's careers in nature study. Women's suffrage reform efforts and the emergence of controversial "new women" with advanced degrees and career aspirations made gender issues prominent in public and professional discussions in the early twentieth century. It was a heady period of change and apparent opportunity but also fraught with tension. The women who went into education were aware, for example, that their proportion of representation in the system decreased in the higher administrative ranks, although the few prominent exceptions sustained the aspirations of some women even as it led to reaction from traditionalists.[123] Nature study had opened midlevel positions for women in supervision, but such participation of women in the natural sciences prompted not-so-subtle arguments from men who believed that women were not suited for the work. The reaction and challenge also came in related fields like forestry and conservation, in which masculinity was asserted as essential. Many male educational leaders emphasized that they were "rational, practical, and—above all—unsentimental." To distance themselves from effeminacy, they often excluded women from their organizations to avoid guilt by association.[124]

Historians have analyzed the sexual divisions of labor in teaching. Statistics reveal that men, on average, taught older rather than younger students, taught more boys then girls, and taught the "harder subjects." Moreover, men were encouraged to teach about ideas and organize the profession, while women were assigned to teach "the ABCs and the virtues of cleanliness, obedience, and respect."[125] An analysis of the membership of the Nature-Study Society (using its one published list) underscores that pattern.[126] The men on the roster tended to be on college or normal school faculties (roughly three-quarters of those identified as men by given name), whereas the women were far more likely to be teaching in schoolrooms and implementing the programs assigned—while adding their own innovations. The differences in responsibility correlated as well with articles published in *Nature-Study Review*, with male faculty and administrators more likely to contribute essays on the definition of nature study, while women were more often authors of reports on classroom practices and results.[127]

This not-so-hidden status hierarchy was shaped by employment preferences that, in turn, reflected perceptions about training and the aptitude of women for science and leadership. Moreover, the high visibility of women like Ella Flagg Young, the nationally known superintendent of schools in Chicago, and the growing proportion of women in education (with a historical peak in 1921–22 when women comprised 87 percent of elementary and 64 percent of secondary teachers) seemed to fuel a backlash. As Kim Tolley and others have pointed out, administrators and male politicians worried in print about the "feminization of education" and the "monopolizing woman teacher."[128] Men in higher education even created an all-male fraternity, Phi Delta Kappa, with the original intention of finding and placing men, rather than women, on the faculties of universities, normal schools, and secondary schools.[129]

Faculty who trained teachers in nature study encountered this issue regularly. Liberty Hyde Bailey of Cornell reassured the parent of a prospective male student in his program that teaching agriculture and nature study had developed through "direct demand" and was a good alternative for "young men" who did not want to go into agriculture or do research.[130] When asked explicitly about whether women were being engaged to teach these subjects, Bailey replied candidly that the "tendency will be to put men into those positions so far as possible" with "an excellent opening for [male] teachers in nature-study who can also give an agricultural turn to the teaching; or who will teach nature-study agriculture."[131] Men more often moved directly into faculty positions at the normal schools, especially as these institutions tried to upgrade their status by offering four-year programs after the turn of the century. The Los Angeles Normal School, for example, deliberately requested a man to teach nature study, seeking a candidate of intellectual stature "whose character and personality are beyond criticism."[132] There were sometimes requests for women, often from women's colleges that offered teaching as part of their mission, as when Lillian W. Johnson, president of Western College for Women in Oxford, Ohio, asked Bailey's recommendation for a woman with an M.A. degree to teach classes in botany and nature study, as well as biology and physiology.[133]

In this relatively new field, the issue of teachers' preparedness for teaching science was a serious one. Although many normal schools with multi-semester programs had introduced basic science into their curriculum by the 1890s, nature study required a different preparation and a method that could not rely simply on learning taxonomic descriptions or basic physical science principles. The assumption that teachers ought to invite unanticipated questions may indeed have daunted young teachers who were "discouraged at the thought of it," according to unnamed administrators quoted in a textbook advertisement.[134] Concern about how to respond to a range of issues of the type children might ask on field trips spurred some teachers to attend institute courses and in Oakland had led teachers themselves to hire Effie B. McFadden to help them prepare for nature study.[135]

Despite such personal initiatives and the myriad of summer institutes, weekend programs, publications, and public lectures, the perception of inadequate teacher preparation persisted.[136] In introducing his textbook, John M. Coulter expressed doubt that even a good manual would be sufficient to help teachers manage a chaotic and ill-defined nature study curriculum thrust upon "unwilling or unprepared teachers." Too often, he lamented, nature study was "the despair of the primary teacher and the joke of the scientific fraternity."[137] Stories, some probably apocryphal, reinforced the presumed ignorance. One recounted a young woman who came into a New Hampshire public library and asked for a book on worms but quickly returned it saying, "I want worms that turn into butterflies." Apparently embarrassed, she then admitted, "I don't know anything about the subject, but I know the proper method of teaching it."[138] Such tales reinforced an ongoing argument about whether teachers, most of them women, were adequately prepared to teach nature study.

Among the most outspoken skeptics was the psychologist Edward Thorndike, who believed that science could only be taught by very "gifted" individuals.[139] While claiming to be neither for nor against nature study, he argued it was too often tainted with the "vice" of "sentimentality." He endorsed a domestic ideology about education that argued that women were particularly appropriate for teaching young children.[140] At the same time, he expressed concerns about "feminization" of male fields and certainly did not think women were suited for the rigors of science.[141] His bifocal lens that could envision women in elementary teaching but did not allow for women to do science kept him from gaining a clear focus on careers of those who were teaching nature study. Fellow psychologist G. Stanley Hall chimed in with a similar bias in his preface to Clifton Hodge's textbook: "Many modern nature books suffer from what might be called effeminization." His endorsement argued that *Nature Study and Life* was "a book written by a man and appeals to boys and girls equally," implying that those written by women did not appeal to boys.[142] The head of the Los Angeles Normal School weighed in with "The Need for More Men in our Public Schools," which argued that boys must get an introduction to science through strong role models.[143] Charles E. Bessey reflected his personal bias when he suggested that nature study had acquired a "namby pambyism which the healthy minded boy so properly hates and despises."[144] Bessey readily encouraged talented women graduate students like Marshall while remaining unclear about what they might do with their degrees—fully aware of the prejudices they faced.[145]

Demeaning comments could be harshly direct or more casual. One critic, C. B. Wilson, posed the issue starkly: "Ask the college professor what he thinks of nature study in our graded schools, and although his inbred courtesy may retain the sneer or the smile, his love for the truth will compel him to reply, as many of them do, that he would prefer to have his students receive no training at all rather than the one they get in graded schools."[146] Wilson went on to suggest that "the teacher"—always referred to as "she"—simply did not have adequate background in science.[147] Sometimes the commentary was sweeping, as when George N. Woods criticized rural schools (nearly all run by women in the early twentieth century), as having an "inefficient and family type management . . . with all the looseness that usually accompanies such management."[148]

Such gendered and sometimes hostile rhetoric was widely used in educational journals and contributed to the attack on the so-called feminization of education in the early twentieth century. But it did not go without challenge. In 1914 Anna Botsford Comstock responded directly to the arguments of two professors who insisted that nature study had "failed" because it was taught by women.[149] The unnamed men, participants in a Chautauqua panel on the question "Has Nature-study in the Schools Been a Failure?" did not publish their comments; but Comstock's editorial response revealed their arguments. One man had claimed that women teachers retarded the growth of nature study arguing: "Women are not primarily interested in science. If they are interested in birds, it is to have them in a cage to show the children; if they are interested in blossoms, it is for dining decorations. Women do not care for science, for the abstraction." The other suggested that "fluffy, moonshiney nature study" was a failure and agreed that women "are fundamentally interested in humanity and romance, not in science," although he qualified his argument by suggesting that "genuine nature study" could be endorsed by scientists. Comstock's retort was that the men's "arraignment of the [female] sex" did not match the reality of women whose work in science and other fields showed an "effectiveness equal to the opposite sex's." If nature study had problems, she suggested, it was that too many [male] school principals, bred and educated in the classics, discouraged nature study and tried to keep "the feminine mind at [only] its ladylike tasks in the schoolroom," rather than encouraging teachers to pursue challenging and outdoor projects.[150] Karen Cruikshank has shown that women teachers typically bore the brunt of criticism directed at the schools in this period, buffeted by administrators, school boards, and parents who all presumed to know ways the job might be done better.[151]

IN THE RANKS

Nature study teachers, women and men, showed remarkable commitment and creativity despite commentary about the quality of their training, the time required for class preparation, and the difficulties faced in the classroom and beyond. Offsetting tales of ineptitude are the reports of teachers who wrote with enthusiasm about their experiences in textbooks, pamphlets, and *Nature-Study Review*. Confident teachers raised school gardens, took their pupils on nature walks, coordinated school museums, and provided classroom terrariums. One historian noted that teachers came from the countryside in disproportionate numbers, making them well aware of nature even when they were not trained scientists.[152] Undoubtedly their skills varied, but teachers' curiosity and commitment are recorded in reports to local supervisors as well as national magazines that described how they guided their students to record daily weather observations, skin and prepare zoological specimens, and pour oil on the top of swampy ponds to limit the mosquito-based threat of yellow fever.[153] Much of the nature study activity required uncompensated effort in terms of preparation beyond the classroom. One teacher wrote, perhaps facetiously or perhaps in frustration, to John Spencer at Cornell that he ought to open a register to "keep a record of all teachers who have participated in this unrequired work" so that the public could know about dedicated teachers whose "first thought is for the welfare of your pupils rather than to hold your position and draw your salary."[154]

Many of the successful teachers rose through the ranks to become leaders in nature study programs across the country, and women were among those moving forward.[155] Indeed, women in the Nature-Study Society represented about a quarter of all members teaching nature study at normal schools and colleges of education, a proportion close to the percentage of women faculty in women's colleges at the turn of the century.[156] Throughout the movement a significant number of women wrote textbooks, readers, manuals, leaflets, and other kinds of materials for teachers and for their pupils. They continued, however, to face prejudices about their work and its marketability. Thus, when David Starr Jordan proposed a collection of animal stories compiled by his wife, the publisher D. C. Heath was candid in his response, recording enthusiasm for the idea but expressing a gendered concern:

> Of course we would much rather have your name on the title page than Mrs. Jordan's, for reasons you both will readily understand. If Mrs. Sheldon Barnes' Histories had been published under the name of a man they would have had a much wider constituency I think, and yet since the book under discussion is of a quite different nature, it might do just as well with a woman's name on the title page. We have no doubt Mrs. Jordan's work will be all right, and yet if we can have your name on the title page, we should prefer it, you giving her credit in the preface for such work as she does in connection with it.[157]

Jordan and many others like Liberty Hyde Bailey, Charles Bessey, and Joseph Grinnell privately encouraged women who taught, wrote textbooks, or supervised nature study. Bailey, for example, frequently recalled that his own interest in nature had been spurred by a woman teacher when he was young. In turn, he offered encouragement to his own advanced students, reassuring one young woman, "You quite underestimate your own knowledge because you feel the responsibility of your work."[158] These men seldom, however, made their support for women explicit and public.

Classroom practice and supervision reveal that the theoretical arguments that undergirded nature study proved less important for many teachers than their very practical concerns about information and resources. Those already teaching found that training through institutes and at normal schools could be of critical importance when nature study was introduced. These supports, along with textbooks and materials from supervisors, were not sufficient to forestall criticism of nature study. Descriptive terms like "soft" and "sentimental" were often code for women's participation or influence and reflected deep prejudices that constrained women's opportunities. Enterprising women turned such charges on their head, however, as when Kate Van Buskirk gave this matter-of-fact advice to teachers: "Don't be sentimental—the bird eats the worm, so dwell on the bird's need for food."[159] Nature study education legitimized women's public discourse about the complicated world of nature as self-confident women taught institute classes, published books and articles, established school gardens, and coordinated programs for entire school systems. Nonetheless, ongoing debates about teacher preparation, appropriate content for nature study classes, and insuring effective classroom methods compelled advocates to establish the Nature-Study Society as a professional base for the national nature study movement.

CHAPTER SEVEN

Forging an Institutional Base

I have always found encouragement rather than dismay in the multitude of opinions which cover the field of nature-study. They all bear testimony to the largeness and vitality of interest in this "problem of problems."
Clifton F. Hodge[1]

THE LAUNCH of *Nature-Study Review* in 1905 by Maurice A. Bigelow, a faculty member at Teachers College in New York City, marked what was perhaps the high point of the nature study movement. Pamphlets, books, journal articles, public addresses, and summer institute and normal school course lists reveal that nature study was by then a staple in progressive education. Teachers in most regions of the country now knew about nature study and many were implementing it in their classrooms. Its ubiquity revealed to advocates and critics alike that, as the term had grown in familiarity, it had also found varied expression, as Clifton Hodge suggests above.

This chapter explores how Bigelow, as editor, provided a forum for discussion that balanced theory and practice and addressed the concerns of teachers, administrators, and educational psychologists. He framed the *Review* and its related Nature-Study Society as mechanisms for stabilizing a definition of the curriculum and appropriate coordinate practices. In the rapidly professionalizing context of education, he turned to "leading men" in universities and normal schools to provide authoritative voices that could quell scientists' skepticism about whether nature study deserved a fundamental role in elementary education. At the same time, the *Review* engaged nature study supervisors and teachers who were implementing the

curriculum in their schools, gave them opportunities to share their classroom practices, updated them on current textbooks, and, not incidentally,
tried to develop a substantial subscription base for the journal.

By 1905, although some teachers still had considerable autonomy in their
classroom, schools were being consolidated, graded classrooms offered
mandated courses of study, and a cadre of specially trained administrative
experts was standardizing education under city authorities, often with reference to state guidelines as legislatures provided educational funds. With
its emphasis on neighborhood nature exercises and its links to international
pedagogical theory, nature study seemed positioned to negotiate some middle position that allowed for creative initiatives on the part of teachers even
as it acknowledged guidance from educational theorists and research scientists. While other publications had provided materials for teachers or for
their pupils, the *Nature-Study Review* was the first focused effort to establish
a professional journal specifically on the topic of nature study.[2] The *Review*
joined a significant number of specialized educational journals, many affiliated with colleges of education and normal schools. Its ambitious young
editor not only established nature study as professional subfield but also
announced plans for a society in the *Review* shortly after its founding, thus
setting up another mechanism intended to direct a movement already well
under way.

Maurice Bigelow, with a Ph.D. in biology and a commitment to elementary and secondary education, was well aware of the growing debate about
whether nature study was an appropriate way to introduce pupils to the
study of the natural world (fig. 22).[3] Biologists, in particular, sometimes
voiced their disappointment that nature study was not more dedicated to
presenting fundamentals of botany and zoology. Educational philosophers
were disturbed by ambiguities in the pedagogical theories undergirding nature study and apparent inconsistencies in practice. It did not help that nature study leaders did not agree among themselves on method and content.
Even strong supporters were uncertain whether the tendency to align (or
coordinate) with such other educational projects as conservation and human hygiene strengthened nature study or diverted its attention from basic
goals. Its breadth and versatility, early lauded for accommodating regional
and school differences, increasingly made nature study inchoate to outside
observers and sometimes to teachers themselves, although initially for the
most committed advocates, like Hodge (quoted above), dissent was a
sign of vitality.

Bigelow was optimistic that inviting the critics, both those within the

FIGURE 22. Harvard-educated Maurice A. Bigelow taught nature study and biology at Teachers College of Columbia University, where he initiated and edited *Nature-Study Review* for its first five years before becoming active in social hygiene activities.

movement and skeptical academics outside it, to state and debate their positions in print would bring greater clarity and strength to the movement. Within two years after starting the *Review*, Bigelow announced the formation of the American Nature-Study Society to provide a more dynamic and immediate forum in which to engage active administrators and teachers in direct discussion. He had a remarkably open editorial policy that, in fact, reveals the myriad of opinions and programs associated with nature study. His ambition was to demonstrate the importance of the subject, establish consensus with regard on methods, and provide clarity for a movement that was becoming controversial among academics and remained a challenge to teachers. In 1905 the prospects for all of this seemed positive.

The lack of coherence was less disturbing because other subjects, including mathematics, geography, and writing, were experiencing similar discussions about their curricular content.[4] Earlier efforts to "capture" the multifaceted nature study movement with state and city curriculum materials had not produced consensus on standards. But the term and programs identified with it were regularly presented and published in the *Journal of Addresses and Proceedings* of the powerful National Educational Association (NEA) as well as Clark University's *Pedagogical Seminary*, the University of Chicago's *Elementary School Teacher*, and other educational journals. Scientific publications, too, had been ready to put their imprimatur on the curriculum that seemed poised to put the natural sciences into public classrooms. The newly revitalized *Science* magazine edited by James McKeen Cattell in conjunction with the American Association for the Advancement of Science (AAAS), for example, published short advocacy pieces by such leading scientists as Stanford's president, the biologist David Starr Jordan.[5] To acquaint scientists and interested colleagues with the movement, the University of Chicago ecologist John M. Coulter described the possibilities in "Nature Study and Intellectual Culture" for *Scientific American* in 1897.[6] Nature study also attracted public attention by its high visibility at the 1904 St. Louis World's Fair, where education was the centerpiece of its exposition. Several educational associations held their annual meetings on the fairgrounds, and entire elementary classrooms were on display.[7] Enterprising nature study supervisors brought exhibit materials from their schools, and Wilbur Jackman gave a keynote address intended to show national and international visitors that nature study was a leading subject in American progressive education.[8] Nature study seemed visible and widely practiced; it only needed to consolidate its program.

Even as Bigelow made plans for the journal, he encountered critics of the curriculum. He reported in his first issue that some "eminent professors . . . were so firmly convinced that nature study is simply a dangerous fad that they counseled against attempting to give the subject recognition in a special journal."[9] Undeterred—and willing to confront critics directly—he published a negative report on American nature study issued by visitors associated with the Moseley Educational Commission in Britain, which had been particularly harsh about the lack of science content.[10] Bigelow probably was not unsympathetic, and he believed that such commentary needed to be aired and addressed to show that the fifteen-year-old nature study curriculum had vitality and staying power.

NATURE STUDY OR SCIENCE STUDY?

Initially many naturalists and social scientists had welcomed the new nature study curriculum, but by the late 1890s, some were asking whether nature study textbooks and classroom practices provided the essentials for later scientific study. Their comments, though typically over generalized, revealed several key concerns. Was the content sufficiently scientific and would it lead pupils toward systematic study in the future? Was it appropriate to use imaginative literature and poetry in conjunction with nature study? Were elementary teachers sufficiently prepared to guide students studying in and beyond their classrooms? The very enthusiasm propelling nature study's rapid dissemination had led to a wide range of practices; the critics wanted a more concrete curriculum and more scientific vocabulary.

Because nature study was described as experiential learning by advocates at major universities, it initially enjoyed support from academic scientists who knew leaders like Liberty Hyde Bailey and Wilbur Jackman personally. At Cornell, Bailey's enthusiasm was contagious, and he correctly assessed that there the "scientific men are closely in touch with elementary education."[11] His colleagues agreed that pedagogical emphasis on observation, comparison, detailed description, and judging the value of facts could prepare pupils well for later scientific study. Several faculty colleagues contributed to his leaflet series. They understood the subtlety in Bailey's widely read book *The Nature-Study Idea*, which declared, "Nature-study is not science," but then emphasized in their contributions precisely the observational skills they knew to be central in field and laboratory research.[12] Nonetheless, I. P. Roberts expressed his unease in the New York Agricultural Experiment Station's *Nature-Study Quarterly* in 1899: "Much that is called nature-study is sugar-coated. This will pass. Some of it is mere sensationalism. This also will pass. With the changes the term 'nature study' may fall into disuse; but the name matters little, so long as we hold to the essence [of using the natural world to teach]."[13] At Chicago, too, the academic faculty offered qualified encouragement. Jackman's colleague, the ecologist John Coulter, warned in *Science* in 1896 that the purpose of nature study must be to emphasize the unity of nature, but "the power of appreciation developed by the humanities must always be tempered by the scientific instincts."[14] Elsewhere the criticism was particularly biting because, Kimberly Perez has pointed out, there was a significant disconnect between scientists who sought a strictly objective outlook and "reluctant modernists" wrestling with scientific complexity and, not incidentally, their own privileged status.[15]

Nature study was a lively topic at meetings of the NEA. In 1900, Charles B. Wilson, president of the NEA Department of Science Instruction that year, expressed sympathy with the "typical college professor" who "feels somehow that this nature work is mere child's play, and he considers it worse than useless, because it inculcates into the child's mind wrong principles, unscientific methods, and inaccurate data at the very period when that mind is most plastic and most receptive to such things."[16] Others had a positive opinion of the techniques of nature study, including Emma G. Olmstead, principal of a training school in Scranton, Pennsylvania, who asserted in 1903 that "nature study must be by the child not the teacher."[17] Still, Otis W. Caldwell, then professor at Eastern Indiana Normal School, warned the following year that nature study should be "less legendary, mythical, imaginary and impersonating." Caldwell, who had a Ph.D. in biology from the University of Chicago and would help develop a national program in general science on the secondary level, insisted that scientists would give their approval only when "elementary science is elementary in a scientific way."[18]

The fiercest critic was the psychologist Edward Thorndike, whose rapier-edged style sliced through the earnest rhetoric of nature study defenders, but his attention to this project suggests that nature study was a territory worth contesting. During his year teaching at Western Reserve College in Cleveland he had observed teachers working with pupils in its practice school. His conclusion was that the teachers indulged in "sentimental conventionality" as they sought to instill a love of animals and plants as a substitution for knowledge of them.[19] Too many elementary teachers, he argued, obscured the fact that nature was "red in tooth and claw." Moreover, Thorndike noted tartly, "St. Francis is no patron saint for science or science teachers."[20] From his point of view, the use of plants and animals for all human purposes was ethical. Children should learn early, Thorndike argued, that the world was a collection of materials to be used by humans. Efforts to teach "love and joy" in the natural world would limit the advance of knowledge derived from dispassionate investigation of living and inanimate objects. Responding to the argument that children were not ready for science, he retorted, "If real science is not fit for children, let them go without it, but let nothing be taught under its banner which is not worthy of the name." After Thorndike joined the Teachers College faculty in 1899, he made no more explicit challenges to nature study, perhaps in deference to his colleagues teaching the subject there, but he continued to raise issues about schools that "feminized" science teaching.[21] The fundamental challenge by the prolific and well-known intellectual echoed criticisms that others made on specific issues, including the use of imaginative literature.

A number of essays in *Science* magazine alerted scientists to practices that incorporated art and literature in nature study. The botanist W. J. Beal, who had worked under Agassiz, was deeply committed to learning through direct examination of natural objects and had passed this idea on to his own early protégé, Liberty Hyde Bailey, at Michigan State Agricultural College. But he diverged from his former student as he reviewed textbooks on nature study. Asking "What is Nature Study?" he compared the answer of Dietrich Lange, that "[n]ature study . . . is understood to be the work in elementary science taught below the high school—in botany, zoology, physics, chemistry and geology," with that of Bailey, who wrote that "[n]ature study is seeing the things which one looks at, and the drawing of proper conclusions from what one sees. Nature study is not the study of science." Beal suggested that Bailey's approach in *The Nature-Study Idea* was dangerous. Beal's essay praised a hypothetical "teacher" who used interest, tact, and enthusiasm to draw out the observations of her pupils, but it disparaged anyone who depended on "marvelous stories" to generate curiosity. His final critique was of a "neat drawing" of honeybees made by a hypothetical normal-school graduate: "The bees are not alike; each has two wings only; the heads and legs are unlike anything ever attached to bees. The apple blossoms are five-lobed (gamopetalous), with three stamens growing from the base of each lobe of the corolla. He has made drawings of imaginary insects seeking imaginary nectar from imaginary flowers."[22] This generalized teacher had not learned basic biology and should not be allowed to discuss nature with pupils.

His concern prompted Beal to ask several colleagues to address the same question—What is nature study?—quite specifically, and *Science* published their responses six months later.[23] Older and established naturalists like A. S. Packard of Brown University and A. E. Verrill of Yale University viewed nature study approvingly as a version of natural history because it required a pupil "to observe experiment and reason from the facts he sees." Jennie M. Arms, head of the nature study program in Boston, reprised Bailey's argument that the study of nature was not the study of books but developed skills of observation, comparison, and inference. David Starr Jordan weighed in briefly, emphasizing the potential of direct observation of nature, but he dismissed the use of "fairy stories of animals and plants, [and] all fantastic stories of creatures more or less imaginary" because, for him, the virtue of nature study lay in its "reality." John Coulter, too, commented, "The association of nature study with poetic literature is probably largely responsible for fostering sentimentality as opposed to knowledge."[24] These published responses seemed to fuel the aversion of academic scientists who were working to establish professional identities for themselves and to make clear the appropriate, if narrow methodology of science.[25] From an empiricist's perspective, imaginative literature was a counterintuitive mode for teaching science.

Nature study advocates shared a different, historically rooted sensibility; they believed that there was no firm distinction or conflict between appreciating nature aesthetically and studying nature systematically.[26] In fact, most leading nature study advocates had been comfortable with using literature alongside nature study. Jackman quoted from Walt Whitman's *Leaves of Grass* as a frontispiece for his *Nature Study* text in 1892; and poetry, ancient myths, popular nature literature, and children's fiction were regularly included in nature study textbooks. Influenced by the Herbartian theory that children are imaginative in their early years, Charles McMurry defended using literature in nature study by quoting John Tyndall: "There are tories even in science who regard imagination as a faculty to be feared and avoided rather than employed. They had observed its action in weak vessels, and were unduly impressed by the disasters. But they might with equal justice point to exploded boilers as an argument against the use of steam."[27] Even William T. Harris, not a fan of nature study but ready to use literature and art in science education, wrote a favorable introduction to one of John W. Troeger's nature study readers, which incorporated ancient myths, folk tales, and poetry (fig. 23).[28] Included in the book are stories about Persephone, who could only spend six months each year with her mother tending to her plants, and about how two girls, one with golden hair and the other with blue eyes became the wild golden rod and purple asters of late spring. The alternative was teaching "strict science lessons," a rigid practice that, according to Ira Meyers, Jackman's colleague in Chicago, had been a "grind" and destroyed children's innate interest in natural objects."[29]

Teachers debated among themselves what kinds of nature-related literature were appropriate. During the nineteenth century a general category of outdoor books developed, which included poetry, natural history, collections of sketches, and short fiction.[30] Nature writing became a popular genre for adults as well as children, sparking a revival of Henry David Thoreau as notable naturalists like John Muir and John Burroughs made pilgrimages to Walden Pond.[31] Alongside literature framed by men who simply enjoyed outdoor life, other outdoor books, many written by women for women and children, broadened interest in nature subjects.[32] Harriet Ritvo has pointed out that the tendency to see the kindred nature of humans and animals was particularly evident after the publication of Darwin's *Origin of Species*, even as the observations of animals became increasingly technical.[33] Anthropomorphism permitted an elision between the objective and the subjective,

APPLETONS' HOME READING BOOKS

NATURE–STUDY READERS

By JOHN W. TROEGER

IV

HAROLD'S
EXPLORATIONS

BY

JOHN W. TROEGER, A. M., B. S.

AND

EDNA BEATRICE TROEGER

NEW YORK

D. APPLETON AND COMPANY

1900

FIGURE 23. John Troeger, principal of Irving Grammar School in Chicago, produced five nature study readers based on the outdoor adventures of Harold; two were coauthored with his wife, Edna Beatrice Troeger. The first three volumes concentrate on things likely to be found close to readers' homes. Author's copy.

giving animals and occasionally even plants some degree of agency.[34] Nature study textbook authors like Lida McMurry, influenced by theories of child development in which children recapitulated the development of the race, used the techniques of animal-human identification as well as myth and poetry in her textbook and her *Songs of Treetops and Meadows.*[35]

Teachers in training might be introduced to nature literature in normal school classes or through their teachers' reading circles. These descriptive books attracted aspiring teachers because, according to Eric Lupfer, such accounts accorded well with their belief that nature could and should inspire religious sensibilities and perhaps even personal reform as well as provide descriptions and examples to be used in their classrooms.[36] There is less evidence that teachers used fiction, although such animal books as *Black Beauty* and *Water Babies* were popular with the general public, and most teachers encouraged pupils to read animal stories and other books increasingly available in Carnegie-sponsored libraries.[37] Teachers believed children could readily distinguish figurative and metaphorical themes in literature from real life and that such material stimulated memory and imagination.

Nature study leaders found it difficult, however, to deal with the controversy that erupted over some best-selling outdoor books at the turn of the century.[38] Critics particularly deplored those nature writers whose descriptions, presented as factual, stretched credulity. The debate about the accuracy of some nature writing began with John Burroughs' essay "Real and Sham Natural History" in the *Atlantic Monthly* in 1903.[39] Soon others joined in challenging William J. Long and even Ernest Thompson Seaton, authors whose first-person accounts were said to create misleading and sometimes sentimental portraits of animal behavior. The charge of "nature faker" could be stretched to encompass children's books that ascribed human behaviors to animals and plants, whether metaphorical or not.[40] Long, who had attended Bridgewater Normal School and was an enthusiastic supporter of nature study, was not intimidated. He dedicated *Fowls of the Air* "to the Teachers of America who are striving to make nature study more vital and attractive by revealing the vast realm of nature outside the realm of science and a world of ideas above and beyond the world of facts."[41]

More than a decade after the debate, Samuel C. Schmucker reflected that it and the term "nature faker" had deepened the scientists' distrust of writers and teachers who presumed to study "animals without chasing them through analytical keys and tacking the Latin names to them."[42] The outcome of these heated debates about imaginative literature and its use in education, according to Priscilla Eccles, was the creation of a "mythology" in the history of science education that the early twentieth-century "nature study program with its sentimental, anthropomorphic subject matter, occupied the time of elementary students."[43]

Within the movement, however, nature study advocates responded, often defensively but sometimes assertively, to the criticism of progressive pedagogy and imaginative literature, quoting Louis Agassiz's exasperated comment, "Pupils study Nature in the Schools, and when they get outdoors they cannot find her."[44] Jackman, too, had pointed out that "nature study is not an exact science, and has been handicapped by those who seek to pretend it is." The controversy, however, prompted nature study leaders to continue to evaluate the use of literary materials and to scrutinize books purporting to provide nature study content.[45]

Nature study had been rhetorically and substantially framed as a reaction to narrow, dry methods of teaching scientific facts, so when poetry and imaginative literature proved effective at awakening pupils' interest and curiosity about the natural world, teachers used them. But were teachers well-equipped to guide nature studies? This often rhetorical question remained another key challenge from scientists; it was associated with doubts about women as teachers and in the sciences. Teachers, especially those already teaching with little or no preparation in nature study or science, also worried about adequate training.[46] The proliferation of textbooks, course guides, manuals, and a new journal helped, but the leading advocates believed that a professional association could guide the thinking and practice of those developing nature study materials or teaching the subject.

NATURE-STUDY REVIEW

Status and authority were important as Maurice Bigelow established an advisory base of "college men . . . interested in the broader problems of science in general education" from prominent institutions across the country for *Nature-Study Review*.[47] With an undergraduate degree in biology from Ohio Wesleyan College and a Ph.D. from Harvard, Bigelow embodied the growing importance of advanced degrees for those who were to train teachers.[48] Teachers College was engaged in ongoing negotiation over its relationship with Columbia University, and faculty members like Bigelow sought to demonstrate that education was an emerging academic discipline with sound research practices and theory comparable to other social sciences. He reflected this agenda in selecting men to help inaugurate the journal.

The core advisory board members were well-known contributors to nature study with academic credentials: Liberty Hyde Bailey at Cornell, H. W. Fairbanks at the University of California at Berkeley, Clifton Hodge at Clark University, and John Woodhull at Teachers College, Columbia.[49] With their help, Bigelow identified nearly seventy geographically diverse "advisors and collaborators," some from teacher training institutions—although no practicing teachers—to "insure that the journal will be entirely independent of local interests and free to become representative of nature-study in all parts of America, the center of the movement."[50] Included in this larger group were nature study authors like Wilbur Jackman, Anna Comstock, and Bertha Chapman Cady; scientists like Charles E. Bessey; and the educational philosopher Charles McMurry and the African American educator George Washington Carver.[51] Bigelow's selection already signaled that he would include articles not only on educational theory and methods but also on classroom practice. Consulted about the proper title for the journal, the board recommended the hyphenated version that Bailey had certified to emphasize the compound term's "unity." Aware of nature study's international spread, Bigelow made the journal broadly North American and in the very first issue reprinted an essay by Dean W. H. Muldrew, the leader of Canadian nature study located at the Macdonald Institute of the Ontario Agricultural College at Guelph.[52]

The *Review*'s masthead made clear that it was intended for active educators and administrators "devoted to all phases of nature-study in elementary schools."[53] Published monthly for the first year, the new journal was to provide a forum to discuss the goals of nature study, the appropriate level for teaching it effectively, and its content and methods.[54] The *Review* gradually included space for teachers to present their own proven classroom techniques. While parents and other readers might be interested in the regular reviews and bibliographies on nature study topics, the announcements of professional meetings and special institutes identified teachers and administrators as the journal's primary readership.

Bigelow was an active manger and editor of the first five volumes, corresponding with potential authors, planning special issues, and underwriting finances (fig. 24).[55] Establishing a new, independent journal was not easy, even though its regular publication represented new stability and visibility for the movement. Bigelow dealt with such practical matters as finding a printer after the first, New Era Printing in Lancaster, Pennsylvania, had difficulty with striking workers and slipped behind schedule.[56] Subscriptions came in slowly, so Bigelow personally subsidized the journal and relied

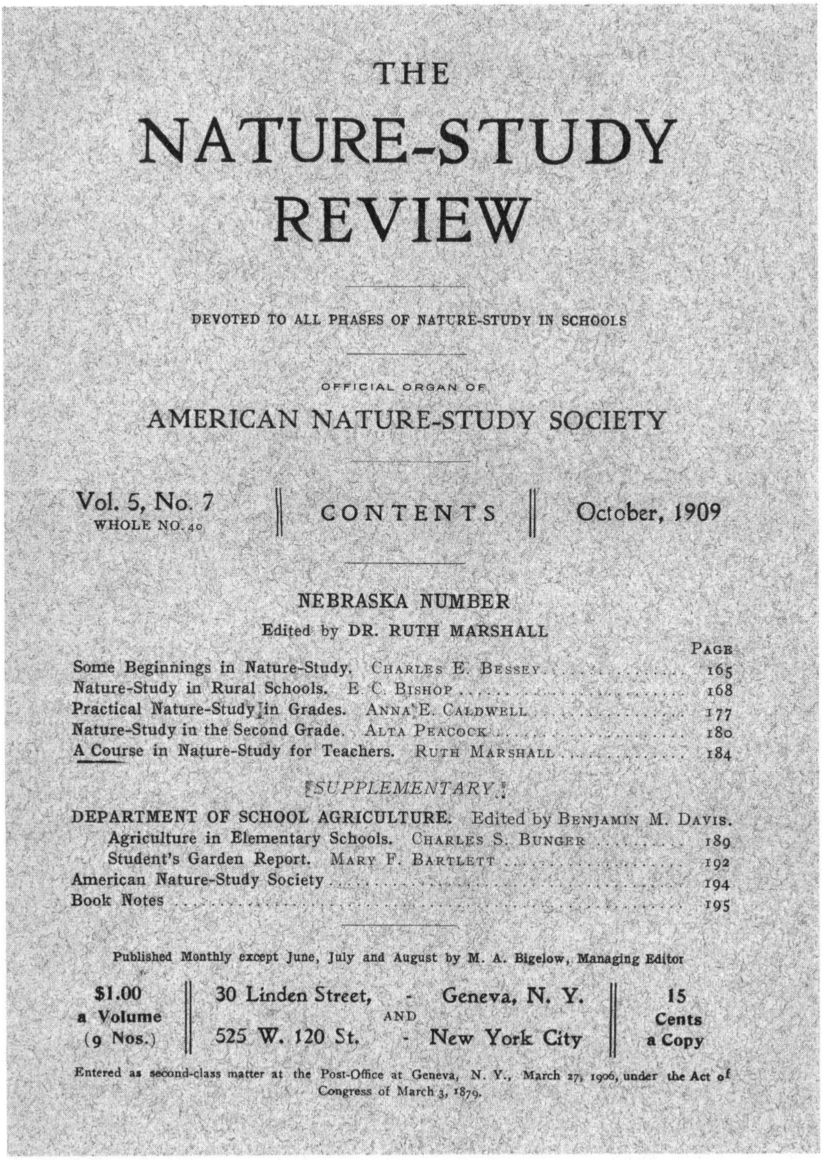

THE
NATURE-STUDY REVIEW

DEVOTED TO ALL PHASES OF NATURE-STUDY IN SCHOOLS

OFFICIAL ORGAN OF

AMERICAN NATURE-STUDY SOCIETY

Vol. 5, No. 7
WHOLE NO. 40 ‖ CONTENTS ‖ October, 1909

NEBRASKA NUMBER
Edited by DR. RUTH MARSHALL

Published Monthly except June, July and August by M. A. Bigelow, Managing Editor

$1.00
a Volume
(9 Nos.) ‖ 30 Linden Street, - Geneva, N. Y.
AND
525 W. 120 St. - New York City ‖ 15
Cents
a Copy

Entered as second-class matter at the Post-Office at Geneva, N. Y., March 27, 1906, under the Act of Congress of March 3, 1879.

FIGURE 24. In an effort to demonstrate the national spread of the nature study curricula and its rich variety of applications, Maurice A. Bigelow sponsored a number of state-focused issues, reflecting programs across the country. Featured in the above *Nature-Study Review* 5 (October, 1909) is the Nebraska program. Courtesy of the Wangensteen Historical Library of Biology and Medicine, University of Minnesota.

on other Teachers College faculty and his wife to help with book reviews and special issues. John F. Woodhull, a former teacher and his colleague in teaching science, contributed physical science essays, some becoming part of his textbooks that linked those topics with nature study.[57] Ada Watterson (later Yerkes), a tutor in biology who had earlier attended Cornell's summer school and also studied at Woods Hole, for years faithfully compiled a "Guide to Periodical Literature" for the *Review*.[58] Such features were intended to help harried elementary teachers sort through the proliferation of texts and teaching aids and find useful advice for their classrooms. The *Review* could also deflect charlatans, as when the editor warned, somewhat tongue in cheek, about a confidence man, "an imposter . . . working among nature-study teachers" in New York and New Jersey, who was selling paid-in-advance subscriptions to a nonexistent nature magazine.[59]

Bigelow encouraged contributors to discuss the theory and pedagogy of nature study as a way of clarifying practice, and early issues featured primarily faculty men who had strong and often diverse views and definitions of nature study. The *Review* was intended to build consensus about nature study principles—but that goal proved elusive. Bigelow's perception of nature study was distinct from that of his colleague Liberty Hyde Bailey, for example, but Bigelow was also well aware of the influence of Bailey's popular book, *The Nature-Study Idea*, and his high credibility among teachers. The child-centered approach that emphasized exploration and hands-on experience in the natural world seemed to Bigelow to be a vague exhortation largely unhelpful to teachers who needed more specific guidance. The journal could provide advice and examples to inform classroom practice. Clearly pragmatic, Bigelow had stretched the movement's reach to include other "phases" of nature study, particularly its geological, physical, and geographical aspects, and also to suggest what links were appropriate to related but distinctive educational programs in agriculture, physiology, and hygiene.

In the inaugural issue of the *Review*, Bigelow created a "symposium" on one core issue of contention among nature study writers, namely, "Nature Study in its Relation to Natural Science." He invited two members of his advisory board, Harold Fairbanks and Clifton Hodge, along with Thomas H. MacBride of Iowa State College, to join him in addressing the issue. The contributors agreed that nature study, by fundamental definition, should encourage pupils to experience nature firsthand. After that, they diverged on its goals, ranging from enhanced appreciation for and keener observation of nature on one hand to a systematic knowledge base for further scientific study on the other. Fairbanks argued approvingly that nature study had moved away from simplistic object teaching over the previous decade to engage the "culture of pupils." Like Bailey, he argued, "Nature-study should lead the child back to this natural intimacy with nature and to delight in her company. This cannot be done by feeding him upon courses of study made up of scientifically arranged facts, but by fitting him in a broad way through the exercise of his observational and reasoning power so that he not only takes pleasure in the world around him but is able to use it more fully to his material advantage."[60] Bailey himself maintained that students should be encouraged to observe and learn about "those [natural] things that are a normal part of one's environment," which meant, for his students, the fields and streams of upstate New York.[61]

Clifton Hodge, aware of the criticism from other scientists, suggested that the greatest risk to nature study came from those who would put back the grinding memorization and tedious repetition that characterized earlier efforts to teach elementary science. Working with teachers in the crowded slums of Worcester, Massachusetts, Hodge sought to present pupils with "those subjects that are most worth knowing" in an industrializing society. His textbook *Nature Study and Life* was intended to challenge critics who dismissed nature study as a "waste of time on fads" and "new-fangled notions" and argued that "they never had to learn such stuff," even though Hodge agreed that some might be "right as to what often goes by the name of nature study." But well taught, nature study had the potential "to bind home and school together as nothing in the curriculum does at present."[62] Influenced by G. Stanley Hall and others in educational psychology who argued that humans from infancy to adulthood recapitulated early stages of human development, he understood history as an account of mastery over nature through domestication of animals and cultivation of the soil. Twentieth-century children needed to learn about a competitive natural world, now very much caught up in industry, and one where they could take responsibility. He had earlier commented bluntly on his peer nature study textbook authors, suggesting that Jackman's detailed curriculum had a "logic-twaddle" method concerned with trivia, while Bailey's attention to detail was a "knot hole method" that began with "a broom splint, a sliver of pine or a knot hole" that arbitrarily filled up time.[63] Hodge's natural world had economic and social elements alongside the botanical and zoological ones.[64]

Bigelow seems to have viewed the difficult Hodge as something of a loyal opposition. Hodge responded by being more circumspect in this *Review* symposium than elsewhere, simply stressing the importance of the local environment in designing nature study activities. In a separate essay, the Chicago botanist Coulter continued his ambivalent support, agreeing with the purposes of nature study but then reminding readers of its "ill-defined beginnings" and noting that it seemed still in an "inchoate stage" that frustrated practicing teachers and exasperated scientists.[65] Bigelow's major goal for the journal symposium was to bring clarity and consistency, and his concluding comments distinguished between nature study as "the simple observation of common natural objects and processes for the sake of personal acquaintance with the things that appeal to human interest directly" and natural sciences as "the close analytical and synthetic study of natural objects and processes primarily for the sake of obtaining knowledge of the general principles which constitute the foundation of modern science." Nature study offered an approach that took into account recent work in educational psychology, which demonstrated that "true elementary science with its very foundation in classification and generalization is not adapted to pupils as young as those in our elementary schools."[66]

The introductory issue of the *Review* fulfilled Bigelow's hope of drawing in established academic leaders. It made clear from the outset that this journal would engage a broad spectrum of university faculty members in education and science and address fundamental issues facing normal school instructors, supervisors within educational systems, and classroom teachers. Early contributors shared a common interest in nature study, but their discrete responsibilities and occupational niches made it difficult for Bigelow to find the exact formula for retaining a professional tone while expanding the content to meet their diverse expectations.[67] His opening editorial disclaimed any intention of providing detailed instructions or classroom plans; instead, his goal was an exchange of ideas among those engaged in preparing teachers, writing articles and books, and conducting school classes.[68] Over the journal's nearly twenty-year existence under three more editors, the balance between theory and content material would fluctuate significantly.

After the introductory issue with its prominent contributors opened discussion on the rationale and scope of nature study, Bigelow accepted more articles by and for teachers. Such contributions moved the journal closer to his promise that it would provide the "combined viewpoints of professional educators with practical acquaintance with the problems of elementary schools—[and] university men who are primarily interested in nature-study as a preliminary phase of science-teaching."[69] That decision to target "progressive teachers of nature-study and elementary education"[70] was also pragmatic, because he sought to increase circulation, explicitly by inviting teachers, in his third issue:

> The first two numbers of the *Review* give prominence to more or less theoretical papers on the educational problems of nature-study, and naturally most of these papers have been contributed by writers who have approached nature-study from the viewpoint of the broader questions of education which appeal especially to college men and school officials. . . . [W]e must give an equally prominent place to the results obtained by teachers who are actually at work in the elementary schools. . . . Teachers who work out even minor points which may interest others are invited to send concise accounts of their results to the editors of this journal.[71]

The invitation to teachers had immediate effect, as elementary teachers and supervisors, primarily women, began sending essays about successful and innovative projects with enough detail to enable their peers to reproduce them.

Because there was so much concern about teacher preparation, Bigelow decided to attack the problem at its source by soliciting reports from "all the best normal schools in the United States and Canada."[72] From detailed descriptions of normal school programs, readers could see the depth and breadth of activities and make comparisons among the leading schools.[73] In early 1906 he compiled the results, identifying distinctions among schools and special programs even as he sought to tease out their commonalities and patterns. Responding colleagues revealed that their teacher training curricula emphasized botany and zoology, but quite a few included additional classes on evolution, hygiene, and advanced biology, as well as mineralogy, meteorology, physics, chemistry, and agriculture.[74] The preparation of students admitted to nature study training programs also varied significantly. It was not uncommon for state-sponsored schools, many in rural districts, to admit students with only eighth-grade schooling, although these might also have had several years of teaching experience. By contrast, 90 percent of those at the Philadelphia Normal School for Girls in Philadelphia, under the watchful eye of Lucy Langdon Williams Wilson, were high school graduates who then undertook four years of teacher training.[75] Teachers College, still just loosely affiliated with Columbia University, boasted that its students typically attended after two years of normal school training or its equivalent. This varied population was the potential subscriber base for Bigelow's journal and for the Nature-Study Society he initiated just two years later.

The new *Review* had the potential to reach beyond the United States to Britain and the Commonwealth countries as well as to continental Europe, where nature study was also introduced.[76] The focus, however, was on North America, where the successful journal could be the catalyst for a "national" society. Bigelow's editorial introduction had been quite explicit: "It is hoped that those interested in nature-study in all the States and Canada will have a personal interest in the development of the journal as though it were the official organ of an American association of nature-study teachers." As the *Review* became "representative of nature-study in all parts of America, the center of the movement," it was appropriate to move to the next step.[77]

AMERICAN NATURE-STUDY SOCIETY

In 1907 an editorial by Bigelow announced an organizing committee for an American Nature-Study Society (ANSS). *Nature-Study Review* would be its official organ, and it included discussion of the organization in all subsequent issues through January 1908, when the society was officially organized at its first annual meeting. After July 1907, all subscribers were automatically enrolled as members. Bigelow, who had been subsidizing the journal, hoped the organization and membership dues would provide additional revenue. Bigelow anticipated membership of at least a thousand "teachers and others who are interested in nature-study as a part of regular school work."[78] The organizing committee included Clifton F. Hodge, Stanley M. Coulter, H. W. Fairbanks, Vernon L. Kellogg, Anna Botsford Comstock, Dietrich Lange, William A. Baldwin, Michael F. Guyer, Frank L. Stevens, William Lochhead, Maurice A. Bigelow, Charles R. Mann, Lucy. L. Wilson, Edward F. Bigelow, and John F. Woodhull. The *Review* soon printed a constitution, announced meeting dates, and reported that local clubs were being established.[79]

The *Review* and the society would have intimately linked fortunes, with ambitious presidents shaping annual meetings and the secretary editing monthly issues of the journal. On January 2, 1908, the ANSS held its first meeting in Chicago with that of the AAAS, which hosted a number of specialized scientific societies during its Convention Week.[80] The announcement of affiliation with the AAAS gave the new society visibility and, Bigelow hoped, would offset the inclination of scientists to "regard nature-study as something of a fad." He was cautiously optimistic, claiming that interest in nature study was "in the ascendancy" among scientists even as he felt compelled to acknowledge that "expressions of approval were commonly coupled with qualifying remarks concerning the kind of nature study" being conducted.[81]

Stanley Coulter, brother of John C. Coulter and soon to be dean at Purdue University, wrote in some dismay after the meeting in Chicago about disagreements between proponents of particular philosophies of nature study at the society's first meeting. He was particularly upset that the acerbic Clifton Hodge distributed his "usual kindly remarks about other workers" and had a large following, despite his "rather ill natured and unfair personal criticisms."[82] However, most participants were enthusiastic about the new organization and did not dwell on pedagogical and personality differences. Liberty Hyde Bailey was not able to attend but was nonetheless elected first president (Otis Caldwell as vice-president stepped in as chair for the meeting), and the *Review*'s editor, Bigelow, became the secretary-treasurer in recognition of his critical role as founder and thus ensured direct communication between the journal and society.[83]

The society's decision to meet initially with scientists rather than with educators was a deliberate attempt to create an informal alliance. Although the society's new president was a scientist, he nonetheless urged Bigelow to plan something with the next NEA meeting in 1908. This would certainly bring in more teachers as members. The society began a pattern of holding additional semiannual meetings with the NEA's summer convention but conducted its annual business meetings during the AAAS convention.[84] In 1908 Bailey gave a plenary talk to the AAAS and conducted a short business meeting.[85] The conference sessions generated essays for the *Review*, as Bigelow requested papers from participants. Having a full-service journal with articles, reviews, and other features was a considerable attraction for potential society members, but Bigelow was nearly overwhelmed as he struggled to put membership apparatus in place, manage finances, and do editorial work.

At first membership grew quite quickly. The only published membership list, for 1908, has useful but incomplete data on nearly eight hundred members' geographical locations and professional positions.[86] Most members came from the Northeast and Midwest, and membership clustered wherever strong nature study teachers' training programs, active nature study courses, or garden projects were located. At the first meeting of the national society, a New York section was announced, with Alice Northrop taking a leadership role.[87] Within two years, sections had been founded in Chicago (north and south side), plus one each in St. Louis, Missouri, and Rockford, Illinois.[88] Another soon sprang up in Grand Rapids, Michigan. Active sections reported holding regular meetings, sharing information, and taking trips to local sites. The relatively few southern members were most often concentrated near state agricultural colleges or institutions where northern philanthropists heavily influenced the curriculum.[89] Teachers in Southern schools had low pay, and school terms were short, resulting in what one historian of education termed "under education" in most of the region; not surprisingly, there is little evidence of sectional groups in the South.[90]

Gender distinctions reflected larger patterns in education at the time. On the 1908 membership list, women (including only those clearly identified by first name) constituted roughly three quarters of those identified as schoolteachers and administrators and about a quarter of those serving as faculty in normal schools and universities. Membership was, of course, self-selecting and reflected only a small proportion of the elementary teachers involved in nature study — consider that Comstock alone had three thousand teachers on her statewide mailing list for New York in 1903.

As editor, Bigelow seemed quite aware of gender distinctions, perhaps because his wife collaborated with him on a biology textbook and other projects. After early issues focused largely on theoretical subjects written by "college men," he turned his attention toward women teachers. Their contributions to the *Review* were "results obtained by teachers who are actually at work in the elementary school." Reflecting the realities of the gendered profession and perhaps for convenience, authors of articles typically referred to the teacher as "she" and the pupil as "he" so that the pronouns become automatic, if misleading, designators for roles.[91]

So long as Bigelow remained editor, from 1905 through 1909, the *Review* gave space to critics of nature study, which perhaps revealed the editor's own increasing ambivalence about the appropriate parameters for nature study. Some of the negative commentary, perhaps meant to be constructive, came from those who had been involved in nature study, such as Charles Bessey, who regularly lectured to teachers' groups in Nebraska, and John Coulter, who coauthored a textbook.[92] They expressed concern about the qualification of teachers and the quality of content presented to elementary pupils. The outspoken director of New York's Zoological Park, William Hornaday, who had trained in engineering and never taught school, contributed "The Weakness in Teaching Nature-Study," and a flurry of comments followed, orchestrated at least in part by Hornaday himself.[93] Hornaday challenged the capacity of teachers to be the "textbook" required by inquisitive children and argued that pupils would, in fact, learn better from books. He urged memorization of important information, suggesting in exasperation that "American children are not such idiots that they cannot learn continents and states from maps, and living things from pictures."[94] The formal responses ranged from mild agreement to Vernon Kellogg's sharp retort that Hornaday's proposal for scientific work was an "old, long-tried, fully tested, and unanimously-rejected one" and his dismissive comment that Hornaday's comments invited "no discussion; there can only be a cross-fire of declarative statements."[95] Bigelow's own position challenged simplistic psychological notions when he asked rhetorically, "Are Children Naturally Naturalists?" in an essay that called for revising pedagogical assumptions about children's innate interest in nature; instead, he called for educational techniques that could arouse interest.[96]

With a limited supply of submitted articles, Bigelow reprinted essays from other educational journals, some identified by his colleague Ada Watterson, who produced a list of articles related to nature study found in both scientific and general periodicals. Sometimes Bigelow summarized material on popular topics, like one on the protective color of gulls taken from the *Atlantic Monthly*.[97] The *Review* earned its title by reporting not only on nature study textbooks but also on a range of scientific books dealing with plants and animals, information on gardening and bee keeping, as well as literary nature study. The naturalist John Burroughs was familiar to New Yorkers through his writing on the landscape and animals of the Catskill and Adirondack mountains, and Bigelow was clearly a fan, frequently citing Burroughs's books. Reviews were combined with articles from scientists like Coulter at Purdue, Kellogg at Stanford, and Hornaday at the New York Zoological Park; from outstanding normal school and agricultural educators like Arthur Boyden at Bridgewater and Dean Muldrew at Guelph in Ontario; and from independent advocates like Edward F. Bigelow (who was not related to Maurice Bigelow), coordinator of the Agassiz Association clubs. The men might comment on theoretical or scientific issues or highlight best practices in training schools while, more typically, women teachers and supervisors wrote about actual classroom practice. Both groups, from their distinctive perspective, discussed adequate preparation for teaching nature study in the normal school curriculum and finding time for it in the classroom.[98]

The society, too, wrestled with managing the multiple facets of nature study. In 1909 the officers appointed a number of committees whose titles identified the admitted if murky connections of nature study to the rest of the curriculum: Committee on Physical Nature-Study, Committee on Industrial Education and Nature-Study, Committee on Nature-Study in Relation to High

School Biology, and Committee on Nature-Study and Agriculture.[99] None of these committees seems to have produced any significant reports, perhaps because their members had to work at a distance. The titles of the committees do reveal a growing emphasis on the "practical or economic side of most biological subjects." That trend paralleled the path taken by some faculty in geographic education, which Susan Schulten notes weakened the place of geography as a natural science and strengthened its place among the commercial and social sciences in the early twentieth century.[100]

EVALUATING NATURE STUDY

When Bigelow stepped down as editor of *Nature-Study Review* after five years, the controversial topics introduced in the first issues were still not resolved. Leaders from the 1890s like Bailey and Comstock remained opposed to strict scientific naturalism. Other academic scientists realized that such resistance limited the scientific outlook they wanted to have inculcated in the youngest future citizens and potential scientists. A comment made at the AAAS in 1910 by John Dewey, who was now at Teachers College, was particularly harsh:

> Visit schools where they have taken nature study conscientiously. This school moves with zealous bustle from leaves to flowers, from flowers to minerals, from minerals to stars, from stars to the raw materials of industry, then back to leaves and stones. At another school you find children energetically striving to keep up with what is happily termed the 'rolling year.' They chart the records of barometer and thermometer; they plot changes and velocities of the winds; they exhaust the possibilities of colored crayons to denote the ration of sunshine and cloud in successive days and weeks. . . . and at the end, the rolling year, like the rolling stone, gathers little moss.[101]

John Dewey's sarcasm may have held a personal edge as he critiqued the rolling year plan of the now-deceased Wilbur Jackman, who had undermined Dewey's wife's position at the Chicago laboratory school. Nonetheless, Dewey's rhetoric at the AAAS meeting, published in *Science*, articulated the reservations held by not a few academic colleagues and educational leaders.

A new generation of educators wanted to make their discipline more scientific. Jacob E. Mayman, completing a Ph.D. at New York University, reviewed the literature on elementary science. He recorded the differences in teaching assumptions, methods, and content, and then concluded that the most effective methods engaged pupils directly rather than relying on books or lectures.[102] This might have seemed quite consistent with nature study methods, but the young academic envisioned laboratory experiments and quantitative assessments, not field studies. A questionnaire circulated by George F. Hunter, a biology teacher at De Witt Clinton High School in New York, to high schools in cities with more than ten thousand people elicited similar opinions. His major question was a leading one: "Assuming that you have nature-study in the grades, to what extent do you correlate grade work in nature-study and human physiology with the biology of the high school?" His results—9 found correlation, 49 slight correlation, and 110 no correlation—reinforced his conviction that nature study was not providing a preparation for advanced scientific work. Based on his own experience, Hunter avowed, "Pupils who have had experimental science . . . come to their biological work with an entirely different mental attitude from pupils who have not had this training. In such pupils the habit of scientific thinking is already forming."[103] By implication the nature study approach did not lead to an experimental outlook or to advanced scientific studies. Thus, even in New York City, home of *Nature-Study Review*, criticism of nature study had not abated.

STRIVING FOR STABILITY

Bigelow stepped down as editor. He was tired by the demands of the position, stung by criticism that he was making money from the journal despite his annual subsidy of about $400, and turning his attention to secondary education and to health and hygiene issues.[104] His successor was Fred L. Charles, an innovative nature study instructor at Northern Illinois Normal in DeKalb under Charles McMurry, who had recently become a faculty member at the University of Illinois. During Charles's short tenure, the *Review* focused on the "civic" development of children and on addressing specific social issues like venereal diseases and vandalism. He planned to "make every issue immediately helpful to teachers," promising that each would be a "teaching monograph" organized around specific topics such as hygiene, which was mandated by many local and state authorities.[105] Clifton Hodge, who was ANSS president—and writing his *Civic Biology*—during the selection process in 1909, was also keen to integrate scientific knowledge with social problems facing pupils in their daily lives and environment. But topics like "dental hygiene" and "pets as disease carriers" were shifting the focus away from the nature subjects envisioned a decade earlier by

Jackman, Comstock, and Bailey, as well as from the more theoretical and philosophical discussion encouraged by Bigelow.[106]

Having worked closely with practicing teachers in DeKalb, Charles was aware of the problems they faced in developing curriculum. He encouraged teachers to submit materials that described practical projects that might be used by others. Articles on maintaining a meteorological journal, observing phenomena like aquatic insects, establishing an ant colony, and preparing cuttings for a window box became staples in the *Review*. These shifts seemed to Bigelow to represent too dramatic a change in the journal he had founded, and he raised his concerns at the next annual meeting of the Nature-Study Society. Otis W. Caldwell, president in 1910, who believed that presentations by teachers about drawings, garden plans, and the nature experiences of pupils made the subject more tangible to their teacher colleagues, had already endorsed Charles's direction.[107]

When Charles died unexpectedly in spring 1911, Benjamin M. Davis, professor of natural history at Ohio State Normal College in Miami and then president of the society, took over for a few issues.[108] Elliot R. Downing, who had a Ph.D. from Chicago, had taught at Northern Michigan Normal School in Marquette, and later studied for a year in Germany, assumed the editorship that October.[109] The ambitious Downing appeared to be a sound choice who could create a link between nature study teachers who still looked to the University of Chicago for leadership and the increasingly professionalized administrators whose interest in pedagogy was framed by practical concerns. At the December meeting in 1911, one member seconded Maurice Bigelow's recommendation that the *Review* return to its initial emphasis with articles on the "same plane" as those in *Elementary School Teacher, School Review*, and the *Journal of Educational Psychology*. This unnamed commentator said that perhaps bimonthly issues of such a *Review* could be interspersed with occasional supplements "for the rank and file."[110] The direction of the journal, however, had permanently shifted.

As editor, Downing continued the nine-month publication cycle and published essays that teachers sent him, but, aware of the criticism that nature study was too informal, he again emphasized method. His own outlook pushed him to make the *Review* directed "primarily to all *scientific* studies of nature in elementary schools," while maintaining that nature study was still a "protest against formalism, bookishness and education by accretion."[111] To assert nature study's intellectual and cultural connections, he cited not only textbooks and specialized articles but also relevant materials from a range of magazines, including *Harper's, Atlantic Monthly, Field and Stream*,

World's Work, and *Outing.*[112] His scientific training led him to encourage research and statistical reports, and he conducted his own survey of normal schools regarding nature study. He found 569 courses offered in 1912, enrolling 142,717 students, which he estimated as about a quarter of the incoming teaching force.[113] A subsequent survey concentrated on how many teacher training schools had practice schools that taught nature study, and the results were significantly higher.[114]

These were encouraging data as Downing and his board struggled to find a formula to attract and maintain subscriptions. [115] Elementary teachers, however, were expected to teach multiple topics. Only a relative few had sufficient salary to subscribe or time to read the specialized journal. Downing worked with Anna Comstock to provide upgrades, like an occasional color plate as frontispiece, and the quality of production did improve. The *Review* continued to solicit advertising, primarily from book publishers; normal school summer programs and seaside laboratories also published announcements of their offerings.[116] Such efforts, however, seemed only to hold the society and journal in place, with annual meetings averaging perhaps a hundred attendees, despite a readership that was announced as over fifteen hundred members. Leadership, moreover, seemed in flux. Bigelow failed to be elected president of the society he had founded, and Hodge moved briefly to the University of Oregon where he taught civic biology before apparently leaving the field altogether. The stalwarts of nature study proved to be Anna Comstock, Liberty Hyde Bailey, and many of the students and colleagues they had inspired.

COMPLETING THE *REVIEW*'S LIFE CYCLE

As Anna Comstock later recounted, the Nature-Study Society maintained itself during the war, perhaps bolstered by the extensive school garden activity, and seemed to be stable as she took over editorship of the *Review* in 1917. In 1920 the membership was nominally twenty-four hundred members, with significant clusters in cities that had active chapters, such as the Webster Grove Nature-Study Society just outside of St. Louis. The *Review* reemphasized teacher involvement and a definition of nature study that reflected early Cornell leaflets, with their focus on discovery of nature out-of-doors. Comstock had signaled her impatience with the vocational trends in "Nature Study as Servant" and, as editor, reasserted the importance of imagination.[117] She unabashedly reintroduced literature, especially poetry, and even a musical play, "Winter Birds: A Bird Masque in Three Acts to

Portray Winter Feeding," in which primary pupils could act out parts of flowers, insects, and snowflakes.[118] The emphasis on what Bigelow had deemed "narrow" nature study that concentrated on out-of-door experiences for children in neighborhoods, parks, and gardens returned.

Nature-Study Society annual meetings continued to draw fifty to a hundred attendees each year. In 1920 ANSS members met in Emmons Blaine Hall at the University of Chicago, coming full circle back to the place where nature study had taken shape under Wilbur Jackman with the support of Anita McCormick Blaine.[119] The following year, in Toronto, there was a lively meeting and evidence that nature study in Canada was prospering.[120] Attendees at the annual meetings, however, were typically a few long-standing members together with local teachers who joined only temporarily when the annual meeting was nearby. Active new members proved difficult to recruit. When the *Review* merged with *Nature Magazine* in 1923, its close affiliation with schools would be largely displaced by a new emphasis on nature study in other venues. The Nature-Study Society, however, continued to publish a newsletter, intermittently, into the twenty-first century, but, lacking a sustaining membership, struggled to survive.

For nearly two decades in the early twentieth century, the *Nature-Study Review* had hosted a national forum for discussions about the theory, practice, and appropriate dimensions of nature study in the schools. Together with annual meetings of the Nature-Study Society, the journal had allowed advocates and administrators authoritative, sometimes contentious, statements about theory and method and encouraged teachers to find and distribute practical ideas about implementation. Elliott Downing, like Bigelow, had never exclusively used the term nature study in his own writing, but he seemed genuinely intrigued by its capacity for nonconformity, observing: "Nature Study should vary its garb, according to locality. Its urban expression is different from its rural expression. . . . It matters not what the predominant subject matter is if it is congruent with the locality, it if be not too narrow, and if it not be dwelt on so persistently that the youthful mind becomes bored."[121] This capacity to be so malleable in the hands of individual teachers and to encourage pupils' own observations of nature in all its intersecting detail may have gradually undermined nature study's position in elementary curricula, but those very features made it attractive in settings beyond formal classrooms.

CHAPTER EIGHT

Reframing and Extending Nature Study

In the nature-study of the past the emphasis was put on observation. . . . The emphasis is now on doing rather than the seeing. It is true that in many instances there was activity in nature-study, but that was not the center of attention. Now the attention centers around what use the child is going to make of his material.
Dora Otis Mitchell (1923)[1]

BY THE TIME Anna Botsford Comstock became editor of *Nature-Study Review* in 1917, nature study activities for children were widespread. Some of the involvement was generated through the school curriculum, but much took place outside formal educational institutions. The ever-optimistic Comstock welcomed these additional paths emerging for nature study by publishing announcements of summer camps, national parks, after-school programs, scouting for girls and boys, and other organizations that seemed, in fact, to revitalize precisely the kind of hands-on programs she and her Cornell colleagues had helped develop within many rural and urban schools in New York and beyond. The *Review* continued to publish material for teachers, even as the Comstock Publishing Company kept her *Handbook of Nature Study* in print and produced a variety of field notebooks, bird identification cards, and other materials suitable for classes or home study. Support for the journal and society nonetheless eroded slowly over the next decade, reflecting the displacement of nature study by elementary science textbooks and new methods in the schools. Contemporary and historical accounts of nature study's demise have, however, underestimated the staying power of the programs introduced under its auspices, despite the painful decline of the organizational framework put in place by Maurice Bigelow and the sometimes derogatory descriptions of nature study education by those advocating change.

The Great War had a sobering effect on the progressive optimism that had fueled educational reform.[2] A conservative mind-set led a few to argue that nature study might be valuable in the Americanization of foreign children, but nature study leaders seldom pursued this line, even in those cities where xenophobic attitudes were prevalent.[3] In the postwar spirit of modernity, science and technology made significant gains in cultural authority, and the vocabulary of objectivity was common.[4] Educational activities that seemed to subvert scientific objectivity lost credibility.[5] From the outset, advocates had struggled to explain why they would not answer the question "Is nature study scientific?" with an unequivocal "yes," and some had felt compelled to provide a simple "no." Equally important, nature study advocates pushed back against the "educational overlords," or school superintendents, whose efforts to standardize curriculum—and emphasis on students' acquisition of knowledge content at the expense of the practices of observation and correlation—constrained the flexibility that had characterized the best of school-based nature study.[6] Determined teachers worked to demonstrate to themselves and others that what critics perceived as "ad hoc" pedagogy lacking quantitative evaluation techniques was, in fact, successful by more qualitative measures of student engagement and classroom energy.[7] Where well-planned nature study programs were in place and positive outcomes perceived, nature study continued as an active program within both urban and rural elementary schools.

By the 1920s, however, the incentives and support for nature study were difficult to maintain in a shifting cultural and educational milieu. Despite all efforts to enhance agriculture and country life, the rural economic crisis had deepened well before the stock market crash of October 1929. Graduates of agricultural colleges no longer intended to be farmers but hoped to become agricultural engineers and agricultural economists. Deborah Fitzgerald argues that such graduates were taught quantitative skills by faculty members who avoided the narrative and anecdotal information used earlier by the extension agents, themselves familiar with farm life, often by having worked with their fathers.[8] In cities, too, interest and enthusiasm for the movement were losing momentum as public attention shifted. One observer commented on the changing membership in the Chicago Women's Club, which had been "an energetic philanthropic organization attacking with courage and fervor the problem of protecting the city's unfortunates.

It was [now] composed of young women imbued with zeal to blaze a trail women had not traversed before. In those early years it even strove with some success to secure membership on the board of education. . . . [but it has since become] a generalizing club, whose members do not pick wild flowers and do advocate the planting of trees."[9]

A changing outlook toward nature meant that making collections of specimens was viewed dimly by a younger generation, for whom conservation was a virtue.[10] Responding to a request from Bertha Chapman Cady, Joseph Grinnell, director of Berkeley's Museum of Vertebrate Zoology, reflected on the shift in attitudes. Cady, then teaching at Chico State Normal School, had asked if Grinnell could provide taxidermy specimens for her classes. His tart response reflected his deep frustration not at her but at the situation facing those who worked with specimens: "The Audubonites have instilled into the youth of the land fear of various punishments of said youth indulging in any sort of 'bird collecting.'"[11] As a result, Grinnell wrote, he knew of no "bird skinners" in the state except for a few "old guard." Ornithology, he felt, was becoming "ornithophily." Grinnell also informed a local teacher that he no longer loaned specimens because the "wear and tear incident to outside handling" could ruin them for further research, and they were no longer easily replaceable.[12] Public attitudes had undergone economic, social, and intellectual reorientation under conservation leaders that resulted in a paternalistic, some would argue a more calculating and exploitative, posture toward the natural world.

Nonetheless, the initial concepts of nature study nurtured by its advocates during the 1890s found expression during and after the 1910s. The painful dismantling of the educational infrastructures established to define and perhaps constrain nature study served, in certain ways, to liberate the vocabulary and methods of the discipline in and beyond the schools. A legacy of window boxes, school gardens, nature walks, classroom terrariums, and hamster cages –whether designated "nature study" or not—remained as evidence of the significant classroom innovation that had occurred in just one generation.[13] Alliances with local museums and botanical gardens remained, offering "field trips" to their facilities and staff members who could guide local educators and their pupils. Anna Comstock directed the *Review* specifically toward active teachers, but she also encouraged contributions from those who led extracurricular and informal programs of nature experiences. Nature study training at Cornell and elsewhere served a new cadre of educators who without hesitation included aesthetic appreciation along with systematic observation as they guided girls, boys, and

adults through city and national parks, summer campgrounds, nature reserves, and local nature trails. The Cornell strand of the movement maintained summer programs for classroom teachers, but the faculty members
attracted additional and quite different constituents during the academic
year as they transformed their program into recreational and conservation
studies.[14]

Although elementary science defined itself in contrast to nature study,
careful scrutiny suggests that the "new" elementary science curriculum endorsed by the National Education Association through the 1920s was not as
original as some sponsors argued. The endorsed curriculum emphasized the
range of topics that had been presented in Wilbur Jackman's *Nature Study
for the Common School* and shrewdly maintained some of the most effective
elements of hands-on learning. There were, of course, visible changes as
nature study texts and vocabulary were replaced in schools and normal
schools (many of which were in the process of being upgraded to colleges)
by those of elementary science. The transition from nature study to elementary science demonstrates that reform in education came gradually, however, and older elements were not fully expunged.

PERSISTENT NATURE STUDY

In the 1890s nature study advocates had been the reformers, directly challenging traditional educational norms. Their efforts, as part of a progressive
agenda, significantly transformed the curriculum of public schools over the
next two decades. But, in turn, the next generation viewed what had been
introduced as dramatic innovation as inadequate, standard fare. The experimental and innovative qualities of nature study that had attracted public
patronage and stimulated classroom change often declined to a prescribed
"course of study" that lacked the dynamic features envisioned by Jackman,
Lange, Hodge, Bailey, Comstock, and others. Thus, in actual practice, nature study in many classrooms was, as Flora Cooke discovered, relegated to
brief lessons with occasional and almost incidental trips to museums and
local parks. Nonetheless, it was deeply embedded and kept authorities like
the Boston School Committee debating just how to implement it: "There
was some difference of opinion in the council whether the [nature study]
course should take the form of projects, introducing a topic related to the
environment of the child and discussing its needs [in] relation to science [,]
or whether it should take a science outline as the basis and carry it out to its
application to the environment."[15]

To administrators, the uncertain status of the term "nature study" had been evident for some time, and they had joined textbook writers in accommodating what were often changes in nomenclature rather than content. In 1918 Gilbert Trafton, whose title was then "Professor of Nature Study" at Mankato Normal School in Minnesota, explained in his introduction to *The Teaching of Science in the Elementary School* that the term "nature study" had become confusing because of its "ever broadening usage to cover all phases of science" in elementary schools.[16] However, when he wrote a classroom textbook nearly a decade later, he brought back the term in *Nature Study and Science for Intermediate Grades* because the phrase was still commonly used in schools around the state.[17] The ambiguity and divided loyalty are clearly demonstrated by the changing identity of the prescribed course in Long Beach, California. In 1925 the course outlines were entitled *Nature Study*, in 1927 they became *Elementary Science (Nature Study)*, and in 1930 they were *Nature Study and Elementary Science*.[18] The term "elementary science" also had multiple meanings, suggesting aspects of nature study in some cases. Those who encouraged its use to describe elementary and middle school courses thought the phrase more clearly labeled such study as "preparatory" for later systematic study of particular sciences. "Elementary science" implied to its advocates that teaching would be didactic and the course work cumulative.

As graduate degrees in education became more common, a new generation of graduate students and faculty conducted research about nature study, particularly in the 1910s. Alice Van De Voort at Teachers College, for example, surveyed forty-five normal schools and teachers colleges and found that thirty-five of them offered nature study. She concluded that nature study was still the dominant approach in elementary training.[19] Professor Elliot Downing in Chicago's College of Education surveyed states to determine the extent to which nature study remained a mandated curriculum and reported that it did in at least twenty-seven states and the District of Columbia. He found that it was "less definitely standardized" than most grade subjects but was surprised that "in many of the city school systems [nature study] is consistently [taught] in excess of the hours prescribed by the state."[20] Downing was in the transitional generation teaching at the University of Chicago, and as part of his responsibility on the faculty he coordinated a *Nature Study* book series with the University of Chicago Press. He was also pragmatic in his support and hardly doctrinaire as he used a variety of terms and phrases to attract potential users to his textbooks, as reflected in the title of his *Field and Laboratory Guide in Biological Nature Study* in 1918.[21] Indeed, the textbook sales of Downing, Comstock, Trafton, and others remained strong through the 1920s, as did sales of classroom books that could be used for pupils in elementary classrooms whatever the title of the topic. The entomology professor Edith M. Patch of Maine, for example, in 1927 produced *First Lessons in Nature* with fifteen stories for third and fourth graders that incorporated many elements of nature study. Her fact-filled book, written in a familiar, informal style, reflected the economic sensibility that had become part of some nature study programs. It started with objects with which children were familiar, like consumable sugar, and then took pupils through an account of sugar cane, its geography, cane processing, and finally human use.[22]

The corporate sensibility of the 1920s made its mark in the schools, and nature study inspired its quota of products. *National Geographic*, for example, hired Jessie L. Burrell in 1919 for its fledgling School Service Division to produce lantern slides and cardboard-backed photographs for classroom use in conjunction with the magazine. She described her work in "Sight-Seeing in School: Taking Twenty Million Children on a Picture Tour of the World."[23] Coca-Cola produced a series, *Nature Study Cards for Visual Instruction*, which was promoted for school use, an advertising gimmick that lasted until 1934.[24] The series was composed of such titles as "Trees and Other Plants Useful to Man" and "Wild Animals."

Schools across the country, large and small, continued to use nature study vocabulary and methods. Dora Mitchell, quoted at the beginning of this chapter, noted in her Harvard thesis (published in *Nature-Study Review* in 1923) that nature study continued to be particularly strong in St. Louis, New York, Illinois, and California; she correlated its ongoing success with the continuing teaching of nature study in related normal schools.[25] New York City schools had implemented nature study early, under the leadership of Teachers College and the sponsorship of private and public organizations, and persisted in using the term and supporting its projects. The American Museum of Natural History (AMNH) and the Brooklyn Botanical Garden were intent on maintaining a "vibrant" collaboration with city schools.[26] Curriculum outlines of "Required Nature Study in the New York City Public Schools" were regularly updated by Clyde Fisher, a staff member of the AMNH, and nature study teacher Marion I. Langham. In response to criticisms of nature study, they opened each of the six grade-level course programs with this warning: "Unguided observation and haphazard questioning result in careless thinking and in a waste of time for both teacher and pupil. But if a child's curiosity is aroused by a thought-provoking question or his attention fixed on an interesting problem, he has a motive for close observation."[27]

Elsewhere, nature study curriculum persisted at sites where enthusiastic teachers and administrators continuously updated and redesigned course outlines in response to local interests and statewide mandates. In Berkeley, California, Clelia Paroni led nature study well into the 1930s, pursuing a program with the outlook of the Cornell program, filtered through Stanford and Berkeley. She emphasized the importance of teachers' initiative and stressed that children, rather than subject matter, should be the focus of attention.[28] Whether they called their work "nature study" as in Sacramento or "elementary science" as in Los Angeles, California normal school programs continued to train teachers to conduct out-of-doors teaching.[29]

Support for nature study could be inconsistent in the decades following World War I.[30] In Minneapolis, where Sarah Louise Arnold had introduced nature study in the 1890s, the subject became subordinated to geography for some years and faded from view before being reinstituted as a distinctive curriculum in the 1920s under supervisor Jennie Hall.[31] At the same time, in neighboring St. Paul, nature study had persisted in the first decades of the century under city coordinator Dietrich Lange. At Winona State Normal School in Minnesota, where the Oswego methods and nature study were integral to the curriculum in the 1890s, nature study was dropped when the object method of teaching went out of currency; but it was reintroduced in 1923.[32] At Peabody College of Education, a leader in southern education owing to its support by the Peabody Foundation (see chap. 6), nature study persisted after its introduction at the turn of the century. It was sustained by teachers like Mary Barry, who taught nature study for elementary and intermediate grades for many years and every summer conducted field botany courses and other out-of-door work.[33] In the 1920s Lucy Gage created a cabin called the "Greenhouse" as part of Peabody's Teacher Laboratory, a name intended to evoke the growth of children as well as plants.[34] Nature study, which never fully overcame its position as optional, persisted best when supported by regional normal schools along with local administrators and school boards.

In its first decade, nature study had been buoyed by the promise of enhancing pupils' acquaintance with the natural sciences and by the general momentum for curricular change. Charles McMurry, who had moved to Nashville from Northern Illinois Normal School, recalled in an unpublished memoir that the early leaders of nature study had been confident, enthusiastic followers of Thomas Huxley, Asa Gray, and Louis Agassiz.

These scientists had not hesitated to insist that "teachers and children alike awaken to the sure things of life, to the beauty and utility everywhere so concretely visible in nature's works.[35] In McMurry's view, nature study lessons had gone through multiple phases and had varying emphases:

> [O]utdoor excursions to study bees and flowers, trees, shrubs, rock and hills, streams and forests, indoor studies in laboratories, with test tubes and apparatus, school gardens and agriculture, scientific inventions like telescopes and thermometers, steam engineers, and electrical machines, school health and hygiene, observations of storms and weather phenomena, the human bodies in health and disease, a scientific dietary, the application of science to industries, and even the sun and solar system.

He ruminated that the range and novelty had constituted a "joy ride into the sciences" that had gradually "dwindled" and "settled down to the commonplace duties of the school." From his perspective at Vanderbilt, he thought nature study remained as one of "six or eight principal studies" in the early grades, but he seemed unsure whether it retained the child-centered outlook he and his colleagues had promoted at Illinois Normal three decades earlier.

Anna Comstock spent her later years trying to revitalize the early spirit of nature study. Even when Elliot Downing had remained nominal editor during the prewar years, the Cornell faulty had picked up considerable responsibility for the *Review* and the Nature-Study Society.[36] Comstock served as president in 1913, followed by Bailey, who had been the society's first president and returned to serve from 1914 to 1917; Comstock Publishing Company continued as publisher. After Comstock herself officially took over the editorship in 1917, she solicited contributions that resemble the Cornell leaflets in their detailed descriptions of animals and plants that school classes might encounter on excursions. She also encouraged accounts of extracurricular nature study by soliciting essays on programs in camps, scouting, and public parks.[37] The Nature Lore School of William Gould Vinal, which educated camp counselors and park guides in basic nature study, particularly attracted her interest, and she actively participated in its summer program (fig. 25).[38]

Ever optimistic, she reported with some pride to her old friend David Starr Jordan in 1920, "For three years I have been editor of the Nature-Study Review and have managed to keep it alive during the War which is more than most editors have been able to do with periodicals of like caliber. I believe the Review has a special place for teachers and am glad to say it is

FIGURE 25. Preparing camp naturalists to do systematic work with their young charges led to the development of special training courses. The Nature Lore staff in 1922 included (*in the back row*) Anna Billings Gallup on the far left, William Gould Vinal in the middle, and Anna Botsford Comstock next to him on the right. *Nature-Study Review* 19 (March 1923): back cover. Courtesy of the Wangensteen Historical Library of Biology and Medicine, University of Minnesota.

on its way to solvency."[39] Her goal was to maintain the educational outlook that nature, rightly studied, had the capacity to uplift children and teach them to appreciate the world around them. Sustaining that ideal, however, proved difficult in the more skeptical culture of the 1920s, and there is little evidence that new members were subscribing to the journal. The stalwart but aging Comstock retired from Cornell in 1922 but remained as secretary/treasurer and editor for another year even as she cared for her husband John, who was seriously ill.[40]

Comstock's retirement marked a significant generational shift. Her faculty position was filled by E. Laurence Palmer, who had been educated in the Ithaca schools and under the supervision of teachers trained in nature study at Cornell before attending New York State Normal School in nearby Cortland. Palmer had subsequently completed a B.S. and Ph.D. in entomology at Cornell.[41] The nature study work had been moved from the Department of Entomology to the Department of Rural Education in 1913 and was staffed by men who tended to have degrees in the social sciences, unlike the team of women and men Bailey had organized fifteen years earlier. Palmer apparently had larger ambitions, but after a brief and awkward negotiation with Teachers College, Columbia, about a position there, he remained in Ithaca, holding the title of Professor of Rural Education and Director of Nature Study.[42] He also began to shape the Cornell program into a B.S. in conservation and recreational nature study.

Perhaps tired of her nearly single-handed efforts as editor, Comstock and her fellow officers of the society responded positively to an offer in 1923 to merge *Nature-Study Review* with the new *Nature Magazine: Illustrated Monthly*, founded in 1923. Comstock had publicly welcomed the "Big Handsome New Sister" as another way to interest children and teachers with its large format, glossy paper, and generous illustrations.[43] But the merger was, in reality, a compromise with a competitor, Arthur Newton Pack, who had launched the rival American Nature Association (ANA) in 1922. Comstock and her board negotiated to keep the society's news as designated column space for nature education and to have its editor elected from the society, but "*Nature Magazine* did not carry out this agreement."[44] For a few issues Comstock edited the Nature-Study Department, providing two or three pages of content and some notices of the society.[45] *Nature Magazine* foundered, and Pack, along with Comstock's successor on the Cornell faculty, E. Laurence Palmer, started *Nature and Science Education Review*, which in 1928 became the new official channel for the persisting Nature-Study Society. But it too failed, and elements were absorbed into *Science Education*.[46] Anna Comstock was apparently surprised at the mixed loyalties of her ambitious young successor, Palmer, but she remained supportive; and her press even published his *Field Book of Nature Study* in 1927.[47]

In 1925 the Nature-Study Society's leaders produced a yearbook of featured articles, imitating those of the NEA and other educational organizations. Its first *Yearbook of the American Nature-Study Society* (fig. 26) was the only one.[48] The goal of the publication was clearly to demonstrate that nature study was established in and beyond public schools. In it Clelia Paroni tabulated responses from 127 of the larger school systems. She reported that twenty-two of them had one or more nature study supervisors, who worked closely with teachers and prepared printed materials for local activities. She also surveyed teacher training institutions with practice schools and found that forty-nine of the fifty-five responding continued to teach nature study and eighteen had school gardens.[49] A somewhat less encouraging report on the preparation in teachers' colleges, conducted by

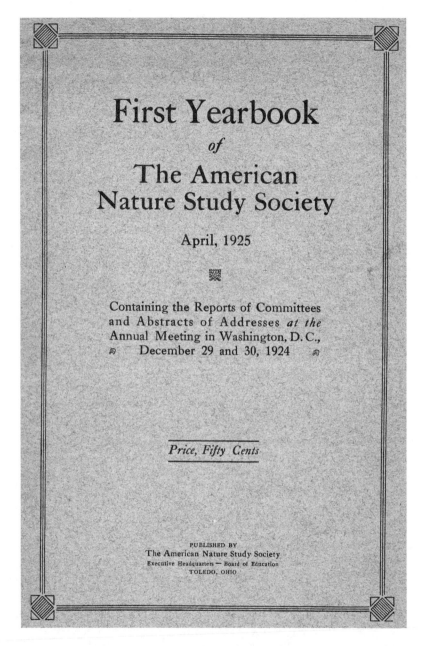

First Yearbook

of

The American Nature Study Society

April, 1925

Containing the Reports of Committees
and Abstracts of Addresses *at the*
Annual Meeting in Washington, D. C.,
December 29 and 30, 1924

Price, Fifty Cents

PUBLISHED BY
The American Nature Study Society
Executive Headquarters — Board of Education
TOLEDO, OHIO

FIGURE 26. The 1925 yearbook, based on papers presented at the annual meeting and research done by academic advocates of nature study, was an effort to demonstrate the professional stature of the Nature-Study Society. Author's copy.

William Gould Vinal, revealed that most instructors held titles that corresponded to high school subjects (e.g., biology, general science, and agriculture); only twelve taught nature study principally.[50] Issues of the *Nature Almanac* reported similar surveys, but those tended to demonstrate that nature study as an educational program under that term was becoming less visible.[51] The competition from the ANA made some American Nature-Study Society (ANSS) members uncomfortable, especially as the association tried to attract teachers rather than simply pursuing its stated goal, "to stimulate public interest in science."[52]

The National Council of Garden Teachers, formed in Cleveland in 1920, was also informal competition to the Nature-Study Society. This group changed its name to the National Council of Nature Study and Gardening in 1922 and then to the Supervisors of Nature Study and Gardening in 1923. Trying to keep pace with curricular changes, members ultimately changed its name in 1932 to the Council of Supervisors of Elementary Science, a move that followed Gerald S. Craig's term in the office of president.[53] Several early members, including Ellen Eddy Shaw, Lenore Conover, and Alice Jean Patterson, were particularly interested in gardening, and thus were comfortably members of both groups.

Without the *Review*, Nature-Study Society members struggled to publicize and sustain their meetings, publish in other educational and scientific journals, and exchange curriculum information. Palmer stated that there were 729 members who pay dues "without material return" but was concerned because the society's annual meeting "is usually rather overwhelmingly represented by Cornell people."[54] They still met with the American Association for the Advancement of Science, and a small group continued to hold regular meetings that were reportedly attended by up to fifty people in the last half of the 1920s. Many of the papers were presented by well-positioned but often aging nature study advocates like Bailey, Comstock, Elliott Downing, Clelia Paroni, Otis Caldwell, and Elizabeth Peeples.[55] Laurence Vail Coleman, prominent in museum studies, came and emphasized that museums had active programs in nature study, and William Gould Vinal described the rapid expansion of nature study in summer camps and national parks.

The 1928 meeting in New York City was a triumph for Bertha Chapman Cady, whose office was at the AMNH. She arranged five days of society activities, including a dinner for nature writers on Friday evening, tours to such local sites featuring nature study as the Brooklyn Children's Museum and the New York Botanical Garden, and special sessions for teach-

ers in the Reptile Room of the AMNH.[56] Still, the membership was slip-
ping, with only about 250 names on a list from the late 1920s, most of
whom had not paid their dues. This number was only about 10 percent of
the membership reported when the ANSS and ANA journals had merged
just five years earlier—although the early figure may well have been in-
flated.[57] In a few cases, energy was shifting to branch chapters: New York
City under the leadership of Ellen Eddy Shaw, curator of Elementary Instruction at the Brooklyn Botanical Garden; Seattle, with Helen T. Bush
as secretary; and Webster Grove, Missouri, with Alfred F. Satterthwait as
secretary.[58] The Webster Grove chapter, with residual members from an
older St. Louis section, had originated when Satterthwait became state entomologist in Missouri. He then coordinated an active group of adults who
regularly took excursions in the area. Although this chapter sponsored a junior after-school group, it seems to have been only loosely connected to the
public schools.[59] In 1931, Satterthwait was elected president of the ANSS,
with Jennie Hall as secretary-treasurer; they discussed ways to revitalize the
membership and made elaborate plans for the New Orleans meeting.[60] With
fewer members, divided loyalties between the society and its sections, and
broad ambitions, the society pulled further from its early focus on formal
school teaching but it could not compete with professional groups that represented museum staff, camp counselors, and other less formal educational
venues. In 1931 the ANSS constitution was revised. It kept dues at one dollar, with active chapters returning only twenty-five cents, and stated that
the society could publish only intermittently. This controversial decision
left few funds for the national organization.[61] In 1933 there was serious
discussion about a name change, with some members arguing that the term
nature study had grown "decidedly out of favor with administrations of
public education," but a poll of members proved so divided that no change
was made.[62]

The organizational momentum of the ANSS had dwindled, although it
was quietly nurtured at Cornell for decades as members turned their attention to local projects. Some of the established advocates continued to present
papers and to publish articles with the words "nature study" in their titles,
or with a nature study outlook, in educational periodicals, including the
NEA's *Journal of Addresses and Proceedings*.[63] The NEA had by now changed
from being a "debating society" of academics to a powerful lobby in Washington. Its leadership now came from the administrative ranks of the state
normal schools, who were engaged in changing educational standards and
consolidating the interests of its membership.[64] Nature study advocates,

however, including some of the leaders from the 1920s, had already been turning their attention toward the broadening inclusion of nature study in other venues.

NATURE STUDY EDUCATION FOR
CHILDREN AND ADULTS

The taste for rational recreation that characterized Victorian life and had fueled natural history clubs, nature writing, and outdoor excursions in the nineteenth century was expressed even more systematically after the turn of the century. Nature study with its close observation of the natural world, hands-on learning, themes of civic and moral uplift, and appreciation for the processes of living things in their environment seemed highly appropriate as comprehensive schools encouraged extracurricular activities and progressive leaders sought to make middle-class activities available to even the poorest children. One result was that while the school curriculum for nature study came to include more health, agriculture, and other industrial topics, the movement's focus on out-of-doors nature gained at summer camps, boy and girl scouting, rural 4-H clubs, and programs in state and national parks. Voluntary youth organizations were particularly active sites for nature study.[65] By the 1920s nature study instruction was often provided by a cadre of specially trained naturalists who worked with parents and youth leaders.[66]

Informal youth activities were not new. The Agassiz Association had loosely coordinated hundreds of them (see chap. 1). The association's Edward Bigelow continued to update his *Guide to Nature* and to hold programs near South Beach through the 1920s.[67] His ambition was to establish a permanent center that would house his publications and serve local residents. John Kennedy Tod, a sympathetic and wealthy neighbor, had provided funds for a small scale "Acadia" that could serve as a center for the Agassiz Association. With its proposed natural history collections, rustic landscape, and an observatory, Acadia was to offer public programs and summer schools along the seashore near Old Greenwich. Tod seems to have anticipated that Acadia would become self-supporting within two years. Although Bigelow devoted himself to the intellectual and the manual labor tasks of building the facility, the informal association did not generate enough income to sustain a permanent organization, even with small donations from local philanthropists in Stamford and a few elsewhere in New England. Lack of progress and perhaps personality clashes led Tod to ask Bigelow to vacate the premises. Ironically, the labor union members to whom Tod then gave the property voted to give Bigelow the building (which he removed), sell the land, and build elsewhere.[68] Other informal nature study efforts were more successful, particularly those affiliated with well-established amateur organizations or newly emerging national programs for youth.

From the 1910s onward, children increasingly encountered nature studies through organized programs. The 4-H movement, in particular, grew rapidly by coordinating the volunteerism of rural families, school teachers, and farm extension agents. The 4-H logo was a four-leaf clover, brought to Washington from Iowa in 1911, along with the slogan, "Head, Heart, Hands, and Health." The leader who played a critical postwar role in building the 4-H club name and its state networks was Gertrude Warren, who had studied with Anna Comstock at Cornell and later taught home economics education at Teachers College before joining the United States Department of Agriculture in 1917 to organize its youth programs.[69] The informal but well-organized 4-H movement was intended to transfer "technological and agricultural advances from the laboratory and the test plot to practicing farmers and homemakers," even as it promoted the advantages of country living for their children.[70] John Spencer's Junior Naturalist Clubs provided an early model, as nature study was an integral part of activities intended to show the value and attractiveness of rural life through gardening, flower arranging, using natural materials for art-and-craft projects, and raising animals for competition. The largely rural club program, with about one hundred thousand members in 1912, had grown to over five hundred thousand by 1918.[71] It would be complemented by scouting programs in small towns as well as in larger cities, the "junior" clubs affiliated with such well-established groups as the Audubon Society and the Garden Association of America, and the youth groups affiliated with natural history museums (fig. 27).[72] Relying on personnel and activities that were tested and proved successful in the nature study movement, these organizations incorporated fundamental elements of the movement's educational outlook.

Summer camps in the 1890s had presented a hardy experience for boys, who were typically housed in tents and given projects that required vigorous exercise in the recreational or marginal rural landscapes.[73] The Worcester Society of Natural History, for example, sponsored a summer camp at nearby Quinsigamond Lake and linked the outdoor experience to its program for school children in its museum.[74] While these camps were for middle-class children, projects for inner-city youngsters became part of the urban progressive reform programs of the next decades. The clear intention

FIGURE 27. The Buffalo Society of Natural History established a Roosevelt Field Club that held excursions with a deliberately conservationist focus, often led by Cornell and other faculty members in the region. *Nature-Study Review* 17 (February 1921). Courtesy of the Wangensteen Historical Library of Biology and Medicine, University of Minnesota.

was to expose children to a living nature that they seldom encountered in their daily lives.[75] By the turn of the century, emphasis on nature characterized most camps for boys and girls.[76] Camp Fire Girls, founded in 1910 by Charlotte and Luther Gulick with the progressive goal of developing the "whole child," incorporated summer camp experiences, as the name suggests. [77] An active reformer, Luther Gulick had helped bring physical education as a "science" into the schools of New York City and to YMCAs across the country; he also encouraged local nature projects. His wife, Charlotte Gulick, extended a program that she had developed for their daughters and friends into summer camps in Vermont and Maine. The organization soon began year-round programs to prepare participants for the camps, and indeed for life itself. [78]

Other youth organizations used nature study as a way to further their goals. James Earl Russell, dean of Teachers College, considered supporting the Camp Fire Girls but concluded they were less interested in citizenship aspects of girls' development than in service and home duties.[79] Instead, he served on the Girl Scouts board and wrote pamphlets for them. Both Boy

Scouts and Girl Scouts celebrated themes of wilderness and camping, particularly as they developed merit badges and a network of outdoor facilities that featured bird and flower walks.[80] The nature topics in these informal organizations could be deeply gendered. Boy Scouts were expected to be out-of-doors "stalking nature" as part of their scouting programs, while Girl Scouts were housed in facilities that more closely approximated conventional domestic settings.[81] Nonetheless, in the Girl Scouts the connection to nature study was reinforced by Bertha Chapman Cady, who held the title of Naturalist of the Girl Scouts in the mid-1920s and worked at its headquarters until her retirement in 1933.[82] She also was secretary of New York City's Coordinating Council on Nature Activities, headquartered at the AMNH.[83] Cady produced a *Girl Scouts Nature Notebook* with detailed instructions on what was needed to earn each badge.[84] The bird badge required major assignments — studying bird anatomy, conducting one full day of field work, building feeders, and monitoring bird behavior, plus five other projects. In both posts, Cady set to work with her accustomed vigor and effectiveness to encourage "out-of-door" work. Under the council's auspices, she and her husband, Vernon Cady, produced a set of nine small pocket *Guides to Nature* for teachers and leaders, identifying their audience as an eclectic group, including "public schools, normal and teacher training schools, camps, Boy Scouts, Girl Scouts, Camp Fire Girls, Girl Guides, Woodcraft League, Girl Reserves, and Nature Study Clubs."[85] Her California colleague Joseph Grinnell assured her that such organizations were important for distributing nature information to the public.[86] He also noted, in passing, that the demand for "nature leaders" was considerable and encouraged Cady to use her organization to ensure an adequate supply of teaching naturalists.

By the 1920s programs were already underway to provide systematic training for those supervising nature study at camps and other sites. Few camp organizers and staff members had the requisite background in botany and zoology or in education to offer extensive instruction, but nature study had developed observational techniques that could be transferred to outdoor settings and provide an introduction to field study.[87] One of the first significant programs was that of William Gould Vinal, who in the 1910s established a Nature Lore School on Cape Cod to offer precamp training for nature leaders; the school became the Nature Guide School at Western Reserve College in Cleveland in 1928.[88] Comstock Publishing Company produced his popular *Nature Guiding*, whose epigraph bowed to nature study's long tradition of incorporating literature by printing Walt Whitman's poem, "There Was a Child Went Forth."[89]

Cornell, which had never established a permanent college in education, turned to the preparation of nature study leaders for youth organizations under Laurence Palmer.[90] Anna Comstock had encouraged these emerging constituencies by dedicating a special issue of *Nature-Study Review* to camping and by publishing essays that related to scouting and Camp Fire Girls. The Comstock Publishing Company had also produced a number of notebooks and worksheets for informal and for outdoor instruction (fig. 28).[91] Palmer maintained nature study classes, intending to train graduates who could both conduct research and translate the natural world to visitors on trips afield "to study the secrets of the hills and forests." These trainee naturalists might also organize and maintain museums, give public lectures, and offer guided tours of natural settings.[92] His Cornell graduates thus pursued careers in informal education and frequently related their work to conservation and preservation activities. Pauline Sauer, for example, received her B.S. in education from Chicago Teachers College and then her Ph.D. at Cornell before going to the Iowa State Normal School, where she pursued conservation interests, encouraged nature study education, and directed the natural history museum.[93] Recognizing that national and state parks sought experienced staff as visitor numbers soared after World War I, Palmer encouraged Cornell students to spend summers working as naturalist interns at Harriman State Park outside New York City.[94]

Throughout the nineteenth century, hunters going into wilderness areas had used guides, but tourism opened up quite different opportunities as families were drawn to the convenience of parks that had roads and facilities. The first trail school for nature guiding in the West was probably at Long's Peak, Colorado.[95] After the National Park Service was established in 1916, its new director, Stephen T. Mather, toured the western parks. At Fallen Leaf Lodge at Lake Tahoe he heard Loye Miller lecturing to guests.[96] Impressed by the overflow crowds gathered to listen and by Miller's uncanny bird whistles, Mather suggested that Miller lecture and give informational tours at Yosemite the following summer. Miller, a graduate of Stanford and Berkeley and an expert on paleontology of the La Brea Tar Pits, was on the faculty of the Los Angeles Normal School, where he taught nature study and natural sciences. He had written *Outline of Nature-Study* with Ernest B. Babcock in 1906, worked closely with Harold C. Bryant, who became educational director at the California Fish and Game Commission in 1914, and for a number of years conducted a summer school for teachers in southern California.[97] Now, he and Bryant shaped a School of Field Natural History to train nature guides and nature study teachers

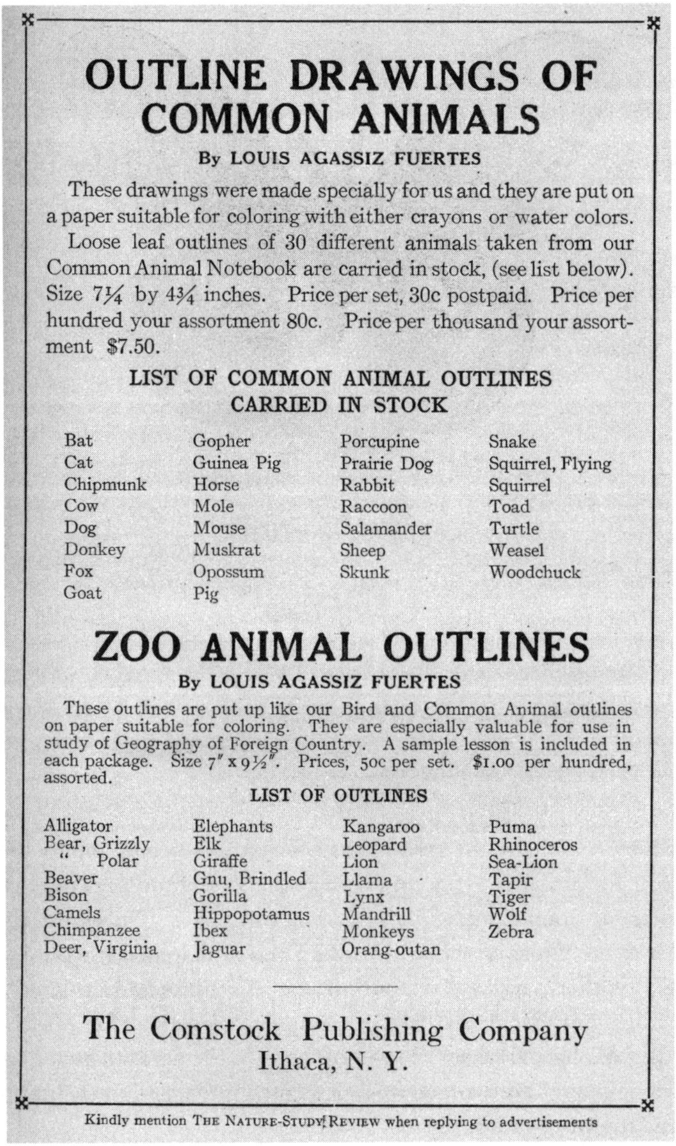

OUTLINE DRAWINGS OF COMMON ANIMALS

By LOUIS AGASSIZ FUERTES

These drawings were made specially for us and they are put on a paper suitable for coloring with either crayons or water colors.

Loose leaf outlines of 30 different animals taken from our Common Animal Notebook are carried in stock, (see list below). Size 7¼ by 4¾ inches. Price per set, 30c postpaid. Price per hundred your assortment 80c. Price per thousand your assortment $7.50.

LIST OF COMMON ANIMAL OUTLINES CARRIED IN STOCK

Bat	Gopher	Porcupine	Snake
Cat	Guinea Pig	Prairie Dog	Squirrel, Flying
Chipmunk	Horse	Rabbit	Squirrel
Cow	Mole	Raccoon	Toad
Dog	Mouse	Salamander	Turtle
Donkey	Muskrat	Sheep	Weasel
Fox	Opossum	Skunk	Woodchuck
Goat	Pig		

ZOO ANIMAL OUTLINES

By LOUIS AGASSIZ FUERTES

These outlines are put up like our Bird and Common Animal outlines on paper suitable for coloring. They are especially valuable for use in study of Geography of Foreign Country. A sample lesson is included in each package. Size 7″ x 9½″. Prices, 50c per set. $1.00 per hundred, assorted.

LIST OF OUTLINES

Alligator	Elephants	Kangaroo	Puma
Bear, Grizzly	Elk	Leopard	Rhinoceros
" Polar	Giraffe	Lion	Sea-Lion
Beaver	Gnu, Brindled	Llama	Tapir
Bison	Gorilla	Lynx	Tiger
Camels	Hippopotamus	Mandrill	Wolf
Chimpanzee	Ibex	Monkeys	Zebra
Deer, Virginia	Jaguar	Orang-outan	

The Comstock Publishing Company
Ithaca, N. Y.

Kindly mention THE NATURE-STUDY REVIEW when replying to advertisements

FIGURE 28. The Comstock Publishing Company, which became the foundation of Cornell University Press, produced its educational materials, like these sketches by illustrator Louis Agassiz Fuertes, at cost, in an effort to encourage users of nature study materials. *Nature-Study Review* 15 (October 1919): advertising insert opposite page 262. Courtesy of the Wangensteen Historical Library of Biology and Medicine, University of Minnesota.

FIGURE 29. Nature guiding offered summer employment and an opportunity for outdoor adventures for nature study teachers, and editor Anna Botsford Comstock described such programs as this one in Yosemite National Park. *Nature-Study Review* 16 (September 1920): 272. Courtesy of the Wangensteen Historical Library of Biology and Medicine, University of Minnesota.

in Yosemite, which by the mid-1920s offered certificates rather than college credit for attendees (fig. 29).[98] Yosemite's program was underwritten by Charles M. Goethe, an active conservationist as well as eugenicist, who viewed it as an extension of nature study and anticipated it would provide a model for other parks.[99] By that time, too, Yosemite had initiated a "trail school" for naturalists who led week-long camping trips for adults. In the early years of the park service, the niche of park naturalists (or rangers) offered employment to women as well as men interested in the natural sciences and out-of-door adventures.[100]

Women's participation had been encouraged by the General Federation of Women's Clubs, which supported the park movement in the 1910s and 1920s and anticipated that educational programs would be necessary as families with children began to visit the parks via railroad and then automobile.[101] Goethe recommended hiring women at Yosemite, reasoning that children might respond better to women naturalists and that they worked for lower salaries. Some women were hired for an increasing full-time staff, while others were teachers who worked in various parks during the summers.[102] The connections between informal and formal learning were evident. Elizabeth Burnell, a teacher from Michigan, was one of the nature educators hired by Enos Mills in Rocky Mountain National Park, where she headed the trail school for twelve summers; she subsequently supervised nature study in the Los Angeles schools.[103] As Polly Welts Kaufman points out, there were obstacles, and women had to challenge a dress code that directed them to wear skirts even when climbing mountains. Although they encountered sexism from some of their male peers, a surprising number enjoyed ranger status until about 1933, when male rangers successfully eliminated women from what became their domain.[104] Men concerned about nature and the environment could be self-conscious about their masculinity, often articulating strong claims about their outdoor work in forest-park management.

In fact, the Depression significantly altered the Department of the Interior and other agencies, and, as staff shrank, staff biologists were pulled from other duties to perform interpretive tasks. But through the 1920s, park employment allowed men and women interested in nature study education to teach the public visitors about nature while visiting state and federal parks. Outdoor and hands-on education in nature study now found a broader base among adults as well as youth even as its position eroded in the schools.

ESTABLISHING ELEMENTARY
SCIENCE AS THE STANDARD

Nature study advocates in schools had, from the start of the movement, debated among themselves and with others about the relationship between elementary science and nature study. Liberty Hyde Bailey had been particularly adamant in drawing a distinction between the two in his *Nature-Study Idea*, while others had wanted to be sure that nature study was scientific. Wilbur Jackman and a continuing line of faculty members and textbook writers presumed that nature study included physics, geology and geography, and astronomy, as well as botany and zoology. Within a decade after nature study had taken center stage as the most likely way to teach the natural sciences, Charles McMurry had already determined to represent nature study as a method rather than a subject; he titled his 1904 textbook *Special Method in Elementary Science for the Common School*.[105] For the next twenty years, while school systems often used the two phrases, "nature study" and "elementary science," simultaneously or interchangeably, especially in local course guides, the balance slowly shifted from the first to the second, toward the term elementary science.[106] Conceding the point, some well-established nature study leaders also mixed the terms in their articles and textbooks, or shifted to elementary science.[107] But this unstable compromise did not hold as an increasingly aggressive position was taken by a few proponents of offering systematic science in elementary education.

In the period after World War I, the relative absence of physics, chemistry, meteorology, and related subjects in American public schools troubled not only scientists but also educational administrators. There was considerable public interest in and respect for discoveries in physics and new applications in chemistry.[108] As the NEA's leadership was taken over by normal school faculty eager to establish their academic credentials, the organization emphasized its scientific approach to education and worked to build better relationships with experts in other academic disciplines. Emerging as the leader of a concerted effort to establish elementary science as the standard for public schools was Gerald S. Craig.

Craig had joined the Nature-Study Society and the School Garden Association but deliberately turned toward "elementary science" education in the mid-1920s.[109] After receiving a bachelor's degree from Baylor University and teaching high school briefly, Craig served in France during World War I. He returned to attend the University of Pennsylvania and taught at nearby Oak Lane Country Day School while occasionally teaching summer school at Pennsylvania State College.[110] In 1924, he moved to Teachers College in New York City, where he completed his Ph.D. dissertation, a multilayered survey of courses and outcomes relating to nature study and elementary science.[111] He then produced an outline, *Tentative Course of Study in Elementary Science for the Horace Mann School*, which specifically called for an alternative to nature study.[112] His research, advanced degree, and affiliation with Teachers College positioned him well as a commentator on the debate. Perhaps because of his experience as a high school teacher, Craig was particularly intent on insuring that the educational ladder had carefully standardized rungs that led toward the upper grades.[113]

Strategically, Craig commented approvingly on Wilbur Jackman's pioneering formulation of nature study but then turned his criticism to the actual practice of nature study that he found in the schools, including the prestigious classrooms of the Horace Mann School, which was affiliated with Teachers College. Jackman, he suggested, "was as American as apple pie. He smacks of Walt Whitman, of John Burroughs, or Theodore Roosevelt. You can smell the frontier in his writing. He always inspired me. I have gone back to his writings again and again."[114] In actual practice Craig sought to establish a "science program in education," and he relied heavily on colleagues like Florence Grace Billig. She had earlier been a nature study supervisor in Kansas but had moved to New York and was fully involved in reformulating the program of study at Teachers College.[115] With his dissertation advisor, S. Ralph Powers, Craig established the National Association for Research in Science Teaching and helped transform the *General Science Quarterly* into *Science Education*, which thus absorbed the former ANSS journal as well.[116] He also began to produce a series of graded textbooks, *Pathways to Science*, with Ginn and Company.

In 1931, when New York State produced its newest recommended curriculum, "Elementary School Science: A Tentative Syllabus for Elementary Schools," the consulting committee included Craig, E. Laurence Palmer, and Van Evrie Kilpatrick, among others. Described as an "experiment" and claiming to include "the essential subject matter heretofore known as nature study," this new course further embodied the "ideals of scientific inquiry, accurate observation, and tested thought." The deprecating commentary then made a deliberate shift in emphasis:

> While for a generation or more nature study has had a place in the curriculum of the elementary school, a three years' search has failed to discover any tangible evidence as to the content or method of instruction for this subject in the schools of the state. Elementary school science attempts to lift nature study out of the incidental, indifferent treatment it has had into a position of respect in the curriculum of the elementary school.[117]

Under the guise of rehabilitation, nature study was in fact dismissed. Craig's position and his determination to eliminate the term nature study seemed initially, and perhaps naively, to Laurence Palmer to be simply "a lot of quibbling."[118]

In practice these activities pushed nature study vocabulary toward the abyss of abandoned curricula.[119] Craig's own expression of surprise at the apparent growing interest in teaching elementary science in his promotional handbook, *A New Science Program for Elementary Schools*, seems disingenuous, given his continuous efforts to establish that phrase and outlook. In the book, he complimented innovative teaching methods among some education colleagues, noting their work as "superior to the type of formal and stereotyped teaching which is so characteristic of high school and college teaching."[120] Publicly, too, he remained conciliatory toward teachers of nature study and did not significantly challenge their independent methods, suggesting that most were "doing creative teaching in the field of elementary science."[121] And, indeed, whether elementary science was dramatically different is contested. A short biographical sketch of Craig points out, for example, that "despite the apparent discontinuity between the nature-study movement of Jackman, Comstock and Palmer, and the new elementary science of Craig, the more one begins to examine Craig's contributions, the continuity is more striking."[122] Craig's initiative and new materials did make him a visible leader in science education and was highly profitable. By the 1950s, Craig's textbook publications supplemented his $8,000 annual salary by nearly $40,000.[123]

The National Society for the Study of Education's *Yearbook* in 1932 has been cited by historian Kim Tolley among others as a critical turning point. Its editors declared that because of the negative "traditions associated with the name," the term "nature study" had been replaced by "science" for the elementary school.[124] By 1940 Craig's challenge to nature study had taken on a negative tone. When he mentioned nature study, it was always to dismiss it. He asserted that nature study had been ad hoc, dependent on "incidents" like a child bringing a frog or flower to school. His claim that there had been no framework of systematically planned lessons or integration of social values does not reflect his own close familiarity with the extensive lesson plans of schools he had studied or the expansive school gardens whose planners had very specific intellectual and civic goals. Looking backward, he chanted a litany of nature study's flaws: "the modern trend seeks to divorce science in elementary schools from all forms of anthropomorphism, sentimentalism, superstition, animism, emotionalism, prejudice, and unfounded opinion."[125] With the age-graded primary school now the norm, the educational ladder needed the testing and standards to support a "return to normalcy" during and after the 1920s.[126] The movement headed by Craig was thus very aligned with larger social and intellectual trends.

Craig's dismissal of his nature study predecessors, however, persisted in textbooks on science education and in histories of it in the 1940s. Robert Hill Lane, for example, in his 1941 textbook wrote that nature study, instead of being based on sound scientific principles, had been "maudlin and fanciful" with "inaccuracies and downright misinformation."[127] Historians made similar arguments. In the preface to Orra Underhill's account of science in elementary education, which became the standard history, Samuel R. Powers offered the following assessment: "Teachers of nature study in the public schools and often in normal schools were, in general, poorly educated in science and, in line with the traditions of the time, tended to become emotional rather than scholarly about natural phenomena."[128] Lost in such accounts was the reality of successful, sustained nature study teaching, and the continuity of many pedagogical techniques from nature study that were used alongside written materials on which pupils might be tested.

INTEGRATED NATURE STUDY

Nature study, never fully defined or completely absorbed into the public schools, had moved beyond them. By the 1930s, the term "nature study" was frequently used to describe programs in parks, summer camps, and other recreational facilities and appeared in headings on bulletin boards, descriptive leaflets, evening lectures series, and park ranger tours. Educators of all types, many of them trained in normal schools and special institutes, took with them pedagogical techniques that promoted participant observation and hands-on learning. Nature study leaders were proud that their flexible educational training allowed them to adopt readily new methods and technologies like phonograph recordings and motion pictures.[129] San Jose State Teachers College, for example, prepared nature study films to be used in classrooms, by scout troops, and at 4-H club meetings; and Cornell produced nature lectures for local radio stations.[130]

Despite the public challenge and new prescription for elementary science by the NEA and outspoken faculty at Teachers College, nature study techniques and vocabulary persisted in many school systems. Local authorities may have been oblivious to the academic debates or perhaps ignored them. Wayne Urban argues that the divide between professors of education — often men — in universities and normal schools and practicing elementary teachers — usually women — widened, at least in part because the pedagogical focus moved the psychology of learning away from effective teaching.[131] So, in practice and in name, nature study persisted without serious challenge in many parts of the country. In Oregon, for example, teachers were encouraged to use Craig's Pathway series along with Comstock's *Handbook of Nature-Study* and other familiar nature study texts, anticipating that school systems and teachers could comfortably integrate these materials.[132]

Nonetheless, the 1920s and early 1930s had been difficult for advocates. In a poignant comment, Jennie Hall reflected that during the first quarter century, nature study leaders and teachers had been characterized by "wonderful ideals, unselfish devotion to a cause, and purposeful building of a society for service" led by leaders of "unusual ability."[133] Her nostalgic view overlooked the contentious debates and ongoing problems that early nature study leaders faced, but her comment captured the sense of loss she felt as nature study was dismissed and even derided by a new generation of educators. Her open letter to Nature-Study Society members in 1932 asked whether in an "urban-minded" world there was time for children to study nature or an educational site where such work could be promoted. Her open-ended question lingered, and it would be much later in the century before curricular attention returned with enthusiasm to nature study's general tenets. The society's efforts "to include not only nature education in the schools but nature education in general" had been overtaken by other organizations with stronger membership or institutional bases.[134]

But the Nature-Study Society and its named curriculum had never been, in the thinking of the early textbook authors, central to their expansive vision for educating youngsters in the natural world. The schools were an important vehicle for establishing model projects and demonstrating the potential in child-centered learning. As educational administrators moved away from nature study, advocates like Anna Comstock, who encouraged her Cornell colleagues to offer courses to address the new developments, welcomed the expanding informal, extracurricular niches for the discipline. As Laurence Palmer watched the vocabulary of elementary science take hold, he noted that it was not, in fact, so very different from the nature study already in place. If the term "nature study" no longer offered inspiration to a generation of teachers, many of them now took for granted that natural sciences in other formats were still embedded in school practice and in programs well beyond classrooms.

CONCLUSION

If facts are the seeds that later produce knowledge and wisdom, then the emotions and the impressions of the senses are the fertile soil in which the seeds must grow.

Rachel Carson[1]

THE ECOLOGIST and nature writer Rachel Carson, writing more than sixty years after the publication of the first nature study manual, echoed the perception of early nature study advocates, including scientists and educational reformers, when she emphasized the importance of both factual knowledge and sense impressions in her popular book, *The Sense of Wonder*, in 1956. The central aim of nature study leaders as they sought to introduce their curriculum to elementary and grammar school pupils was to acquaint children with nature firsthand, equipping each individual to be a keen observer and to appreciate the complexity and significance of the natural environment. The potential tension among those who variously emphasized elementary science, a sympathetic approach to nature, practical implications of knowing the natural world, or moral responsibility to conserve and preserve it became explicit and sometimes heated in the 1910s. By the 1920s, the authority of rational science and a deliberately modern outlook had steadily gained ground, but this emphasis never completely undermined those nature study leaders who identified an intimate connection between nature and human experience.[2]

Nature study's initial sponsors included men and women with scientific training, several of whom also had experience teaching school. They were joined by progressive educational reformers who shared a scientific outlook. Both groups agreed on the proposition, whether from personal philosophy or pedagogical theory, that children needed to be taught in ways that were age-specific and intimately related to their personal experience. Responsive teachers took such theories into their classrooms. Particularly in the early years, when there was considerable autonomy in local schools, they adapted nature study's content and methods in the innovative ways described in previous chapters. The result was a new curricular space for the natural sciences in the public schools — an effervescence of progressive curricula that incorporated nature study into urban, rural, and small town schools across North America in the first three decades of the twentieth century. In narrating and analyzing the extent of this activity and the responses to it, this history offers a deliberate corrective to dismissive descriptions of the nature study movement. Too many casual descriptions in histories of science, education, environmentalism, and pubic culture have labeled nature study as naive in outlook, marginal within the school curriculum, or ineffective in educational practice.[3] Although nature study could be controversial and more demanding to teach than some other subjects, evidence demonstrates that it enjoyed widespread and sustained support from administrators, teachers, and communities where it was implemented. Moreover, nature study was intimately related to the social outlook and political activism that undergird a broad interest in the natural environment and generated conservationist and preservationist activities in the early decades of the twentieth century.

Academic scientists and science educators at normal schools built the essential infrastructure that made it possible to introduce nature study into public schools in most parts of North America.[4] Their enthusiasm was reflected in commentary from civic-minded university presidents like Charles Eliot of Harvard and David Starr Jordan of Stanford, both trained in the sciences, who became active in the increasingly influential National Education Association (NEA) and made known their views about the importance of science education in *Science* and *Scientific American,* and at meetings of the American Association for the Advancement of Science, the Society of American Naturalists, and specialized societies. Eliot and Jordan shared with other scientists an aspiration to recruit more and better-prepared future college and graduate students, and both promoted textbooks and other materials that could help fill that pipeline. Scientists' enthusiasm and support waned over time, in part because the methods being used in the early grades were aimed more at advancing childhood development than at introducing the fundamental principles of science. This later generation of scientists with advanced degrees was also less often in the public eye and more withdrawn from public discussion about science than their predecessors; and those who stayed engaged with education were primarily involved with sciences in the high school curriculum.[5]

Actual leadership in the nature study movement thus concentrated in the domain of professional educators, especially those teaching in the normal schools, where faculty members were actively engaged in discussions about methods, content, and essential resources. Whether couched in the pedagogical terms of progressive educators like Wilbur Jackman or the inspirational rhetoric of nature study evangelists like Liberty Hyde Bailey, their stated goal was to awaken and enhance the capacities of children to understand their natural environment. Using nature in elementary classrooms coincided with the basic ideas of child psychology at the turn of the century, including G. Stanley Hall's developmentalism, John Dewey's constructivism, and even Edward Thorndike's associationism. Barbara Beatty argues that, although fundamentally distinct and even competing in certain ways, these psychological approaches shared the common premise that educators needed to attend to the learning capacity of "the child" and that introducing mathematics and science early was important.[6] Indeed, the ideas of all three of these educational psychologists were invoked by various nature study leaders.

Expressing strong confidence in the capacities of elementary classroom teachers even as requirements for teaching certification grew steadily, Francis Parker and his normal school faculty did not tightly codify course content in nature study but rather emphasized its expansive capacities and introduced potentially useful leaflets and textbooks. The Chicago educators encouraged teachers to promote student initiative and use local objects and sites for study, an approach that could cause practice to vary significantly from school to school, even in the same system. Classroom teachers, most of them women, responded to the invitation to use their immediate environment, buttressed by their formal instruction in normal schools and summer institutes, course of study handbooks, and, in urban areas, supervisors who supplied materials. Nature study was promoted at certain leading preparatory schools for teachers and administrators—the University of Chicago, New York's Teachers College, Clark University, Peabody College of Education, and Harris Normal School in St. Louis—as well as at state normal schools across the country, most notably in New England; in the states of New York, Illinois, and Nebraska; and in the cities of New York, Chicago,

St. Paul, Los Angeles, Nashville, and Oakland. Faculty members in these institutions presented nature study in theory and practice at professional meetings, published in the numerous national and state educational journals of the period, wrote textbooks, and collaborated with school systems that provided courses of study for teachers. With support from active colleagues, in 1905 Maurice Bigelow started *Nature-Study Review* to provide a forum for discussing theory, publish examples of good school practices, and establish visibility for a subsequent Nature-Study Society. Such professional tools reflected the widespread establishment of nature study as well as ongoing debates about its goals and implementation.

Nature study curricula could not, however, have taken hold without the teachers, school administrators, textbook authors and editors, voluntary local sponsors, and educational program leaders who contributed to the reformulation and extension of nature study education and extracurricular activities. Enthusiastic advocates found nature study expansive, flexible enough to be implemented in different geographical settings, and readily aligned with other social and political agendas. Although teachers were never uniformly enthusiastic about the prospects of teaching nature study, those who found it an effective pedagogical program remained committed to its principles and practice and enjoyed the flexibility and autonomy it promoted in their classrooms. As Anna Comstock commented, the teacher herself became the "center of influence" in nature study, for many of them a daunting but exhilarating position.[7] Alice Patterson made a related point, noting that one adept teacher in a system could influence others, sparking their initiative by her own.[8]

Despite opportunities for young women to gain higher education in the sciences by the end of the nineteenth century, their job prospects were not always good. A few of the most enterprising, as Margaret Rossiter has demonstrated, created new niches of "women's work" where they could use their expertise. The emerging field of nature study opened opportunities for women (as well as men) to work as supervisors of nature study in urban school systems.[9] They, in turn, encouraged and supported teachers by not only implementing the theoretical insights of educational philosophers but also introducing practical methods that gained and sustained the interests of restless pupils and providing materials for their classrooms. These coordinators came from the ranks of the best trained and most ambitious women teachers. Those who had advanced degrees, like Ruth Marshall and Lucy Wilson, sometimes attained positions on normal schools faculties and in women's colleges.[10] By the 1920s, the niche opportunities for nature study teachers in schools, city educational administrations, and normal schools shrank, fading with other professional opportunities for women in science. Nonetheless, in the first two or three decades of the twentieth century, nature study provided a space created at the intersection of science, still dominated by men, and public school education, where women were the clear majority. Moreover, their nature study activities out-of-doors required similar work from girls and boys. Many would agree with Edward Bigelow's assertion that terms like "tomboy" and "sissy" were "old time vernacular" when it came to nature study and that each sex could gain from having some of the qualities traditionally associated with the other.[11]

However, in the unclear, overlapping domains of teaching and science, nature study had opened what was also a controversial space for women. Their expertise took them into leadership positions in public schools and related careers in museums, botanical gardens, public parks, and national youth organizations. Undoubtedly women's success meant that gender was implicated when a new science education curriculum displaced nature study, as Kim Tolley argues. But there were multiple factors that challenged a nature study curriculum and help explain why the first generation was so successful and the later generations were not.[12] One was the lack of support in the interwar period from a generation of social, natural, and physical scientists with public authority. Critics like Edward Thorndike stigmatized nature study as soft and sentimental and argued that there was a lack of "rigor" in its content. From their university positions, some academics blamed normal schools for providing inadequate training and the teachers for not insisting on systematic memorization of basic scientific details. Increasingly, too, state normal schools focused on teaching high school teachers and educators, adding a fourth year of study that transformed them into four-year colleges offering B.A. and B.S. degrees; in that process many minimized their original emphasis on training elementary school teachers.[13] As state and national standards were put in place under these shifting circumstances, the latitude earlier given to teachers in nature study did not suit "professionals with cosmopolitan rather than local perspectives."[14] The best of nature study practices were gradually absorbed into another, more hierarchical elementary science program.

In the 1920s, too, leaders of the NEA came from the administrative ranks of the state normal schools that now acknowledged the role of schools as "social agencies." These executives wanted more measurable evaluation of outcomes and emphasized civic values, but they simultaneously wanted to delete from the curriculum such topics imposed by outside advocates as hygiene, vocational education, agriculture, and conservation.[15] Nature study was attacked by a new generation of educational leaders concerned about standardization and core curricula. The malleability of nature study in terms of locality, pedagogical philosophy, and adaptation to other trends in the progressive educational enterprise, once its peculiar strength, now became its greatest vulnerability. Given that educational reforms move in successive waves of enthusiasm, and, as David Tyack and Wayne Urban have argued, fundamental transformation has eluded most reformers in the twentieth century. In this context, the evident persistence of nature study for three or four decades in many public schools stands out for its remarkable strength and resilience.[16]

Nature study was an innovation that initially appealed to educators and a wider public who defended the place of science in the curriculum because of a conviction that no one should "separate science from culture or culture from science."[17] Its advocates were not prepared for the twist in thinking that not only viewed science positively but also placed it in a superior position and distinct from humanities, as C. P. Snow noted when he discussed "two cultures" a half century later. The integrative outlook of some nature study teachers and leaders put them on a course that ran counter to centralizing tendencies in education and a strict progression of subject matter favored by academic scientists. Nature study advocates had stressed child development over subject matter, although never ignoring the importance of systematic understanding of the natural world. They typically demonstrated cooperation rather than competition both in their approach to nature and in project development. They also argued that the empirical approach, allowing children to go from data to generalization, was a more satisfactory teaching method than rote learning. Nature study enthusiasts pushed against the scientific tide of what they viewed as narrow objectivity as they adhered to Louis Agassiz's invitation to "study nature, not books." The fact that nature study was aggressively challenged by prominent social and natural scientists was evidence of its perceived influence and, perhaps, its determinedly independent stance.

Nature study reformers had hoped to provide the architecture for a dramatically transformed and truly innovative curriculum, a large ambition that is probably rarely realized and seldom sustained in education. But educational reform and institutional change have never been the same thing.[18] Nature study did contribute to a significant reform in the schools, although as in most reforms, the outcome was not controlled by the initiators but took its own path as other players and the rapidly changing school environment affected it. Nature study educators invented versatile programs that took pupils out-of-doors for nature walks, school gardening, and hikes in parks and nature centers, even as they provided classroom terrariums and window boxes—practices that continued well after the phrase nature study had become a descriptive phrase for outdoor activities rather than a deliberate course of study. They adopted ideas from child psychology and German educational philosophy. They made use of current scientific investigation, as when Jackman acquainted his teachers with the ecology of the dunes along Lake Michigan or when Bailey incorporated up-to-date theories of botany and experimental agriculture into his leaflets. Many of the successful strategies that had been introduced with nature study—seasonal approaches, recognition of age-specific skills in introducing content, and emphasis on learning by observation and experience rather than rote memorization—were incorporated into the courses of study termed elementary science. Moreover, the legacy of eggshell gardens, ant farms, and gerbil cages remains permanently in classrooms from elementary grades through high school.

Nature study teachers and supervisors collaborated with a broad spectrum of educational leaders, civic boosters, enterprising parents, conservation activists, and women's clubs. In many ways these partners helped extend nature study content and techniques beyond the school setting to summer camps, museum programs, national parks and preservation areas, nature trails, and the scouting and outdoor organizations.[19] All these dimensions added to (and sometimes distracted from) the relatively specific educational goals of the early leaders—but they do not, in fact, provide any final assessment of nature study's long-term influence.

Ultimately, education is highly individual. Arguably the most important impact of nature study would have been the direct and personal one experienced by those pupils who were educated in its spirit through activities led by effective teachers. Most of that influence can never be known, but an example may suggest something of the significance nature study could exercise in one life. A child born in 1907 into a relatively poor rural family had a dedicated mother who read avidly in the nature literature of the day and took her children outside to use the nature study leaflets that they brought home from their local country school. Growing up surrounded by sixty-four acres of farm and woodlands in western Pennsylvania, the growing girl emulated Anna Botsford Comstock, whose personal experience led her to argue that early and direct involvement would bring a "sense of companionship with life out-of-doors, and an abiding love of nature." The young adolescent loved to read, and among her favorites were such lively animal books as Beatrix Potter's *Wind in the Will*ows and, later, the novels of Gene Stratton Porter, especially *Girl of the Limberlost,* whose active hero deliberately eschewed social life in order to spend more time in the fascinating outdoor world around her. A subscription to *St. Nicholas* magazine reinforced the girl's enthusiasm for stories that stressed "protection of the oppressed, whether human or dumb creatures" and for illustrations that reflected accurately the native environments of the wild animals portrayed. The magazine's pages were filled with stories from nature, some of them published by young readers. It was here that Rachel Carson published her own first story.[20]

Although Carson wrote about a military hero, inspired by a letter from her brother then serving in World War I, by age fourteen she would produce her first nature account, "My Favorite Recreation," reflecting on her walk with camera, dog, and notebook (fig. 30). It was published in 1922. While a scholarship student at the Pennsylvania College for Women (later Chatham College), she temporarily turned from a literary career toward science after taking a biology course taught by Mary Scott Skinker, a former school teacher who had a master's degree from Teachers College in New York City. Encouraged by her mentor, Carson went on for an M.A. in zoology from Johns Hopkins and there acquired a lifelong fascination with the sea. The emerging ecologist combined her writing and scientific skills, and eventually was appointed chief editor at what became the United States Fish and Wildlife Service. Well-remembered for her best-selling book, *Silent Spring,* which exposed the effect of pesticides on the environment, Carson also wrote *The Edge of the Sea, The Sea Around Us,* and other nonfiction books and articles filled with rich ecological detail that she hoped adults would use with children to acquaint them with the wonders of nature.[21] Carson traced her commitment to writing and her fascination with plants, animals, and their environment to those childhood influences near Springdale, Pennsylvania, where, with guidance, she learned to love the study of nature, and indeed nature itself.

Rachel Carson represents the ideal outcome of the formal and informal nature education that would have pleased the advocates who defined nature study. She grew up in the rapidly changing semirural borderland region near Pittsburgh on landscape similar to the Finger Lakes region of upstate New York, which had been the inspiration and provided the material culture for the programs elaborated by Comstock and Liberty Hyde Bailey at Cornell. Carson's pursuit of lifelong and systematic studies in biology was a

FIGURE 30. Rachel Carson grew up in rural western Pennsylvania where she relished the out-of-doors and seemed always to have a dog by her side, as in this picture from the 1910s. Collection of American Literature, Beinecke Rare Book and Manuscript Library, Yale University.

vindication of Wilbur Jackman's assertion that awakening children's powers of observation would open them to the possibilities of advanced work in science. Her integration of civic, social, and moral sensibilities about the nature she knew so intimately also aligned with Clifton Hodge's goal of putting his pupils into the modern natural world as participants in its sustainability. Rachel Carson's multifaceted career thus glosses the significant and serious tensions expressed by advocates within the nature study movement, even as her life reflects the shared goals of educators who caught and sustained the vision of nature study teaching in and beyond progressive school classrooms. Her personal and professional life history is significant because nature study advocates had always stressed the importance of individual experience even as they built the capacity for advanced systematic study of the natural world including its human inhabitants.

Nature study, envisioned as a way of promoting all the sciences of nature into the public schools, flowed along multiple tributaries into the mainstream movement as advocates across the country implemented a curriculum that allowed for considerable innovation on the part of teachers and school systems. In the process, nature study became an integral part of the transformation of education in the United States between the 1890s and 1930s. By the latter date, systematic study of the natural world, now more explicitly identified as science, was presumed to be an essential part of the public school experience of every child from the earliest grades through high school.

In the years after that, elementary science framed a different approach, more standardized and didactic, that dominated educational theory and practice through the middle decades of the twentieth century. Starting in the 1970s, in the context of growing concerns about the environment and a renewed sensibility about nature itself, conservationist leaders and educators once again sought to encourage young children to appreciate and respond to their natural world and, equally important, to think about their role in shaping it.[22] This new generation of advocates rediscovered a wide range of facilities—nature preserves and nature centers, state and local parks, museums and botanical gardens, as well as school spaces—with significant resources for nature exploration by school children. Many of these sites, established during the early twentieth century, had adapted their educational programs to broad public concerns while retaining much of the outlook toward nature that motivated their founders.

Today, many teachers echo Louis Agassiz's admonition to "study nature, not books," or pursue Anna Botsford Comstock's to go "through books to nature." They may, unwittingly, follow the example of Clifton Hodge and organize projects in line with community needs, identify museum and institutional resources like those provided by Anna Billings Gallup, or lead expeditions into nearby parks or woods. In the early twenty-first century, educators are once again pursuing new forms of nature study in and beyond school classrooms, and their goal remains that of establishing connections between civic life and study of the natural world. A contemporary injunction to "think globally and act locally" reconfigures nature study and helps us imagine the power of the historical nature study outlook as it awakened the sensibility of children in relationship to the natural world, for their sake and for that of the entire planet.

American Nature-Study Society Officers to 1940

PRESIDENTS
(AFFILIATION AT TIME OFFICE WAS HELD)

The records are incomplete and secondary sources inconsistent for the later years. Where they are available, I have relied on primary source materials.

1908	Liberty Hyde Bailey (1858–1954), Cornell University, Ithaca, New York
1909	Clifton F. Hodge (1859–?), Clark University, Worcester, Massachusetts
1910	Otis W. Caldwell (1869–1947), University of Chicago, Chicago, Illinois
1911–12	Benjamin M. Davis (1867–1953), Miami University, Oxford, Ohio
1913–14	Anna Botsford Comstock (1854–1930), Cornell University, Ithaca, New York
1915–17	Liberty Hyde Bailey (1858–1954), Cornell University, Ithaca, New York
1918–19	Samuel Christian Schmucker (1860–1942?), West Chester State Normal School, West Chester, Pennsylvania
1920–21	J. Andrew Drushel (1872–1940), Harris Teachers College, St. Louis, Missouri
1922–23	William Gould Vinal (1881–1973), Rhode Island College of Education, Providence, Rhode Island
1924–25	Morrison R. Van Cleve, Toledo Public Schools, Toledo, Ohio

1926-27	George Rex Green (1884-1947), Pennsylvania State College, State College, Pennsylvania
1928	Alice Jean Patterson (1862-1929), Illinois State College, Normal, Illinois
1928-30	Bertha Chapman Cady (1873-1954), Girl Scout Naturalist, New York, New York
1931-32	Alfred Fellenberg Satterthwait (1879-1954), State Entomologist, St. Louis, Missouri
1933-34	Ellis C. Persing (1885-1947), Cleveland Public Schools and Western Reserve University, Cleveland, Ohio
1935-36	E. Laurence Palmer (1888-1970), Cornell University, Ithaca, New York
1937-38	Edith Marion Patch (1876-1954), State Entomologist, Orono, Maine
1939-40	Ellen Eddy Shaw (1874-1960), New York Botanic Garden, Brooklyn, New York

SECRETARY-TREASURER (WHO ALSO HELD OFFICE OF EDITOR OF *NATURE-STUDY REVIEW*, 1907-23)

1907-10	Maurice A. Bigelow (1872-1955), Teachers College, Columbia University, New York, New York
1910-11	Frederick L. Charles (1872-1911), University of Illinois, Urbana-Champaign, Illinois
1911	(After Charles's death, President Benjamin Davis ran the journal for several months.)
1911-17	Elliot R. Downing (1868-1944), Northern Michigan Normal School at Marquette, Michigan; University of Chicago, Chicago, Illinois
1917-24	Anna Botsford Comstock (1854-1930), Cornell University, Ithaca, New York
1925	E. Laurence Palmer (1888-1970), Cornell University, Ithaca, New York
1926	Clara M. Cheatham
1927	J. Andrew Drushel (d. 1940), New York University, New York, New York
1928-29	E. Laurence Palmer (1888-1970), Cornell University, Ithaca, New York
1930	Ellen Eddy Shaw (1874-1960), New York Botanical Garden, Brooklyn, New York
1931-34	Jennie Hall (1874-1949), Minneapolis Public Schools, Minneapolis, Minnesota
1937-40	Nellie F. Matlock (-1941), St. Louis Public Schools, St. Louis, Missouri

Anecdotal List of Nature-Study Supervisors

The following alphabetical list is incomplete but includes individuals noted in a wide range of ephemeral sources as supervisors of nature study (the title varied) during the period 1890–1932. It lists the cities in which they did the supervision and decades during which they were known to be supervising (with 1900s indicating 1900–1909); this list is meant as a snapshot and not a definitive description of nature study supervision. Their individual stories are often fascinating and could be explored using local archives.

Sarah Louise Arnold, Minneapolis, Minnesota (1890s); Boston, Massachusetts (1890s)

Jennie M. Arms [later Sheldon], Boston, Massachusetts (1900s)

Florence G. Billig, Emporia, Kansas (1910s, 1920s)

Sarah E. Brassil, Quincy, Massachusetts (1890s)

Elizabeth Burnell, Los Angeles, California (1920s)

Bertha Chapman Cady, Oakland, California (1900s); Girl Scout Naturalist (1920s)

Lenore Conover, Detroit, Michigan (1920s)

Clara Dietz, Glencoe, Illinois (1910s)

Henry L. Drummer, Steuben County, New York (1900s, 1910s)

Charles Lincoln Edwards, Los Angeles, California (1910s, 1920s, 1930s)

Violet Finley, Wilmington, Delaware (1920s)

Mattie B. Fry, Anderson, Indiana (1920s)

Delia I. Griffin, North Attleborough (1890s) and Newton, Massachusetts (1890s, 1900s)

Jennie Hall, Minneapolis, Minnesota (1920s, 1930s)

Harriet S. Hayward, Haverhill, Massachusetts (1890s)

John Hollinger, Pittsburgh, Pennsylvania (1920s)

Edward Hughes, Stockton, California (1900s)

Dietrich Lange, St. Paul, Minnesota (1890s, 1900s)

Alice M. Macomber, Newton, Massachusetts (1890s, 1900s)

Alice G. McCloskey, Ithaca, New York (1890s, 1900s)

Effie B. McFadden, Oakland, California (1890s)

Lida Brown McMurry, Normal (1890s) and DeKalb, Illinois (1900s)

Louise Miller, Detroit, Michigan (1900s)

Persia K. Miller, Baltimore, Maryland (1900s)

Ruby Minor, Anderson, Indiana (1920s)

Edna Mosher, Hampton, Virginia (1900s)

Minnie J. Nielson, North Dakota (1920s) (Superintendent of Public Instruction)

Clelia A. Paroni, Berkeley, California (1920s, 1930s)

Elizabeth Peebles, Washington, DC (1920s)

Ellis C. Persing, Cleveland, Ohio (1920s)

Mary Payne, Winnetka, Illinois (1910s)

Hattie Rainwater, Atlanta, Georgia (1920s, 1930s)

Kate Sessions, San Diego, California (1910s)

Fannie A. Stebbins, Springfield, Massachusetts (1900s, 1910s, 1920s)

Gustave Straubenmueller, New York City, New York (1900s, 1910s)

Helen Swett, Alameda, California (1900s)

W. W. Thomas, Wyandotte, Kansas (1910s)

Gilbert H. Trafton, Passaic, New Jersey (1900s)

Annie T. Washburn, Princeton, New Jersey (1920s)

Carolyn Wood, New Bedford, Massachusetts (1900s)

Laura Embree Woodward, City Training School, Trenton, New Jersey (1900s)

Emilie Yunker, Louisville, Kentucky (1910s)

PARTIAL LIST OF NATURE STUDY SUPERVISORS IN MUSEUMS, 1890–1932

This is a list of some of the more noted staff members who specialized in working with children in natural history and other museums with scientific collections.

Katharine Blake, Erie Public Museum, Erie, Pennsylvania

Cornelia James Cannon, Museum Extension Work in Cambridge, Massachusetts

Louise Connolly, Newark Museum, New Jersey

Anna Billings Gallup, Brooklyn Children's Museum, New York

Kathryn B. Greywacz, New Jersey State Museum, Trenton

Delia Griffin, St. Johnsbury Museum, Vermont; Boston Children's Museum, Massachusetts; Museum of Hartford, Connecticut

Elizabeth Letsom, Buffalo Society of Natural History, New York

Harold Madison, Museum of Natural History, Cleveland, Ohio

Amelia Meissner, St. Louis Public Schools Educational Museum, Missouri

Helen Howell Neal, Director of Nature Lore in Gulick Camps

Ellen Eddy Shaw, Curator of Elementary Education, Brooklyn Botanical Garden, New York

Ann E. Thompson, American Museum of Natural History, New York

Agnes L. Vaughan, American Museum of Natural History, New York

Alice Wilcox, St. Johnsbury Museum, Vermont

Flarida A. Wiley, American Museum of Natural History, New York

William S. Wright, San Diego Natural History Museum, California

PARTIAL LIST OF NATURE-STUDY INSTRUCTORS AND SUPERVISORS IN NORMAL SCHOOL TRAINING OR PRACTICE SCHOOLS

This list references those individuals who played a significant role in the activities mentioned in the book; also see list of over four hundred active instructors in nature study (defined as nature study, elementary science, and elementary agriculture) in *Nature-Study Review* 12 (February 1916): 69–78.

Mary Pearle Anderson, Horace Mann School, Teachers College, Columbia University, New York

Guy A. Bailey, State Normal School at Geneseo, New York

Mary Barry, Peabody College of Education, Nashville, Tennessee

Maurice A. Bigelow, Teachers College, Columbia University, New York

Bertha Chapman Cady, State Normal School at Chico, California

Elizabeth Carss, Horace Mann School, Teachers College, Columbia University, New York

Breta W. Childs, State Normal School at Worcester, Massachusetts

Anna M. Clark, New York Training School for Teachers, New York

Lenore Conover, State Normal School at Detroit, Michigan

J. A. Drushel, Harris Teachers College, St. Louis; New York University, New York

Lucy Gage, Peabody College of Education, Nashville, Tennessee

Gertrude A. Gillmore, City Normal School, Detroit, Michigan

Gertrude B. Goldsmith, Normal School, Salem, Massachusetts

George Rex Green, State College of Pennsylvania, State College, Pennsylvania

Roland Guss, State Normal School, North Adams, Massachusetts

Frederick Holtz, Brooklyn Training School for Teachers, New York

Joseph W. Hungate, State Normal School, Cheney, Washington

Harriet A. Luddington, Brooklyn Training School for Teachers, New York

Grace G. Lyman, New York Training School for Teachers, New York

Pearl McCoy, State Normal School, Bridgewater, Massachusetts

Effie Belle McFadden, San Francisco State Normal School, California

Anna E. McGovern, Iowa State Normal School, Cedar Falls, Iowa

Lida Brown McMurry, Illinois State Normal School (at Normal); Northern Illinois Normal School, DeKalb, Illinois

Ruth Marshall, Rockford College; Northern Illinois Normal School, DeKalb, Illinois

Loye Holmes Miller, Normal School, Los Angeles, California

J. P. Munson, State Normal School, Ellensburg, Washington

Alice J. Patterson, Illinois State Normal School, Normal, Illinois

Charles S. Preble, Normal School at Farmington, Maine

Samuel C. Schmucker, West Chester State Normal School, Pennsylvania

Ellen Eddy Shaw, Normal School at New Paltz, New York

William H. Sherzer, Normal School at Ypsilanti, Michigan

Adeline F. Schivley, Normal School, Philadelphia, Pennsylvania

Susan B. Sipe, J. Ormond Wilson Normal School, Washington, DC

Maud J. Staber, Teachers Training School, Brooklyn, New York

Charles A. Stebbins, State Normal School, Chico, California

Gilbert Trafton, Normal School at Mankato, Minnesota

W. G. Vinal, State School of Education, Providence, Rhode Island

Lucy L. W. Wilson, Girls Normal School of Philadelphia, Pennsylvania

Notes

ACKNOWLEDGMENTS

1. Significant support for time and travel came from National Science Foundation grants nos. 0115772 and 9123719. I appreciated the dedicated leadership of the section on history and philosophy of science while Ronald Overman was in charge and learned from the thoughtful comments of reviewers. Research was also facilitated by a summer research grant from the University of Minnesota.

INTRODUCTION

1. The term nature study was often hyphenated, most notably in the *Nature-Study Review*, founded in 1905, where its editor suggested that they would use this form because they sought to emphasize the unity implied in the compound form; Bigelow, "Editorial," *Nature-Study Review* 1 (May 1905): 140. Over time, the authors tended to drop the hyphen in journal articles, although not in its title.

2. Tolley's *Science Education of American Girls* is based on her dissertation also cited in this book; also see Doris, "The Practice of Nature-Study." Both attend to some of the women who taught and supervised nature study in the schools, and emphasize the positive, empowering impact of nature study education for the women and their pupils. Armitage in "Knowing Nature: Nature Study and American Life, 1873–1923" describes a complex antimodernism in nature study. Amy Susan Green argues that nature study teaching reinforced class and ethnic distinctions in her dissertation, "Savage Childhood," esp. chap. 4.

3. On the struggle for authority in this period, see the sources cited in Linn, Songer, and Eylon, "Shifts and Convergences in Science Learning and Instruction," 445–47. For a discussion of the ambivalence of women about professionalization in general during this period, see Brumberg and Tomes, "Women in the Professions," 275–95.

4. Kohlstedt, "Collections, Cabinets and Summer Camp," 10–23.

5. Most of these accounts rely heavily on the dismissive perspective in Underhill's *The Origin and Development of Elementary School Science* and its preface by Samuel R. Powers who noted, with clear disdain, "Teachers of nature study in the public schools and often in normal schools were, in general, poorly educated in science and, in line with the traditions of the time, tended to become emotional rather than scholarly about natural phenomenon" (vii). Several

older dissertations have a defensive tone and concentrate almost exclusively on the educational philosophers and textbook authors who articulated the arguments for nature study; they deal little with women educators and the institutional development of the movement. See Olmsted, "Nature-Study Movement in American Education"; Minton, "History of the Nature-Study Movement and Its Role in the Development of Environmental Education"; Kraus, "Science Education in the National Parks of the United States"; Bellisario, "E. Lawrence Palmer: His Contribution to Nature, Conservation, and Science"; Nutt, "History and Development of Nature Education in California with Emphasis on the Period up to 1935"; Arnold, "Nature Study: An Analytical History of Its Demise"; and Corcoran, "Formative Evaluation of Science and Natural History Curriculum Materials." A thoughtful contemporary summary is presented in a thesis at Harvard, later published, by Mitchell, "History of Nature Study."

6. The thread of direct criticism included Kuslan's suggestion that nature study was a "craze of the nineties" foisted on reluctant teachers in his "Elementary Science in Connecticut," 286-89. An account of the British movement is similarly critical, following the revisionist themes common in the history of education in the 1970s and early 1980s, as suggested by the title "Science, Sentimentalism, or Social Control" by E. W. Jenkins. These authors relied on a few prescriptive publications, perhaps because the evidence of classroom activity is not easy to find. Records of teaching practice are often ephemeral and, until recently, few libraries sought to collect the letters and diaries of women teachers, course handouts and homework, as well as tactile objects used in classrooms. The Internet has made the scope, scale, range, and persistence of the nature study movement much more evident, providing the capacity to find more easily the elusive books, periodicals, courses of study, leaflets, and myriad publications that were produced across North America (and beyond) that I had begun to identify on a library-to-library search in the 1980s.

7. The best source I found for the materials produced to assist teachers in their classroom work, covering nature study for the period 1921-34, was in the Teachers College Special Collections Library, which had file drawers with mimeographed and typed curriculum files arranged by subject. Among the nearly forty outlines for nature study were materials for city schools in Anderson, Indiana; Fresno, California; Grand Rapids, Michigan; Buffalo, New York; Manitowoc, Wisconsin; and New Bedford, Massachusetts, as well as recommended state outlines for the rural schools of California and statewide programs for New Jersey, North Dakota, Arizona, Wisconsin, and Delaware, in addition to Washington, DC, and Hawaii. By the 1920s, recommendations for field trips and nature materials in the classroom were common, as were references to hygiene and, in rural states, to agriculture. Typically, the course guides specified plants, animals, and other materials that would be of particular value, often by seasons, for students in each locale. The Special Collections holdings of Gutman Library at Harvard University include a fine collection of textbooks for pupils and for normal school students for the pre-1950 period in American education.

8. Frontispiece quotation in Parker, "Syllabus of a Course of Lectures," 16.

9. See Rossiter's chap. 3, with that title, in *Women Scientists in America: Struggles and Strategies*, 51-72.

10. The erosion is evident in Rossiter, *Women Scientists in America: Struggles and Strategies*, esp. chap. 5 on "Growth, Containment, and Overqualification," 129-59; and the strong outlook is described in Shapin, *Scientific Life*, 27-46.

11. Very few records of teachers, most of whom were women, exist in established archives. Many taught for just a few years, and their years of teaching may not even be noted by obituaries. There is useful information, based significantly on published examples, about teachers' practices in Doris, "Practice of Nature-Study." Also see the opening remarks and papers edited by Warren, *American Teachers;* Finklestein, "Governing the Young: Teacher Behavior"; Biklen, *School Work: Gender and the Cultural Construction of Teaching;* and, for the antebellum period, Kohlstedt, "Parlors, Primers, and Public Schooling," 424-45.

12. For a thoughtful discussion of the contentious but deeply contentious historiographical discussion on this issue in American education, see Ravitch, *Revisionists Revised,* which argues for a balanced approach to the complicated factors that create these central and culturally powerful institutions.

13. The literature on public schools from 1890 to 1920 is substantial, but surprisingly few scholars have sought to recover the basic curriculum taught in the elementary or even the high schools. The best overview of this period in education remains Tyack, *One Best System.*

14. Attention to public outdoor spaces, urban and rural, expanded dramatically in the period discussed here; see Schuyler, *New Urban Landscape;* and Cranz, *Politics of Park Design;* and on national parks, see Mark Daniel Barringer, *Selling Yellowstone.* Peter Schmitt in his classic account of attitudes in the nineteenth century, *Back to Nature,* explains the nature study movement narrowly and aligns it with the thesis of his book, but his discussion of nature enthusiasm as a reaction to urbanization and industrialization is important if insufficient to fully appreciate the phenomenon; also see Norwood, *Made from This Earth,* 79-80.

CHAPTER ONE

1. Agassiz, "Methods of Study in Natural History," 336. This essay was part of a series by Agassiz that traced the development of key ideas in natural history from the time of Aristotle.

2. Natural history held a particular place in the commonwealth of learning in the seventeenth century, a way of interpreting both human nature and the nature that surrounded humans. By the nineteenth century, changing ways of approaching nature led to the use of the term "natural sciences," which suggested both the multiplicity and the systematics of what would become disciplines at the end of the period. For general discussion and examples, see Jardine, Secord, and Spary, eds., *Cultures of Natural History.* Thomas Huxley in "Natural History" explained to educators that "natural history is the name familiarly applied to the study of the properties of such natural bodies as minerals, plants, and animals; the sciences which embody the knowledge man has acquired upon these subjects are commonly termed Natural Sciences, in contradistinctions to the other so called 'physical sciences'" (473).

3. See Warner, "Science Education for Women in Antebellum America," 58-67; and Tolley, *Science Education of American Girls.*

4. See Blum, *Picturing Nature;* and Porter, *Eagle's Nest.*

5. Kohlstedt, "Parlors, Primers, and Public Schooling," 424-45.

6. For the complex attitudes of the British, often influential in North America in the nineteenth century, see Secord, *Victorian Sensation.*

7. A classic account of this fascination was Perry Miller's *The Life of the Mind in America.* Less positive was E. Branch, *The Sentimental Years, 1835-1860,* which took a whimsical view of "nature addicts," including the Transcendentalists. Porter's *The Eagle's Nest* offers the most systematic view of active naturalists, focusing particularly on those in Philadelphia.

8. See Elizabeth C. Agassiz to Mother, n.d., Agassiz Papers, Schlesinger Library, Radcliffe Institute, Harvard University, Cambridge, MA. Louis Agassiz advocated public support for the Boston Society of Natural History in 1861 because it agreed to provide access to its holdings by local teachers; see the anonymously published *Boston Society of Natural History, 1830-1930,* 13-14.

9. John Lee Comstock, who wrote books for the Sabbath schools in the 1840s, suggested, "There is hardly any subject which more interests the mind of youth than that of Natural History, particularly when it is illustrated by pictures." Quoted in Welch, *Book of Nature,* 137.

10. In the nineteenth century, many elementary teachers simply went from being pupils to teaching, while others attended an array of pedagogical seminaries, academies, teacher institutes, high schools, normal schools, preparatory departments, and colleges. See Fraser, *Preparing America's Teachers.*

11. See Wells, *A Graded Course of Instruction;* and Foote, *Teachers' Institute.*

12. This theme is elaborated in Jones's classic *O Strange New World.*

13. Stearns in *Science in the British Colonies of America* details the role of North American collectors who were, in varying degrees of formality, affiliated with the Royal Society.

14. The argument that Americans were particularly interested in Baconian empiricism is found throughout Daniels, *American Science in the Age of Jackson.* Norwood in *Made from This Earth* describes the work of women as writers, illustrators, landscape designers, and conservationists to make the point of the popularity of nature topics and activities and the active engagement of women.

15. Greene, *American Science in the Age of Jefferson;* and Beadie and Tolley, eds., *Chartered Schools.*

16. Monroe, *History of the Pestalozzian Movement.*

17. On the prescriptive nature of these arguments, see Beaver, "Altruism, Patriotism, and Science"; and Kuritz, "The Popularization of Science," 259-74.

18. A. A. Gould, "On the Introduction of Natural History as a Study to Common Schools"; and W. Channing, "On the Moral Uses of the Study of Natural History"; both in American Institute of Instruction, *Lectures* (Boston: Carter, Hindee, and Co., 1835). Other advocates are cited in Van De Voort, *Teaching of Science in Normal Schools,* 1-10.

19. Gould, "On the Introduction of Natural History," 239.

20. Phelps's *Familiar Lectures on Botany* went through multiple editions and sold over one hundred thousand copies. Her books were addressed to mothers and teachers, those most clearly engaged the instruction of younger children.

21. Baym, *American Women of Letters and the Nineteenth-Century Sciences,* chap. 2; and Houghton, *Country Walks of a Naturalist.*

22. Welch, *Book of Nature,* 44-54. Welch makes the specific claim that North American naturalists built on their proximity to create a distinctively North American way of linking species and their geographical range.

23. On the prevalence and style of art representing living animals in nature, see the detailed discussion in Blum, *Picturing Nature.*

24. Barry to Baird, July 9, 1855, RU 52, Smithsonian Institution Archives, Washington, DC.

25. Hill, "The True Order of [Natural History] Studies," 273-84.

26. According to the census for 1870, for example, there were 141,629 schools and 220,022 teachers, indicating that while some were in schools with several teachers, many must have been in charge of rural one- and two-room schools. See "Progressive Development of Schools and Other Institutions of Public Education," *American Journal of Education* 32 (1882): 818-71 (table on p. 871); and Fuller, *The Old Country School.*

27. Doris, "The Practice of Nature-Study," probes this issue and relies largely on the self-reporting of teachers for the early twentieth century.

28. Rossiter, *Women Scientists in America: Struggles and Strategies to 1940,* 4-6.

29. Cushing, *Reminiscences of School Teachers in Dorchester and Boston* (c. 1884); found at the American Antiquarian Society, Worchester, MA.

30. Mann, "Reminiscences of School Life and Teaching," 751. Also see Mann's entry by Jonathan Messerli in James and James, eds., *Notable American Women,* 2:488-90.

31. A particularly charming example is in Margaret Emerson Bailey's autobiographical account of the school her mother conducted in their home to add to the family income and which capitalized on the knowledge of the botanist father who taught at Brown University. See Bailey, *Good-bye, Proud World,* 72-78.

32. Abbott, *The Teacher,* 237. Abbott's use of a woman as negative example here is unusual because most of his references to "the teacher" indicate a man.

33. Teachers themselves also were concerned about such issues as evidenced in Finklestein, "Governing the Young"; and Smallwood, *And Gladly Teach.*

34. Robin, "Science for Children and the Untutored in Eighteenth-Century England."

35. Pond, *Science in Nineteenth-Century Children's Books.*

36. Anna Letitia Barbauld, *Lessons for Children from Four to Five Years Old.* This was, according to Pond's exhibition catalogue, based on an English reading text from 1778, but took its examples of natural history and scientific objects from the American scene.

37. Samuel Goodrich urged children to keep a scrapbook of pressed plants as a healthy and educational pastime in "About the Leaves of Trees," 41-42. On rhetorical devices that signal intended audiences, see Shteir, *Cultivating Women, Cultivating Science,* chap. 2.

38. Kohlstedt, "Parlors, Primers, and Public Education," 434-39.

39. Abbott, *Rollo's Museum,* 104, 110.

40. Welch, *Book of Nature,* 138-39.

41. An annotated list of juvenile books from this period is found in Underhill, *Origins and Development of Elementary School Science,* 259-85.

42. Gray, *Botany for Young People and Common Schools.* Note that Gray intended the book for the public as well as private school use.

43. Worthington Hooker's *The Child's Book of Nature* was to be used with children aged six to nine years. For science books intended for secondary schools, see the series published by Steele, including *Fourteen Weeks in Zoology,* which included a significant number of line drawings.

44. Women nature writers, and Phelps in particular, are discussed in Baym, *American Women of Letters,* chap. 2; and Norwood, *Made from This Earth,* 19-23.

45. Kohlstedt, "Collections and Cabinets" and "Museums on Campus: A Tradition of Inquiry and Teaching."

46. Preparation for teaching before the 1830s often required only that candidates showed aptitude for school and had graduated with some endorsement, and this practice would continue in some communities through the century. Both city and rural schools also hired college students and graduates, but these were gradually displaced by normal school graduates; in some cases the normal schools also functioned as preparatory schools for students who then attended college. See Herbst, *And Sadly Teach,* 141.

47. Lurie, *Louis Agassiz,* 200-201. His other commonly cited aphorism: "The pupil studies nature in the class-room, and when he goes out of doors he cannot find her."

48. Green, "Lucretia Crocker," 407-9.

49. Johonnot, "The Story of a School," 496-508. Frederick W. Putnam was hired to teach zoology at the Salem State Normal School, for example, and had students handling scalpels and forceps nearly as well as "a set of medical students." Putnam to Albert Gunther, December 5, 1872, Gunther Papers, British Museum, London.

50. At Eastern Michigan Normal School, for example, students learned sciences in a general way, using laboratories during the 1870s and 1880s; and the emphasis was on acquiring scientific knowledge with apparently little attention to teaching it specifically in schools. Isbell, *A History of Eastern Michigan University,* 253.

51. Burstyn, "Early Women in Education," 50-64. Among the future nature study educators who studied with Agassiz were Lucretia Crocker of Boston, Susan Blow of St. Louis, and Henry A. Straight of Oswego and later Cook County Normal School. Also see the scrapbook of Mary Beaman Jozaleman commemorating her summer at the Penikese; she subsequently taught natural science at the normal school in Binghamton, New York. Scrapbook of newspapers clippings at Boston Public Library, Boston.

52. Quotation of Louis Agassiz, written to Burt G. Wilder, is found in Tolley's doctoral dissertation, "Science Education of American Girls," 216.

53. The graduates were a precocious lot who carried Agassiz's vision with them, including Lydia Maria Shattuck of Mount Holyoke and David Starr Jordan, subsequently president of Indiana University and then Stanford University.

54. Jordan, *Scientific Sketches*, 146–47.

55. Ibid.

56. Louis Agassiz to Alfred M. Mayer, November 25, 1873, Hyatt and Mayer Papers, Manuscripts Division, Princeton University Firestone Library. For an excellent summary of the opportunities for women, see Burstyn, "Early Women in Education," 50–64; she points out that Agassiz, open to admitting women to the school, envisioned the young men as producing science while the women were seen and often viewed themselves as the distributors of concepts developed by others (61).

57. In the early years, women gained access to these sites, as suggested in Sloan's "The Founding of the Naples Table Association," 208–16. Most of the seaside laboratories that came later catered to academic male faculty; see Benson, "From Museum Research to Laboratory Research," 49–86.

58. See correspondence with Forbes from a number of people who had taken the class or intended to do so in the future, including A. E. Bourne, October 5 and 24 and November 10, 1874; G. W. Mason, January 31, 1876; and E. A. Gastman, June 21, 1876; Natural History Survey, Chief's Office, 1871–1909, University of Illinois Archives (hereafter UIA). Marshall, *Grandest of Enterprises*, 361–72. The amphioxus materials from Naples were put in order by Burt G. Wilder of Cornell, who came to teach in the summer program.

59. B. H. Van Vleck to Forbes, Cambridge, April 15, 1876, Natural History Survey, Chief's Office, 1871–1909, UIA. Forbes was actively interested in educational issues, as indicated by his contribution to a series of *Educational Papers, 1889–1890* on science education, produced by the Illinois Education Association; papers in UIA. See also Ilerbaig, "Pride of Place," chap. 5.

60. Ralph W. Dexter, "Views of Alpheus Hyatt's Sea-side Laboratory and Excerpts from his Expeditional Correspondence," *The Biologist* 39 (1956–57): 5–11.

61. Ralph Dexter discussed the seaside laboratories in a number of short essays, including "The Annisquam Sea-side Laboratory of Alpheus Hyatt," 112–16. Because Hyatt also worked for the United States Fish Commission, much of the direction of the Annisquam laboratory was under B. H. Van Vleck.

62. A few applications for participation in the school and descriptions of some programs are in the Boston Society of Natural History Papers at the Boston Museum of Science Archives.

63. Quoted in Dexter, "An Early Movement to Promote Field Study in the Public Schools," 345. However, it is clear that this was not always an easy combination.

64. He hired a number of women, including Jennie Arms [later Sheldon], who continued to produce literature for science teachers and who gratefully dedicated one of her later books to Hyatt, who provided "my opportunity for scientific study and whose suggestive teaching and incisive criticism always pointed to the broadest scientific view." In Sheldon, *Observation Lessons on Animals*, dedication page.

65. Agassiz, *A First Lesson in Natural History*. The four-by-six-inch little pamphlet has a number of lovely sketches.

66. Lindsay, "Intimate Inmates," discusses Elizabeth's role as partner, as well as her intellectual achievements.

67. One pamphlet described Woods Hole as a place for study "free from the inconveniences and distractions of fashionable summer resorts"; see *Nature Study, A Course Designed Especially for Teachers and Students*, 11; copy found in the Woods Hole Archives.

68. Maienschein, "Whitman at Chicago: Establishing a Chicago Style of Biology?" 151–84.

69. In 1900, the Woods Hole laboratory ran a nearly all-female class for teachers of nature study, an acknowledgment, perhaps, of the visibility of the nature study curriculum. Its flyer boasted an impressive list of permanent and visiting faculty, including Liberty Hyde Bailey, John C. Coulter, Clifton Hodge, and Alpheus Hyatt. See *Nature Study: A Course Designed for Teachers and Students, Thursday, July 5 to Wednesday August 15, 1900* (copy found in the Woods Hole Archives). That the program narrowly designed for teachers had little outdoor and laboratory

work was apparently disappointing to the Wisconsin teacher Mary Bradford, who found when she attended in 1903 that "in most lines of pedagogy the West was fully abreast of what the lecturers presented there." See her *Memoirs of Mary D. Bradford*, 255.

70. Crocker's former student and well-known author of children's books on geography, Jane Andrews, remembered her as "one of those remarkable persons who can act as a catalyst, bringing order and direction to the thinking of others, inspiring them to their best work." Quoted in Green, *A Forgotten Chapter in American Education*, 32. Andrews taught nature study and had an Agassiz Association chapter at her school.

71. Kaufman, *Boston Women and City School Politics*.

72. See the biographical sketch by Green in *Notable American Women* (1971), 1:407-9. There are few records of this remarkable woman, well known among her contemporaries, but see Cheney, *Memoirs of Lucretia Crocker and Abby W. May*. Crocker published *Methods of Teaching Geography* "printed at the request of teachers in attendance," and a building at Framingham Normal School was named in her honor. Also see Boyden, *Nature Study in the Public Schools of Boston*.

73. Shaler, a student of Agassiz, helped formulate the summer school program and remained interested in education, according to Livingstone, *Nathaniel Southgate Shaler*, chap. 9.

74. Wright was a prolific author on travel, religious, and temperance topics; her *Seaside and Wayside* trio of books identified and described objects that might be found in the Boston area. Health also distributed British books, including Ricks, *Natural History Object Lessons: A Manual for Teachers*.

75. Hoyt, "The Home of Nature at the Root of Teaching and Learning the Sciences," 61.

76. Boston School Committee, *Annual Report for 1891* (1892): 27-28. Their report observed simply that teachers needed to be given resources because they "cannot make bricks without straw" (28).

77. Report of the Committee on Instruction and Lectures, May 30, 1882; item 289, Academy of Natural Sciences of Philadelphia.

78. The 1871 report by Harris was later published as *How to Teach Natural Science in Public Schools*. Also see Troen, *Public and the Schools: Shaping the St. Louis System*.

79. He later edited an educational series for which Howe wrote *Systematic Science Teaching*. Harris's introduction uses the term nature study essentially as a substitute for traditional natural science, and he emphasizes giving teachers the latitude to introduce topics from nature.

80. Harris, "The Study of Natural Science—Its Uses and Dangers," 277-87.

81. Phillips, *Modern Methods and the Elementary Curriculum*, vii.

82. Many colleges provided summer programs for teachers at the end of the century, including Harvard University; see Livingstone, *Nathaniel Southgate Shaler*, 253-60.

83. Geitz, Heideking, and Herbst, eds., *German Influences on Education in the United States*.

84. Daum, *Wissenchaftpopularisierung im 19. Jahrhundert* [Popularizing Science in the Nineteenth Century].

85. Shapiro, *Child's Garden: The Kindergarten Movement from Froebel to Dewey*.

86. "The Iowa Course of Study and Manual for the Ungraded Schools" is provided in Cordier, *Schoolwomen of the Prairies and Plains*, 122.

87. Bullough, *Cities and Schools in the Gilded Age*.

88. Tebbel, *History of Publishing in the United States*, 2:583.

89. Quoted in Downing, *Teaching Science in the Schools*, 8.

90. This volume, by Mayo, *Lessons on Shells as Given in a Pestalozzian School*.

91. Möbius's title in German is *Naturgeschichte in der Volkschule. Der Dorfteich als Lebensgemeinschaft*. See discussion of this education in W. Zenz, "Naturgeschichte," in *Enzyklopadisches handbuch der Erziehungskunde* [Encyclopedic Handbook on Pedagogy], ed. Joseph Loos (Vienna and Liepzig: A Pichlers Witwe & Sohn, 1908), 2:103-6; also see Nyhart, "Economic and Civic Zoology in Late Nineteenth-Century Germany," 605-30. I am grateful to students in my classes

who helped me translate German publications while I taught for one semester as visiting professor at the Amerika-Institut at the University of Munich.

92. Friedrich Paulsen, *German Education: Past and Present*, trans. T. Lorenz (London: T. Fisher Unwin, 1908), 252-53. When an administrator from St. Paul, Minnesota, visited Germany, he reported "no educational fire-works, no parading of theories or methods, and no educational experiments. There is no trifling with the child's mind, but it is guided step by step according to well-known and long established principles." Handwritten notes, "Comparison of German and American Schools" (ca. 1894), Hiram W. Slack and Family Papers, Minnesota Historical Society (hereafter MHS).

93. The title in German is *Natur und Schule;* the journal only existed until 1907. See Scheele, *Vone Lueben bis Schmeil,* 202, 249 ff.

94. See W. Zenz, "Naturgeschichte," 103-6.

95. Applegate, *A Nation of Provincials,* 62-78; Confino, *Nation as a Local Metaphor,* 108-15; and Hermand and Steakley, eds., *Heimat, Nation, and Fatherland.* Although there is no simple definition of the movement and most accounts concentrate on its political and patriotic implications, another important aspect relates to the sense of place and knowledge of location that encouraged hiking and other outdoor activities brought with German immigrants to North America.

96. Geitz, Heideking, and Herbst, eds., *German Influences on Education,* passim. Many scholars went to Germany for advanced degrees, but a number of educators also went there to see first hand how schools were run. See, for example, Hiram W. Slack's journal of visits to schools in Leipzig [*sic*] May 2 to June 2, 1894, where he also attended a field trip with practice teachers. Slack was born in Oswego County, New York, in 1843 and moved to Hudson, Wisconsin, in 1872, later becoming principal of Edmund Rice School in St. Paul. Hiram Slack Papers, MHS.

97. See Ross, *G. Stanley Hall.*

98. The single best account of this movement is Cruikshank, "The Rise and Fall of American Herbartianism." Charles DeGarmo was one of the leading proponents and quickly went on the lecture circuit as, for example, when Stephen Forbes invited Charles DeGarmo to give a talk about his recent work in Germany at a program for Illinois teachers. Stephen Forbes to Charles DeGarmo, June 2, 1890; Forbes sent a copy of DeGarmo's paper to Wilbur Jackman, October 14, 1893, and offered extras to Charles A. McMurry for the Pedagogical Club, December 12, 1893, all in Natural History Survey, Chief's Office, 1871-1909, UIA. Also see R. A. Croker, *Stephen A. Forbes and the Rise of American Ecology.*

99. Scheele, *Von Lueben bis Schmeil,* 98-110 and 145-66. I thank Tom Haakenson for his detailed translation of this material.

100. Sheldon's *Lessons on Objects,* published in 1863, was the first significant description of this method written by an American educator, identifying his name with the movement.

101. See Rogers, *Oswego: Fountainhead of Teacher Education;* and Ned Harland Dearborn, *The Oswego Movement.* Sheldon's manual of instruction in 1863 devoted nearly two-thirds of his object lessons to physical, biological, and earth sciences. Oswego provided, in effect, a new model training school for teachers, eventually creating a practice school to provide supervised experience to teachers in training and a place to experiment with educational methods.

102. Ballard, "History of the Agassiz Association," 93-96.

103. Mary Mapes Dodge's father was a freethinker and scientist, and his daughter created a companionable and amusing magazine for juveniles that captured their imagination and interest with stories of both fact and fiction. See Burlingame, *Of Making Many Books,* 199-201; and Chew, *Fruits among the Leaves,* 131-41.

104. Klassen, "The School of Nature."

105. Ballard, *Hand-book of the St. Nicholas Agassiz Association,* 2d ed.; later editions were titled *Three Kingdoms: A Handbook of the Agassiz Association.* Ballard also wrote texts for writing and had several other publishing activities.

106. In addition to reports in *St. Nicholas,* see the handwritten minute book for 1885–86 of Manhattan Chapter B (whose list of members had predominantly German surnames) and two printed listings of programs for Manhattan Chapter Z. Agassiz Association, Special Collections, American Museum of Natural History, New York.

107. Agassiz Association, *The First General Convention of the Agassiz Association. The Naturalists' Journal* had only one issue (March 1884), published by the Association in Philadelphia.

108. Robert Welker, *Birds and Men,* 179–81.

109. A single bound volume of minutes and a few handwritten papers with student presentations point to the active years, 1883–87, of St. Paul Chapter C, when five to ten boys met weekly during the academic year, elected officers, and heard presentations on topics from barnacles to iron pyrite. They used their weekly dues of five cents to subscribe collectively to *Youth's Companion.* Agassiz Association Papers, MHS.

110. Massachusetts State Assembly of the Agassiz Association, *Proceedings of Meeting held in Fitchburg, 30 May 1890* (published by the Assembly).

111. Comment by Harlan H. Ballard, "The Agassiz Association," *Science* 13 (1889): 5–6. If the reporting chapters in the magazine were typical, most were small, with five to thirty members and often with many adults as well as children involved in the collecting and field work. *The Swiss Cross* 3 (1888); the publisher, Hodges, also edited *Science;* see Kohlstedt, "*Science:* The Struggle for Survival, 1880–1894," 33–42.

112. For additional discussion of the clubs, see Keeney, *The Botanists: Amateur Scientists in Nineteenth-Century America,* 140–45; and Mark V. Barrow, Jr., *A Passion for Birds,* 14 and 165.

113. *The Agassiz Association* (Stamford; n.d.) contains a summary history and claims there are 1,200 local societies and 20,000 members in chapters that range from family groups and classroom chapters to adult groups.

114. Other magazines were established to serve the association or publish its material, including *Swiss Cross, Agassiz Companion, Santa Claus, Observer,* and *Popular Science News,* as well as various chapter periodicals by active groups, including the Wilson Ornithological (which became the active amateur Wilson Ornithological Society), the Linnaean Fern, and the Gray Memorial Botanical chapters.

115. *The Observer* was made the official organ of the Agassiz Association in 1894, and its original subheading was "For Popularizing Knowledge and Love of Nature Among Young and Old." For some account of Bigelow's initiatives, see a privately printed booklet (untitled) of correspondence of J. Kennedy Tod of Innis Arden House, South Beach (Old Greenwich) Connecticut. Bigelow and others who were involved in a failed project to create a new center for the association from 1909 to 1911. A copy of the correspondence is in the Stanford University Libraries. An announcement of this initiative is found in *Nature-Study Review* 5 (January 1909): 31.

116. There had been a long-standing interest in natural theology for children as in Gallaudet, *The Class Book of Natural Theology for Common Schools and Academies,* which defined natural theology as all that can be learned about God merely from examining the things He has made, without the aid of revealed theology. Adults, too, persisted in this interest; see Robert Welker, *Birds and Men,* 179–81.

117. *Nature Study Review* 3 (April 1907): 118–19. On Bigelow's multifaceted career as printer, teacher, lecturer, and leader of the Agassiz Association, see his *New York Times* obituary (July 15, 1948), 17.

118. Cutts, *Index to Youth's Companion,* 1:xii–xiii.

119. Cutts, *Index to Youth's Companion,* 2:1061. The contributors were primarily literary figures, most of whom were not well known, but William J. Long was a frequent contributor, and there were also several short articles by Nathaniel Southgate Shaler.

120. Kohler, *All Creatures,* chap. 2. The classic discussions are in Huth, *Nature and the Americans;* Nash, *Wilderness and the American Mind;* and Shi, *The Simple Life.*

121. The complex scientific, aesthetic, and cultural significance of illustrations are explored

in many sources, including Blum, *Picturing Nature;* Rindge, "The Painted Desert"; and Sandweiss, *Print the Legend: Photography and the American West.*

122. Barrow, Jr., *A Passion for Birds,* 54–73.

123. Philippon, *Conserving Words: How American Nature Writers Shaped the Environmental Movement;* and Wilkins, *Clarence King: A Biography.*

124. Burt's initiative is discussed in Lutts, *Nature Fakers: Wildlife, Science, and Sentiment,* 28–29. Burroughs's school readers, produced by Houghton, Mifflin and Company and by Ginn and Company, sold about three hundred thousand copies between 1889 and 1906.

125. Kohlstedt, Lewenstein, and Sokal, *Establishment of Science in America,* 33. J. T. Kessler's address on "Nature Study and Biology" noted that the Society had gone on record in 1888 that the study of plants and animals should begin in the primary grades. Published as a Baylor University *Bulletin* 9 (January 1906), 10.

126. Rice, *Science-Teaching in the Schools.*

127. See, for examples, *Education: A Monthly Magazine Devoted to the Science, Art, Philosophy and Literature of Education,* including A. Tovel, "Plea for the Study of Nature in School," 375; B. B. Russell, "Nature Study and Intellectual Culture in Our Schools," 345; and H. L. Clapp, "Nature and Purpose of Nature Study and Intellectual Culture," 597.

128. Edward Hyatt, "Teacher Enthusiasms" (handwritten notes, about 1915, but reflecting on the 1890s), Hyatt Papers, University of California at Los Angeles. He suggested it was "better than anything before" because they were "trying to plant interest, alertness toward nature, not information."

129. B. M. Davis, teaching at the State Normal School at Chico, suggested these kinds of experiences accounted for a "conservative attitude" among some teachers toward the early introduction of nature study in "Progress of Nature-Study in California," *Nature-Study Review* 2 (November 1906): 257.

130. Bigelow, *Spirit of Nature Study,* 145–46.

131. On education as a "civic religion," see Reese, *Power and the Promise of School Reform,* 150; and his "Between Home and School: Organized Parents, Clubwomen, and Urban Education," 3–28.

132. Cubberley, *Changing Conceptions of Education,* 40–45.

133. Quoted in Worster, *Nature's Economy,* 185.

134. Rudolph estimated that at least eighteen hundred women actively pursued botanical studies in his "Women in Nineteenth Century American Botany," 1346–55.

CHAPTER TWO

1. Stephen Forbes to Wilbur Jackman, December 31, 1891, Natural History Survey, Chief's Office, 1871–1909, University of Illinois Archives (hereafter UIA).

2. Croker, *Stephen Forbes,* 64–74; and Lovely, "Mastering Nature's Harmony."

3. Forbes to Jackman, December 31, 1891, Natural History Survey, Chief's Office, 1871–1909, UIA.

4. Croker, *Stephen Forbes,* 64–74.

5. G. Stanley Hall quotes K. Lange in Hall's "On Entering School," *Pedagogical Seminary* 11 (1891): 145.

6. Cruikshank, "Rise and Fall of American Herbartianism," but see the ambiguous role of Ellwood P. Cubberley at Stanford University, as described in Tyack and Hansot, *Managers of Virtue,* 125.

7. Quoted in Solberg, *University of Illinois,* 71.

8. Kohlstedt, "Nature not Books," 324–52.

9. Commentary on the goals and the varieties of theoretical approaches appeared in Parker,

Textbook in the History of Modern Elementary Education; a discussion of the array of experiments in education is in Lazerson, *Origins of the Urban School.*

10. On the pressures and aspirations, see Tyack and Cuban, *Tinkering toward Utopia.*

11. Cuban, *How Teachers Taught.*

12. Bullough, *Cities and Schools in the Gilded Age,* 22-23. Pressures on the system, with over six hundred thousand pupils and over five hundred public schools, were enormous. Some schools reported that nearly half their pupils were unable to speak English; truancy was such a serious distraction that there was a special school for truants, and many classes were housed in outdated or converted facilities.

13. Goals varied among those who wanted to extend public education. The urban middle class sought to free children from work and make mothers responsible for child nurture so that children gained a broadened experience. Working class and farm families were more interested in having children gain skills that would help them contribute to the home economy. See Clement, *Growing Pains.*

14. Van De Voort, *Teaching of Science in Normal Schools and Teachers Colleges,* 16. There was also considerable debate about whether the teaching knowledge should be equivalent to academic knowledge taught in colleges during the 1880s and 1890s, but few normal schools could provide such education, and increasingly administrators wanted to stress the professional nature of subject classes.

15. Tyack, *One Best System;* also see Mattingly, *Classless Profession: American Schoolmen.*

16. The phrase is that of Bullough, *Cities and Schools in the Gilded Age,* 99. On this and related themes, see Worster, *Nature's Economy;* Leo Marx, *Machine in the Garden;* Nash, *Wilderness and the American Mind;* Schmitt, *Back to Nature;* and Cremin, *Transformation of the Schools.*

17. Quoted in Bullough, *Cities and Schools in the Gilded Age,* 7.

18. Jordan presented his arguments before the Buffalo meeting of the NEA on "Nature Study and Moral Life," *Annual Proceedings of the National Educational Association* (1896): 130; reprinted in his *The Story of Innumerable Company and Other Sketches* (San Francisco: Whitaker and Ray Co., 1896). Also see Lazerson, *Origins of the Urban School,* 241-52.

19. Quoted in Bullough, *Cities and Schools in the Gilded Age,* 100. Also see Troen, *Public and the Schools.*

20. Boyden, *History of Bridgewater Normal School,* 46-50.

21. Boyden did not produce a major textbook but did write for educational journals about teaching from nature and picked up on the term "nature study" when he published his curriculum ideas in the 1890s. See *Journal of Education* 43 (1896): 238; 44 (1896), 253-55 and 390-91; 46 (1898): 375; 47 (1898): 116, 166, 212, 277, and 294. Also see his perspective on the history of the movement in "Nature Study: Then and Now," *Nature-Study Review* 19 (March 1923): 93-96.

22. Boyden, "Nature Study," 93-108; and Daum, *Wissenschaftpopularisierung im 19. Jahrhundert.* Lynn Nyhart's book, *Modern Nature,* came just as I completed the final editorial revisions and thus was not able to incorporate any of her important insights into natural history and the new biology into this volume.

23. Massachusetts Board of Education, *Annual Report for 1890-1891* (1892): 80-88; also see the *Report for 1889-1890,* 144-49. The local activity in Plymouth was documented in 1889 through a questionnaire, which demonstrated the effect of having children collect and report on their observations rather than using recitations. In fall 1890, teachers brought a wide-ranging set of school collections of natural objects and drawing and language work based on them to an annual meeting of the Plymouth County teachers association as an argument for more support for such work. Liberty Hyde Bailey notes this early effort in *Nature-Study Idea,* 1-26. Tyree Minton suggests that Frank Owen Payne also used the phrase nature study in 1889 in conjunction with work in New York State in his "History of the Nature-Study Movement," 105.

24. Sarah Louise Arnold Papers, Simmons College Archives, Boston, Massachusetts.

25. Tebbel, *History of Book Publishing*, 580-82. Kellogg edited *The School Journal* and *The Teachers' Institute* and also published books on the history, philosophy, and practices of education.

26. A perceptive summary of Parker's work is in Katz, "'New Departure' in Quincy 1873-1881," 3-30; and Cremin, *Transformation of the School*. G. Stanley Hall said of the Chicago program, "I come here every year to set my educational watch" and took his enthusiasm back to Clark University where he served as president; see Parker, "Francis Wayland Parker, 1837-1902," 120-32.

27. On Parker, see Wilbur Jackman, *In Memoriam* (Memorial Number of *The School Reporter*) and "Francis W. Parker," *American Monthly Review of Reviews* (April 1902), n.p.; both copies in the College of Education Papers, University of Chicago Archives (hereafter UCA); also Parker, "Francis Wayland Parker," 120-32.

28. Francis Parker on "Herbartianism" in *First Yearbook of the Herbartian Society for the Scientific Study of Teaching* (1897), cited in Cohen, *Education in the United States*, 3:1888-91.

29. Quoted in a report of the Committee of Sixty assigned to promote field work and nature study in the public schools of Chicago, Scrapbook 13 (1896), Francis Wayland Parker Papers, UCA.

30. Katherine M. Stilwell, "Colonel Parker as I Knew Him," in Flora Juliette Cooke Papers, Chicago Historical Society (hereafter CHS). Parker's confidence in the intuitive knowledge of teachers and his persistent demand for the student-teacher relationship to be a primary factor in education brought him incredible loyalty, but it flew in the face of increasing centralization of educational curriculum and hierarchical authority. On the loss of control by teachers, see Herbst, *And Sadly Teach*.

31. The persistence and level of this potentially debilitating commentary is well documented in Cruikshank, "Rise and Fall of American Herbartianism," chap. 2.

32. Mitchell, "History of Nature Study," 296.

33. See the discussion of Straight in Lavender, "History of Nature Study in Texas," 35-37; and McLaughlin, "Interpretation of the Environmental Movement within Manual Arts."

34. In his "History of Nature-Study," Mitchell points out that Straight did advanced work at Harvard with Nathaniel S. Shaler and at Cornell before joining the Oswego program. The critique of object teaching concentrated on a growing rigidity in the use of objects according to Parker, *History of Modern Elementary Education*.

35. Livingstone, *Nathaniel Southgate Shaler*, 253-60. Although Shaler's initiatives hardly were unique, given the tradition of summer institutes for teachers, the availability of Harvard classes to local teachers until the 1890s was significant. Boyden, *Nature Study in the Public Schools of Massachusetts*, 6.

36. Champagne and Klopfer, "Pioneers of Elementary School Science: I. Wilbur Samuel Jackman," 145-75.

37. These contributions are noted in the obituary by Maurice Bigelow, "Wilbur S. Jackman," *Nature-Study Review* 3 (March 1907): 65-67. Nature study was also featured at the Chicago World's Fair; see the *Catalogue of the Educational Exhibition of the Commonwealth of Massachusetts of the World's Columbian Exposition* (Chicago, 1893). Minton, "History of the Nature-Study Movement," portrays Jackman as the embodiment of the new connection made between science teaching and educational psychology.

38. After considerable effort to find the origin of the term stated in the singular (as distinguished from the fairly common reference to "nature studies") Liberty Hyde Bailey concluded that the idea and the term had no single or simple origin but was introduced sometime in the late 1880s; see Bailey, *Nature-Study Idea*, 16-26. Two significant parallel developments may or may not have been related to Jackman's use of the phrase. One came from the Boston area where Arthur G. Boyden taught summer institutes and where organizers at the educational exhibits

in Boston in 1890 and 1891 thought that the term "nature-study" seemed a good equivalent of the German *Naturkunde*. The second came from Amos M. Kellogg, editor of *The School Journal*, who asked a regular contributor, Frank Owen Payne, to prepare a number of specific lessons for teachers about the natural world with that heading about 1889.

39. Jackman, *Nature Study for the Common Schools*. Bigelow suggests that there was considerable discussion about the title with Henry Holt and Company before they settled on hyphenated "Nature-Study." Bigelow, "Wilbur Samuel Jackman," 66.

40. Jackman, *Nature Study for the Common Schools*, iv.

41. Ibid., 8.

42. On the use of dialogue in nature writing especially for young children, see Gates, *Kindred Nature*, esp. 34-44.

43. The idea of using the calendar year was deployed with considerable success in adult books at midcentury by Susan Fenimore Cooper in *Rural Hours* (1850) and in Henry David Thoreau's *Walden* (1848).

44. Quotation taken from an obituary for Jackman in *Nature Study Review* 3 (March 1907): 65. This idea was picked up by other authors such as Warren in *From September to June with Nature*.

45. Jackman produced *Nature Study for the Grammar Grades*, with its intentional subtitle, *A Manual for Teachers and Pupils below the High School in the Study of Nature*.

46. Wilbur Jackman, "Natural Science of the Common Schools," *Journal of the Proceedings of the National Educational Association* 30 (1891): 581.

47. Stephen Forbes to Wilbur Jackman, December 31, 1891. Within a year, Forbes suggested that he might collaborate with others on a course of practical work and study in zoology more appropriate for country schools; letter to B. P. Colton, April 7, 1892. Both in Natural History Survey, Chief's Office, 1871-1909, UIA. Forbes was a leader in formulating field sciences for practical ends; see Juan Ilerbaig, "Allied Sciences and Fundamental Problems," *Journal of the History of Biology* 32 (December 1999): 439-63.

48. Quoted in Troeger, *Harold's Rambles*, xiii. Coulter also produced his own series of books for beginning botany for Appleton publishing house, which were billed as twentieth-century textbooks.

49. "Recent Publications," *Garden and Forest* 5 (September 14, 1892): 443. The reviewer noted the breadth of subjects and the organization of the topics so that the subjects coincided with the materials available throughout the school year.

50. Cruikshank, "Rise and Fall of American Herbartianism." Jackman served as secretary of the Herbart Society, founded by him and Illinois colleagues.

51. For a detailed study of one aspect of this work, with an emphasis on student projects, see Doris, "Practice of Nature-Study," 137-70.

52. Wilbur Jackman, *Number Work in Nature Study, Part I*.

53. See, e.g., *Nature Study for the Common Schools* (New York: Holt and Company, 1896); *Number Work in Nature Study* (1893); *Field Work in Nature Study* (1894); *Nature Study for the Grammar Grades* (1899); and *Nature Study Record* (1895); plus numerous presentations and articles in educational journals.

54. Wilbur Jackman, "Representative Expression in Nature Study," *Educational Review* 10 (October 1895): 248-61.

55. Scrapbook 13 (1896), Francis W. Parker Papers, UCA.

56. Comment is in Cuban, "Persistence of Reform in American Schools," in *American Teachers; Histories of a Profession at Work*, ed. Warren, 370.

57. There were nine subject categories discussed by such subcommittees meeting at various places at the same time, December 28-30, 1892, across the United States. The prestigious Committee of Ten then reviewed, summarized, and had their results published by the United States Bureau of Education as *Report of the Committee of Ten on Secondary Studies* (Washington, DC,

1893). On the issue of secondary curriculum, also see Heffron, "Knowledge Most Worth Having: Otis W. Caldwell," 227-52.

58. It appears that Merritt did not actually attend the meeting; see the official list on page 7 and the report of deliberations in *Report of the Committee of Ten*, 138-61.

59. *Report of the Committee of Ten*, 139.

60. Ibid., 142-51.

61. The description of the NEA leaders is in Tyack and Hansot, *Managers of Virtue*, 137; on the report's reception see Sizer, *Secondary Schools at the Turn of the Century*, 192-93.

62. Clapp, "The Nature and Purpose of Nature Study," *Education* 15 (June 1895): 600; quoted in Olmsted, "Nature-Study Movement," 71.

63. Joseph M. Rice, "Public Schools of Minneapolis and Others," 376. Although Rice uncovered much that was wrong in the public schools, he was impressed with nature study and its correlated art and diary writing projects. See Rice, *Public School System of the United States*, 213-15.

64. Wilbur Jackman to Anita McCormick, November 23, 1896, McCormick Family Papers, Wisconsin Historical Society, Madison, Wisconsin (hereafter WHS).

65. Stephen Forbes to Wilbur Jackman, October 10, 1895, Natural History Survey, UIA.

66. Stephen Forbes to Wilbur Jackman, May 16, 1896, Natural History Survey, UIA.

67. The politics of the Chicago schools at the turn of the century were very difficult. Although University of Chicago's President Harper and his faculty studied local education and produced reports, Chicago remained one of the most stressed of all the large urban systems, with low expenditures per student and a student-to-teacher ratio that grew out of hand as the population grew by the tens of thousands each year. See Herrick, *Chicago Schools*, 82-118; and Schiltz, "Dunne School Board: Reform in Chicago, 1905-1908."

68. Parker had also been through a positive but bruising review based on a political battle over his management of the normal school; see Selzer, "History of the Chicago Normal School," 62-73.

69. Blaine's diaries for 1898 and 1899 record her growing excitement as she met with President Harper and Colonel Parker, as well as with such visiting dignitaries as Patrick Geddes. She "agreed delightedly" to start university work for teachers and found talk of a new school a "plan with some hope" and was "all enthusiasm" after a discussion with John Dewey. See August 1, 1898, November 13, 1898, and January 21, 1899. Blaine diaries, McCormick Family Papers, WHS.

70. Russell's response was to note that top faculty might earn $3,500, much more than the lesser class of "adjunct professors and the women of the faculty." Russell to Blaine, June 28, 1900, Russell Papers, Teachers College Library, Columbia University.

71. Records of the Chicago Institute, Academic and Pedagogic, constitute boxes 157-59 in the extensive McCormick Family Papers, WHS.

72. *Catalogue of the Chicago Institute: Academic and Pedagogic, 1900-1901* (Chicago: Chicago Institute, 1900), 4. In the Department of Science and Nature Study, Jackman taught biology; Charles W. Carmen, physics and chemistry; Alice P. Norton, elementary chemistry; Willard Streeter Bass, elementary physics; Ira Benton Meyers, zoology; and Harriet T. Bradley, botany.

73. Chicago Institute, *Catalogue*, 7-8.

74. Ibid.

75. "Preliminary Announcement," *Catalogue*, 5. The sixteen-page brochure outlined the three major components of the school: (1) Academic—a school for children 4-18 years old, (2) Pedagogic—a course of study for teachers in training, and (3) Summer School—intended largely for teachers seeking to enhance their skills. The Chicago program also intended to include physical sciences, often subsumed under the nature study rubric for the elementary level. Physical sciences were presumed to be an appropriate part of various subjects; see Olsen, "Nature Study: Some Physical Laws Necessary to the Study of Geography and Agriculture."

76. Kohlstedt, "'A Better Crop of Boys and Girls.'"

77. The Institute's *The Course of Study: A Monthly Publication for Teachers and Parents* 1 (July 1900), contains Parker's "Plan and Purpose of the Chicago Institute" (9–16). This plan, under the editorship of Jackman, initiated the *Elementary School Teacher.*

78. Henry Sabin to Parker, quoted in Parker to Faculty of the Chicago Institute, January 14, 1901, Chicago Institute Papers, UCA.

79. Clipping from the *Chronicle*, October 7, 1896, in Scrapbook 13; also see undated clipping in Scrapbook 13 (1896), Parker Papers, UCA.

80. Quoted in a report of the Committee of Sixty concerning field work and nature study in the public schools of Chicago. Scrapbook 13 (1896), Parker Papers, UCA.

81. "Morning Exercise," Box 1, Folder 3, Chicago Institute Papers, UCA.

82. The expenses ($400,000 for the property on North Park Avenue between Belden and Webster Streets; $100,000 for staff and supplies, plus the costs of construction) began to overwhelm even the wealthy Anita McCormick Blaine, as indicated in the Minutes of the Board of Trustees, Box 160, McCormick Papers, WHS. The Board thought of requiring teachers to live in the northside neighborhood near the school and interviewed Jane Addams about her settlement house, but Parker was concerned that it would be hard to recruit teachers if they had to live in "vicious wards."

83. Cited in Cremin, *Transformation of the School,* 129.

84. Biography of Dewey by Judy Survatt, *Notable American Women,* 1:467–69.

85. Mayhew and Edwards, *Dewey School,* esp. 1, 64, 78, and 153.

86. Young was a dynamic administrator, who had been a student of Nathaniel Allen and earned her Ph.D. at Chicago. She left the University of Chicago in 1904 to head the Chicago Normal School, before becoming superintendent of the Chicago schools from 1909 to 1915. Ellen Condliffe Langemann, "Experimenting with Education: John Dewey and Ella Flagg Young at the University of Chicago," *American Journal of Education* 104 (May 1996): 171–85; and her sketch in *Biographical Dictionary of American Educators,* 3:1453–54.

87. The contract of merger is dated April 15, 1901, and included several key provisions. The School of Education was to be under Parker. John Dewey headed the Department of Pedagogy and was also in charge of the secondary school. The summer school was to operate independently and be self supporting, with faculty to be paid only from its income. The building was to come from a grant and to be equipped by the University. School of Education Papers, UCA.

88. College of Education, Records 1900–1926, UCA. Also see Griffith, *A History of the Origins of the Laboratory School at the University of Chicago;* and, on the controversies, DePencier, *History of the Laboratory Schools,* 26–52.

89. Jackman to Parker, February 14, 1901, Parker Papers, UCA. Parker envisioned a program parallel to Teachers College, Columbia, which drew students primarily from the East; Chicago would, he thought, serve teachers and administrators in the West. School of Education Papers, UCA.

90. Eventually Jackman retreated from the University faculty to become principal of the University Experimental School in 1904, a position he held until his death in 1907. According to rumors that made it to California, Dewey was having a difficult time and "his practice school had been taken away from him . . . [because] he was making too wide generalizations from his experimental work." David Starr Jordan of Stanford was nonetheless interested in hiring Dewey. See Helen Swett to Charles Schwartz, April 25, 1901, Swett Papers, Bancroft Library, University of California at Berkeley.

91. Jackman found Alice Dewey, principal of the elementary school, to be critical and autocratic, an ineffective administrator who could be rude to pupils as well as visitors from other normal schools. See Jackman to Trustees of the Chicago Institute, April 23, 1904, McCormick Papers, WHS.

92. He died in 1907, and at his funeral, selections from Walt Whitman's "Song of the Open Road" and "Song of Myself" were read, favorites among many of the nature study teachers. Obituary clipping, McCormick Papers, WHS.

93. Based on the extensive correspondence in the Cooke Papers, CHS. She also produced a series of articles suggesting nature projects that could be done near Lake Michigan.

94. Kimberley D. Finley traces the influential networks of women intent on encompassing culture and particularly art (sometimes in relationship to nature) within the Chicago public schools in "Cultural Monitors: Club Women and Public Art Instruction in Chicago," 144-47.

95. Chicago Board of Education, *Bulletin*, for January 5, 1903, November 2, 1903, and October 4, 1904.

96. Curtis, "Vacation Schools, Playgrounds, and Settlements," 7.

97. Klunk, "Chicago Academy of Sciences," esp. 127-28, 147-51, 154, and 157-60. Because the Field Museum was so much larger and included the entire world in its purview, the Academy with more limited resources chose to sponsor fieldwork in the greater Chicago area and to present habitat displays relating to that region.

98. Charles T. Chamberlain, "John M. Coulter," reprint from the *University of Chicago Magazine* (n.d.), Faculty Obituary Files, UCA.

99. Mitman, *State of Nature: Ecology, Community, and American Social Thought*, 16-17; and Kingsland, *Evolution of American Ecology*.

100. Jackman's *Third Yearbook of the National Society for the Scientific Study of Education, part 2, Nature Study*, particularly emphasized the landscape and the relationship of elements within it; the dunes of Lake Michigan had been one of the early sites of ecological research. Recent studies in the history of biology also identify both the persistence of this contextual outlook and some of its new dimensions, including Ilerbaig, "Pride in Place"; Kohler, *Landscapes and Labscapes;* and Pauly, *Biologists and the Promise of American Life* and *Fruits and Plains.*

101. Patterson was a graduate of Illinois Normal School in 1890 and attended the University of Chicago in 1896-97 plus several subsequent summers before completing her B.S. degree. Her obituary in the campus newspaper made clear her active professional participation, public lectures in a number of other states from Tennessee to Pennsylvania, presentations at summer camps for youth, and local initiatives to teach and improve the natural history of Normal, Illinois. *The Vidette* [student paper of Illinois Normal School] 49 (June 17, 1929); accessed at Illinois State University Archives.

102. Ange O. Milner, Library of Illinois State Normal School, to L. H. Bailey, September 9, 1904, requesting multiple copies of the leaflet series, Bailey Papers, CUA.

103. Parker, introduction to Lucy Wilson, *Nature Study in Elementary Schools*, vii. Wilson, whose title page always indicated her Ph.D. degree, presented topics by month, following Jackman. She offered more illustrations, helpful advice to teachers about finding material, and descriptions of projects enjoyed by pupils. The manual also presented material by grade or a combination of grades, which also distinguished it from Jackman's textual *Nature Study and Related Subjects for Common Schools.*

104. Francis Parker on "Herbartianism" in *First Yearbook of the Herbartian Society for the Scientific Study of Teaching* (1897); see Cohen, *Education in the United States*, 3:1888-91.

105. Finley, "Cultural Monitors: Club Women and Public Art Institutions in Chicago," 144-47.

CHAPTER THREE

1. Shaw, "The Place of Children's Gardens," *Nature-Study Review* 6 (February 1910): 43. Shaw taught at the State Normal School at New Paltz, New York, before taking a long-term position working with school visitors at the Brooklyn Botanical Garden.

2. Bullough, *Cities and Schools in the Gilded Age*, 7, 22–23.

3. New York City Board of Education, *Annual Report for 1874*, 33:254–55; and *Annual Report for 1870*, 29:158.

4. Columbia College became Columbia University and a formal agreement to link the two institutions officially was finally signed in 1915.

5. James Earl Russell taught briefly at Cascadilla School in Ithaca and subsequently traveled to Germany, where he wrote a dissertation on secondary and extension education. He was deeply committed to the idea of universities, rather than normal schools, as the best place for professional education of teachers who would wok in model schools and serve as school administrators. A good summary of Russell's life is the guide to the Russell Papers and the archives, Teachers College Library, Columbia University (hereafter TC).

6. The tension between liberal arts and professional education for women, especially those becoming teachers, is described in Weneck, "Social and Cultural Stratification in Women's Higher Education," 1–25.

7. By the 1920s, Teachers College concentrated on philosophical foundations and focused on graduate education rather than undergraduate training. Cremin, Shannon, and Townsend, *History of Teachers College, Columbia University*, 16.

8. The issue was enjoined by the decision to allow the Normal School of the City of New York to offer liberal arts degrees in 1895, which slowly eased the institution away from its role in education. Grunfeld, "Purpose and Ambiguity: The Feminine World of Hunter College."

9. Russell, "The Function of the University in the Training of Teachers," *Teachers College Record* 1 (1900): 6.

10. Careers proved very fluid for some nature study educators, as shown in the career of Lloyd who left Columbia to go to the Carnegie Institute's Desert Botanical Garden in Tucson and became managing editor for *Plant World*. Lloyd to Russell, February 9, 1907, Russell Papers, TC. Lloyd also coauthored an article, "Courses in Biology in the Horace Mann High School," with Maurice Bigelow, and their colleague Elizabeth Carss contributed a "Course on Nature Study in the Horace Mann School."

11. Russell, *Founding Teachers: Reminiscences of the Dean Emeritus*, 57–58. Russell observed, "The time was ripe for development of the field which we afterwards labeled educational psychology. Stanley Hall had opened it up by his child study. Colonel Parker and Doctor Sheldon were trying out practical uses of the new ideas. Professor Hinsdale had dared challenge the authority of the doctrine of formal discipline," and so he investigated the work of a promising young man, Edward Thorndike, and brought him to Teachers College (52). Mark Largent generously shared a letter on this matter from M. A. Bigelow to C. A. Davenport, April 19, 1899, Davenport Papers, American Philosophical Society, Philadelphia (hereafter APS).

12. M. A. Bigelow to C. B. Davenport, July 19, 1907, Davenport Papers, APS.

13. M. A. Bigelow to L. H. Bailey, December 17, 1906, Bailey Papers, Cornell University Archives (hereafter CUA). Bigelow thought it would be wise to add agriculture to the program and suggested a cooperative program with Cornell in which Teachers College students would attend summer classes in Ithaca.

14. See Ravitch, *Great School Wars*, 107–88.

15. Superintendent of Schools, New York City, *Annual Report* 1 (1898–99), 1. For the pervasive ways reform sensibilities intersected in politics and education, see Recchiuti, "Origins of American Progressivism: New York's Social Science Community"; this lively account follows the political framing of changes in the New York system, eventually consolidated under a central board in 1901.

16. Ravitch, *Great School Wars*, 169–70.

17. Biographical sketch by Robert J. Fridlington in *Notable American Women*, 2:239–40.

18. Requests for Cornell nature study leaflets came from administrative officers, principals, and practicing teachers in New York to John Spencer; for example, from Jenny Merrill, Decem-

ber 14, 1897, and Catherine. H. Murphy, January 28, 1898, Spencer Education Extension Papers, CUA.

19. Pauly, *Biologists and the Promise of American Life*, 174.

20. New York normal schools educated over a thousand student teachers a year in the early twentieth century and the overwhelming majority were women. Men entering the New York City schools tended to come from the City College of New York, and in 1906 the College officially established a Department of Education. Gorelick, *City College and the Jewish Poor*, 104 and 108-9.

21. New York City Superintendent of Schools, *Annual Report* (1902-3): 68-73.

22. *Course of Study in Nature Study, Elementary Science, and Geography* (New York City Department of Education, 1903).

23. New York City Superintendent of Schools, *Annual Report* 5 (1902-3): 71. In addition to Straubenmueller, committee members were superintendents Edward B. Shallow and Grace Strachan (the only woman on the Board of Supervisors), as well as James E. Peabody, chairman of Morris High School, Bronx; John P. Conroy of P.S. 179, Manhattan; John Doty, P.S. 21, Manhattan; Rosina J. Rennert, Girls' Technical High School, Manhattan; Josephine E. Rogers, P.S. 75, Manhattan; and Georgiana Brown, P.S. 91, Brooklyn. This committee had the largest number of women of any of the curriculum committees.

24. The "Dr." in his title was unexplained, since he had an undergraduate degree from the College of the City of New York (1880). The *New York Times* obituary of May 14, 1934, noted his fifty years of service and credited him with "introducing many of the features which give to the New York public school system its distinctive character" (17).

25. New York State Department of Education, *Annual Report*, Supplemental Volume (1905): 424.

26. These included "specimens of plants, animals, and minerals used for science teaching including life-sized models and colored representations." New York City Superintendent of Schools, *Annual Report*, Appendix F (1903-4): 241; New York, *Journal of the Board of Education* (1906): 1236-37. Also see John E. Findling, ed., *Historical Dictionary of World's Fairs and Expositions, 1851-1988* (Westport: Greenwood Press, 1990), 178-86.

27. New York City Superintendent of Schools, *Annual Report* 5 (1902-3): 174; and *Annual Report* 11 (1908-9): 194.

28. Whitney, "Vacation Schools, Playgrounds, and Recreation Centers," 298-303.

29. Fannie Griscom Parsons directed the program and described it in *The First Children's Farm School in New York City* and "A Day in the Children's School Farm in New York City," *Nature-Study Review* 1 (November 1905): 255-61. On the school gardening movement, see Kohlstedt, "'A Better Crop of Boys and Girls,'" 1-36.

30. Parsons to Liberty Hyde Bailey, Bailey Papers, CUA. She concluded that "we not only have to train teachers in our Normal Schools but should give simple practical lessons to the teachers who are already in the schools."

31. *New York Times*, January 17, 1905, quoted in Ravitch, *Great School Wars*, 169.

32. Albert Bickmore's vision of an educational museum is evident in his unpublished typescript "Autobiography with a Historical Sketch of the Founding and Early Development of the American Museum of Natural History," as well as the minutes of the Board of Trustees, both in the archives of the American Museum of Natural History, Special Collections, New York.

33. The state funding was $18,000 in 1884 and persisted (with the exception of 1890 and 1892) until 1904. Sherwood, *Free Nature Education by the American Museum of Natural History*, 9.

34. New York City Superintendent of Schools, *Annual Report* 8 (1905-6): 431-34.

35. New York City Superintendent of Schools, *Annual Report* 12 (1909-10): 200-201.

36. New York City Superintendent of Schools, *Annual Report* 12 (1909-10): 20. The Museum also added a room for the blind, understanding how important objects would be for this group.

37. Brochures in the papers of Alice Rich Northrop, Schlesinger Library, Radcliffe Institute, Harvard University.

38. These letters are to Ira S. Wile, Chairman of the Committee on Studies and Textbooks, from Holtz, February 26, 1913, and Knox, March 10, 1913, Wile Papers, Rush Rhees Library, Special Collections, University of Rochester (hereafter URSC).

39. These letters are all to Ira S. Wile, Chairman of the Committee on Studies and Text-Books, from Charles Townsend, February 10, 1913; from M. A. Bigelow, February 17, 1913; from T. Gilbert Pearson, February 17, 1913; from C. Stuart Gager, February 17 and 19, 1913; and from Anna Clark (and written as the New York representative of the Nature-Study Society), March 21, 1913, Wile Papers, URSC. Bigelow's letter implies that Frank McMurry was on a committee that recommended dropping the subject.

40. This is a separate, undated typescript entitled "A Brief on the Teaching of Natural History in the Schools of New York," Wile Papers, URSC.

41. E. Esther Pritchett to Ira S. Wile, March 15, 1913, and an extended report (June 13, 1913) that recommended the course of nature study (June 20, 1913); all found in Wile Papers, URSC.

42. Shaw, "The Place of Children's Gardens," *Nature-Study Review* 6 (February 1910): 43–45.

43. "Anna Billings Gallup," in *Who Was Who in America* (Chicago: Marquis, 1960), 3:310. Also see Alexander, *Museum in America: Innovators and Pioneers*, 133–46.

44. Eastman, *Story of the Brooklyn Institute of Arts and Sciences*.

45. Phillip Mershon, "Brooklyn Children's Museum: The Story of a Pioneer" (typescript, d. 1959, provided by the Brooklyn Institute Library), 1–11.

46. Gallup, "Work of a Children's Museum," reprinted in American Association of Museums, *Proceedings* 1 (1907): 144–47; also see her "The Children's Museum as an Educator," 371–79.

47. Superintendent of Schools to New York City Board of Education, *Annual Report* 13 (1910–11).

48. On the problems, see B. Ellen Burke, April 4, 1897, Spencer Extension Education Papers, CUA; also the pamphlet *Program of Study and Syllabus in Physical Culture, Physiology and Hygiene, Nature Study at the Archdiocese of New York* (New York: Catholic School Board, 1911), found in the Andover Cambridge Divinity School Library, Cambridge.

49. *Nature Study* (Philadelphia: Archdiocese of the Catholic Church, 1932), TC. There is miscellaneous evidence that nature study was pursued by individual teachers and schools in systems that did not provide any support, as evidenced by letters to John Spencer from Ellen Burke, April 4, 1898; Dominican Sister Superior, September 12, 1898; M. Theresa Elder, March 27, 1899, and others in the Spencer Papers, CUA.

50. *Nature Study* (Cleveland: Diocesan School Board, 1927). The handbook was unabashed about introducing religious materials, including Biblical references to birds and lilies; found in TC collection.

51. Kelly described his program for "teachers to study nature first hand" through twelve excursions in the New York City Superintendent of Public Schools, *Annual Report* 1 (1899): 75–76; also Hine, "School Camera."

52. Tyack, *One Best System*.

53. One example is discussed in Schiltz, "Dunne School Board."

54. Janet Miller, "Urban Education and the New City: Cincinnati's Elementary Schools," 385–87.

55. A biographical sketch by Rachel Arnold Hefler describes Arnold's outdoor childhood on a farm and years of teaching before moving west, and a series of scrapbooks have early writings by Arnold in a number of educational journals, sometimes without the source but typically dated. This material was kindly provided to me by Claire Goodwin of the Simmons College Archives, Boston (hereafter SCA).

56. *Annual Report of the Board of Education and of the Superintendent of Schools of the City of St. Paul, Minnesota, for the Year 1893–1894*, 70–73.

57. Elizabeth Williams [History of Minneapolis Public Schools, c. 1902], typescript docu-

ment 127.K17.12 (F), Minnesota Historical Society, St. Paul (hereafter MHS); an attached report by Jean Goudy waxes eloquent on nature study, pointing out that nature study had replaced map memorization and field lessons and provided important preparation for later studies in geography.

58. S. L. Arnold to L. H. Bailey, March 31, 1906, Bailey Papers, CUA.

59. Clippings of her publications are in scrapbooks; the earliest are from 1891, and she reported in some detail on examples of children's correlated work in *Primary Education* (May 1894) in Scrapbook 4, Arnold Papers, SCA.

60. C. E. Bessey to S. L. Arnold, March 31, 1894, Bessey Papers, University of Nebraska Archives.

61. Massachusetts Board of Education, *Annual Report for 1890-1891* (1892), esp. 27 and 270. Boyden took considerable pride in his Bridgewater graduates, claiming that they created special apparatus while those of other normal schools "make the mistake of teaching by means of lecturing or of closely following the book and going over much ground" (205).

62. These annual course outlines became increasingly detailed as part of the effort to establish a uniform curriculum across the state, by grades. Although there was no state mandate to implement them, a system of official state "visitors" to the schools reported to the board about each classroom and thus created pressure to conform.

63. Massachusetts Board of Education, *Annual Report for 1891-1892* (1893), 84, 117-19, 225.

64. Walton had been an official school visitor for twenty-five years when he wrote his "reminiscences" about the changes in schools for the Massachusetts Board of Education; see its *Annual Report for 1895-1896* (1897), 250-51. He was an advocate of the new nature studies, believing that nothing was equal to the "tendency to elevate the taste of the children as nature herself" (251).

65. Report of Arnold in Boston School Committee, *Annual Report for 1897* (1898), 125-41; and *Annual Report for 1898* (1899), 147. She included nature study in her *Waymarks for Teachers . . . with Illustrative Lessons* (New York: Silver, Burdett, and Co., 1894).

66. "Nature Study in City Schools," *American Primary Teacher* (June 1899), in Scrapbook 4, Arnold Papers, SCA. In 1902 Arnold joined the faculty of newly opened Simmons College as dean of students and head of home economics; she also later played an active role in promoting the Girl Scouts.

67. Massachusetts Board of Education, *Annual Report for 1897-1898* (1899), 214-24.

68. Boston School Committee, *Annual Report for 1890* (appendices), 10.

69. He also published *Our Native Birds: How to Protect Them and Attract Them to Our Homes* to encourage children to work with their families on nature study.

70. See *Report of the Teachers Training Schools of Minnesota*, which recorded a statewide attendance of 6,276 teachers the previous year.

71. Clippings of his newspaper articles and drafts of his books including his guide to birds, together with sporadically maintained journals, are in the Dietrich Lange Papers, MHS.

72. Kliebard, *Forging the American Curriculum*, 51-67. Hall had declared, "As our methods of teaching grow natural, we realize that city life is unnatural, and that those who grow up without knowing the country are defrauded" (54). For a slightly different view of Hall's interest in schooling, see Cruikshank, "Rise and Fall of Herbartianism."

73. "Nature-Study and Moral Education," *Harper's Weekly* reported July 11, 1896, that this meeting was "scholarly, thoughtful, public spirited, and progressive."

74. Ross, *G. Stanley Hall: The Psychologist as Prophet*, 286-308, 299. Also see correspondence to and from Hodge between 1889 and 1913 in the G. Stanley Hall Papers, Clark University Archives, Worcester. Hodge had spent the 1891-92 academic year teaching biology at the University of Wisconsin; *Wisconsin Alumni Directory, 1846-1919* (Madison: University of Wisconsin, 1921), 617. On Hodge, see *Who Was Who in America*, vol. 4 (Chicago: Marquis, 1961-68).

75. From the outset Hodge took an independent stance on nature study; see the two-part

essay on "Foundations of Nature Study," *Pedagogical Seminary* 6 (1899): 536-63 and 7 (1900): 208-28.

76. Hodge had initially worked on brain cell fatigue and seems to have turned his attention fully to nature study after about 1896, publishing articles until about 1912, when he focused on high school biology.

77. Hodge, *Nature Study and Life*. Not all teachers appreciated it, however, and the Ohio Teachers Circle teachers, who used it as one of their basic readers in 1909-10, were quite critical of its organization. Lupfer, "Reading Nature Writing: Houghton Mifflin Company," 49.

78. It was, for example, enthusiastically reviewed in *Science* 16 (November 1902): 739-41.

79. Hodge, "The Bird Census and Food Chart," in *Nature Study and Life*, 319-25.

80. Hodge, *Nature Study and Life*, ix-x. Also see Kohlstedt, "Collectors, Cabinets, and Summer Camp," 10-23.

81. Hodge, *Nature Study and Life*, v.

82. It was the one required books for the Nebraska Teachers' Reading Circle; see *Biennial Report of the Nebraska State Superintendent of Public Instruction for 1901-1903* (Lincoln, 1903), 125-27. Participation was not required, but teachers were awarded certificates for their participation. Also see Cordier, *Schoolwomen of the Prairies and Plain*, 61.

83. Merrick Bemis in President's Report, May 23, 1894, Worcester Natural History Society Papers, American Antiquarian Society. He also noted with regret that the Society did not have resources to be as helpful as the museums in New York and Boston, but over the next two years it did loan seven hundred items. Also see Kohlstedt, "Collectors, Cabinets, and Summer Camp," 10-23.

84. Hodge to Spencer, March 30, 1898, Spencer Extension Education Papers, CUA. Hodge relished his independent stance and was highly enthusiastic about his work with local teachers.

85. Hodge, "Practical Work with Mosquitoes," *Nature-Study Review* 3 (January 1907): 33-36. Doris, "The Practice of Nature-Study," 80-85, notes that a number of other schools also did projects with mosquitoes and uses this as an opportunity to show how different teachers might stress quite different aspects of such study. A rural version dealt with agricultural pests, such as "Blister Beetles" in Missouri, described in Dewey, *New Schools for Old*, 280-81.

86. See Cohen and Mohl, *Paradox of Progressive Education: The Gary Plan and Urban Schooling*; and Mohl and Betten, *Steel City: Urban and Ethnic Patterns in Gary Indiana*.

87. Flexner and Bachman, *Gary Schools: A General Account*, 108-9.

88. A Cornell graduate who taught briefly in Gary during its years of experimentation recalled that her pupils were a mix of upper-class students with a strong educational background and others from southern Europe whose fathers worked in the hot steel mills pouring the molten steel into forms. Clara Koepke Trump to Edward H. Smith, May 8, 1986 and to Editor of *Alumni News*, July 31, 1986, both in Trump Papers, CUA.

89. Flexner and Bachman, *Gary Schools*, 110-12. The Rockefeller Foundation-funded analysts were obviously uncertain about nature study since they found the work "formless and discontinuous in character" but also noted the enthusiasm and activity that surrounded the children's observational and practical activity.

90. Cohen and Mohl, *Steel City*, 133. Also see Ahearne, "Nature Study in the Gary Schools," *Nature-Study Review* 11 (February 1915): 59-63.

91. Cremin, *Transformation of the School*, 148.

92. Cohen and Mohl, *Steel City*, 11. The authors point out that Gary, with a fluid population and a determined board with adequate funding, provided an unusual opportunity to experiment with the possibilities of education that included all the elements of the progressive agenda but provided in an authoritarian context.

93. Such in-service training programs were prevalent from the 1880s into the 1920s. They were free or low fee, sometimes mandatory, and could vary from three days to six weeks in

length. For an insightful overview of their dynamics but with little attention to curriculum, see Spearman, "Peripatetic Normal School: Teachers' Institutes in Five Southwestern Cities." These in-service programs followed national trends and targeted local issues as a way of enhancing the education of practicing teachers. Those that featured nature study were, after 1905, regularly advertised in the "News Notes," as, for example, in *Nature-Study Review* 2 (May 1906): 192.

94. *Report of the Teachers Training Schools of Minnesota*, 8, 28, 33, 35, 42, 52. Nature study was offered in about half of the programs, with the report from the southern Minnesota town of Austin suggesting, "Perhaps no work taken in summer school was more interesting to the students and attracted the attention of outsiders more than this [nature study]" (52); nature study was sometimes taught in conjunction with physical geography, elementary science, and natural philosophy.

CHAPTER FOUR

1. Quoted in Minton, "History of the Nature-Study Movement," 122.

2. The best general discussion of these activities remains Colman, *Education and Agriculture*; and also useful is Smith, *People's Colleges*.

3. Liberty Hyde Bailey, *Nature-Study Idea*, 65-66.

4. Henson, "Evolution and Taxonomy: J. H. Comstock's Research School in Evolutionary Entomology" and "Comstocks of Cornell: A Marriage of Interests," 112-25.

5. Extension lectures could be on campuses but were also given in counties across the state coordinated by a state supervisor, often in the major state university. Colman, *Education and Agriculture*.

6. Anna Botsford had attended Cornell in 1875-76 but left after marrying John Comstock and accompanying him to Washington. She resumed her studies in 1883 and received a B.S. in 1886; see *Ten Year Book of Cornell University* 4 (1898-1908), 164. Also see Champagne and Klopfer, "Pioneers of Elementary School Science: II. Anna Botsford Comstock," 299-322.

7. Comstock and Comstock, *Manual for the Study of Insects*. Anna's sense of humor was evident when she reported to a mutual Cornell friend, "Do you know that Henry and I have a new wife? My cousin Helen Willis is the victim. She cooks our dinner, sews our buttons, mends our gloves, and in short is the goddess of our domestic life . . . I can still work with Harry and we have our home life too." Comstock to William Trelease, January 16, 1881, Comstock Papers, Cornell University Archives (hereafter CUA).

8. Students and others recalled her lectures and striking presence in their reminiscences, such as that of Edwin Emerson, undated typescript, "Groves or Academe"; and Lydia [no last name] to Anna Botsford Comstock, February 28, 1876, Comstock Papers, CUA.

9. Lind, *Methods of Teaching in a Country School*, 159.

10. The term "abandoned farm problem" was used by Densmore, "Nature-Study at Cornell," 101. For a discussion of the intensifying nineteenth-century problem of "worn out soil" in the northeastern states, see Rossiter, *Emergence of Agricultural Science*.

11. Comstock, *Comstocks of Cornell*, 89-197.

12. For a discussion of her controversial appointment and its gender implications, see Comstock, *Comstocks of Cornell*, 254; and Pamela Henson, "Comstocks of Cornell."

13. Charles B. Scott to J. H. Comstock, May 4, 1895, Comstock Papers, CUA. Scott's textbook, *Nature Study and the Child*, detailed the pedagogical philosophy of the program to teachers.

14. Scott to John Spencer, March 25 and May 28, 1898, Spencer Extension Education Papers, CUA.

15. This theme of the artistic and poetic qualities that Comstock brought to her work is highlighted in Champagne and Klopfer, "Pioneers of Elementary-School Science: II. Anna Botsford Comstock," 299-322.

16. Copies of some of these ephemeral leaflets, along with descriptive and encouraging letters from Anna B. Comstock to Alila Miller are in the Grace Miller Papers, Suffolk County Historical Society, Rivershead, New York. Each leaflet was topical, typically four pages, and described and traced the life history of a particular group like ferns or butterflies; and these were accompanied by a "supplementary pamphlet" with questions to be answered by the student. Some of these series produced between 1899 and 1911 are now available on microform at the CUA. Comstock's assistant was Miss Georgia, a former teacher, who helped respond to the literally hundreds of teachers who participated; L. H. Bailey to C. D. Bostwick, December 21, 1907, Bailey Papers, CUA.

17. This pamphlet was found at Ohio State University, with no date except a postal permission of July 16, 1896; *Home Nature-Study Course,* n.s., vol. 1, no. 4, p. 4.

18. *Home Nature-Study Course,* n.s., vol. 1, no. 4, p. 6.

19. Comstock produced over twenty illustrations for Samuel Scudder's major text on butterflies, earning about nine dollars each; but she concluded, "I think it is a mistake for a woman who is at the head of an establishment—and who has many home duties—to try and be a business woman at the same time." Comstock to Scudder, June 14, 1899, Samuel Scudder Papers, Boston Society of Natural History.

20. The complete title was quite clear about its educational goal, *Insect Life: An Introduction to Nature-Study and a Guide for Teachers, Students, and Others Interested in Out of Door Life.* Appleton recognized a good market and paid $1,000 for her engravings, making the cash an attractive alternative to producing and distributing this book through their own press, according to Comstock, *Comstocks of Cornell,* 189.

21. Anna Botsford Comstock, *Ways of the Six-Footed.*

22. Effie [original family name unknown] married another Comstock-trained entomologist, Mark V. Slingerland, and signed her slides as "Mrs. M. V. Slingerland," even after her husband died and until she remarried. See Effie E. Yantis to J. G. Needham, February 7, 1947, Needham Papers, CUA.

23. The attention to social problems is subtle and rarely political as writers of the leaflets sought to resonate with a diverse audience feeling economic and social pressures. Bailey reflects this compassionate perspective in his *Outlook to Nature,* 168.

24. An early account of the Comstock press is found in the Cornell University Press and Comstock Publishing Company Papers, CUA; and in Comstock, *Comstocks of Cornell,* 187–89. In early 1915, the productive press was able to build a "chalet" in which it could be housed, based in part on the income from Anna's popular *Handbook of Nature Study.*

25. Comstock, *Comstocks of Cornell,* 234. The *Handbook* went through twenty-four editions between 1911 and 1939 and is still published by Cornell University Press.

26. Typescript of an oral interview with L.H. Bailey, apparently by E. L. Palmer (n.d.), CUA.

27. Ruth Sawyer, "What Makes Mrs. Comstock Great," in *The Woman Citizen* (September 20, 1924), 26–28; copy in Comstock Papers, CUA. Although little suggests that she was outspoken on women's rights, she was deeply committed to women's education and helped establish William Smith College for women in Geneva, New York.

28. Comstock to L. H. Bailey, August 28, 1905, Bailey Papers, CUA.

29. Her undated "Syllabus of Lectures: Nature Study" pointed out the advantages to each pupil and the teacher, arguing that such work also created a new relationships between them and relief from the routine of traditional classroom studies. Comstock Papers, CUA.

30. Comstock, *Comstocks of Cornell,* 193. Although the Chautauqua is most remembered for the famous lecturers who visited, an important attraction had always been the green and rolling natural landscape of western New York; see MacCaughey, *Natural History of Chautauqua.*

31. Her national contributions are described in an oral interview of Laurence Palmer, Comstock's successor at Cornell; transcript of an interview by James E. Rice, CUA.

32. Typescript of an interview with L. H. Bailey, apparently by E. L. Palmer (undated), CUA.

33. Bailey's first book, *Talks Afield about Plants and the Science of Plants*, had established his interest in working with public audiences and demonstrated his liberal Protestant view that the systematic study of nature would reveal the hand of the Creator. For an account of Bailey's philosophy of education, see Azelvandre, "Forging the Bonds of Sympathy," 99-203.

34. The organizational structure was unstable in the agricultural college, with names and appointment structures varying during this period. In general, the Bureau of Nature Study remained part of the University Extension of Agricultural Knowledge division within the College of Agriculture and thus was eligible for various state and federal funding programs that not only underwrote tuition for students in the agricultural college but also provided most of the funding for education extension programs. The best general discussion of these departments remains Colman, *Education and Agriculture*.

35. The book's subtitle indicated his outlook: *Being an Interpretation of the New School Movement to Put the Child in Sympathy with Nature*. Also see LaBud, "Liberty Hyde Bailey: His Impact on Science Education."

36. Scott taught nature study to about fifty teachers each year and oversaw the public school program for nearly five hundred pupils in Oswego. He was also chair of a committee of the state teachers association that sought to introduce "real" nature study and offset the "tendency to be satisfied with mere book work and cramming for examination." Scott to John Spencer, March 25 and May 28, 1898, Spencer Extension Education Papers, CUA.

37. Bailey urged teachers to begin with a child's interests and with objects taken from the local environment, taking advantage of children's natural tendency to investigate. See his report for the Cornell University College of Agriculture, *Fourteenth Annual Report* (1901): 219-20; and in "Nature Study Movement," 109-16.

38. Bailey to George Brett of Macmillan Publishing Company, May 9, 1897, Bailey Papers, CUA.

39. "How the Squash Gets Out of the Seed," 1. Thirty thousand copies of this leaflet were distributed.

40. The leaflets were produced under a number of series titles and, depending on funding, by various divisions within Cornell Agricultural College; primarily these were the *Nature-Study Leaflets* (1896-1904), *Cornell Rural School Leaflets* (1907 to at least 1944), and *Teacher's Leaflets*; other related pamphlet publications were the *Home Nature-Study Course*, *Junior Naturalist Monthly* (1900-1907), and the *Nature-Study Quarterly* (the latter produced in conjunction with the New York Agricultural Experiment Station).

41. Bailey, *Outlook to Nature*, 178-79.

42. Bailey, *Nature-Study Idea*, 5. The second part of the book posed questions that teachers might have about implementing nature study, perhaps gleaned from his discussions with lecture audiences. Bailey was impatient with "abstract theories of pedagogy" and indicated that he worked out his teaching from the point of view of the child and his environment and "let others parallel it with theories of pedagogy." L. H. Bailey to D. H. Roberts in Ypsilanti, February 5, 1908, Bailey Papers, CUA.

43. Meine, *Aldo Leopold: His Life and Work*, 296.

44. Bailey to George P. Brett, January 6, 1898, Bailey Papers, CUA.

45. Typescript of an interview with L. H. Bailey, apparently by E. L. Palmer (n.d.), CUA.

46. New York Department of Education, *First Annual Report, Supplemental Volume* (Albany, 1905), 424. The recommended texts for teachers were Bailey, *Nature-Study Idea*; Hodge, *Nature Study and Life*; and Scott, *Nature Study and the Child*. Also see DeAntoni, "Coming-of-Age in the Industrial State"; and Bailey to W. L. French, 7 December 1906, Bailey Papers, CUA.

47. The *Catalogue of the State Normal and Training School in Cortland for 1906-1907*, for example, indicated that its students devoted five of fifty-two required course hours to "Methods in

Science and Nature Study." See also Corby, "Centralization of Educational Administration in the New York State Normal Schools."

48. Copy of letter from L. H. Bailey to Thomas O. Baker of Brooklyn P.S. 128, October 31, 1905, Bailey Papers, CUA.

49. Olive E. Weston, for the Chicago Women's Club, to Liberty Hyde Bailey, August 21, 1905, Bailey Papers, CUA.

50. Bailey to Macmillan Company, June 11, 1897, Bailey Papers, CUA.

51. Bailey seemed to have recognized the potential conflict of interest even as he sent his publisher lists of farmers who attended Farmers' Institutes, secretaries of local Grange chapters, and coordinators of regional Teachers' Institutes. He asked that his name not be used in advertisements because he wanted "to keep these schools perfectly free of all personal interests." Bailey to Macmillan Company, August 13, 1896, Bailey Papers, CUA. Bailey recruited other authors and he personally revised Asa Gray's 1868 volume *Field, Forest, and Garden Botany*.

52. Colman, *Education and Agriculture*, 129–31. Also see Eva L. Gordon's history, "Cornell Nature-Study Leaflets, 1896–1956," *Cornell Rural School Leaflet* 50 (Fall 1956). Quantities were significant; in 1904 W. F. Humphrey, publisher in Geneva, New York, published twenty thousand leaflets and eighteen thousand bulletins; Humphrey to Bailey, December 1905, Bailey Papers, CUA.

53. Maintaining lists and distributing leaflets were time-consuming tasks, and the administrators debated how to be most effective. A. B. Comstock reported to L. H. Bailey that she distributed 2,000 leaflets each month, 1,750 to people who had recently requested them, and the rest to experiment stations and libraries. Moreover, 700 teachers had responded with their lesson plans. See letter of May 6, 1905, Bailey Papers, CUA. Materials were also sent, when requested, out of state, as to Mary Catherine Thomson who taught in a private school in Charlottesville, Virginia; Thomson to L. H. Bailey, April 14, 1905, Bailey Papers, CUA.

54. See, for example, Alice G. McCloskey to L. H. Bailey, July 7, 1905, Bailey Papers, CUA.

55. The summer school course in the Cornell *Catalogue* for 1899–1900 offered free tuition to New York state teachers who devoted full time to study of three distinct sections, one "on the farm" (taught by Professor Isaac Roberts), another "on insect life" (with two sections, one taught by John and the other by Anna Comstock, here listed as Assistant Professor), and a third "on plant life" (with Professor Bailey), 336–37, held at CUA.

56. Marital partnerships in nature study publications were not uncommon, typically with wives collaborating with their faculty member husbands. Needham wrote several texts on nature study, some of them illustrated by his wife, Anna Taylor Needham; see Needham, *Outdoor Studies: A Reading Book of Nature Studies*. Also see Pycior, Slack and Abir-Am, *Creative Couples in the Sciences*.

57. McCulloch to John Spencer, March 25, 1901, Bailey Papers, CUA.

58. The course included Henry David Thoreau, Gilbert White, and Robert Jefferies. John G. Coulter, *The Dean [Stanley Coulter]*, 105–6. The Coulters lived that summer with the Baileys at Bailiwick, an attractive cottage "high above Cayuga's Waters."

59. Purdue University published a series of twenty-four short *Leaflets on Nature Study*, several by Coulter and one reprinted from Liberty Hyde Bailey, "designed for the use of teachers in the public schools." These were modest, lacked illustrations, and included the usual range of articles from wild and domestic animals, to children's gardens, to conservation as in Lillian Snyder's "Our Friends, the Birds" (Purdue University, n.d.); copies found at the main library of the University of Illinois. Stanley and his brother John M. Coulter were the sons of missionaries to China; Stanley founded the Indiana Academy of Sciences and was interested in forestry and conservation in his later years; see *National Cyclopedia of American Biography*, 33:315.

60. The two-year course earned a certificate, not a degree. L. H. Bailey to Mary Whitson, 25 August 1903, Bailey Papers, CUA.

61. Comstock, *Comstocks of Cornell*, 192-93.

62. Anna Comstock to David Starr Jordan, November 8, 1902; her husband John reported similarly that she had been "preaching about the teaching of Nature Study" from Lake Erie to the eastern end of Long Island, December 12, 1902, both letters in Jordan Papers, Stanford University Archives (SUA).

63. Dorf, *Liberty Hyde Bailey*, 53; Margaret W. Rossiter kindly brought this citation to my attention. Reference to opposition from faculty and trustees is recorded in an interview with Liberty Hyde Bailey, October 21, 1951, CUA.

64. Margaret Rossiter documents and discusses salary disparities and gender-based task assignments in her classic account *Women Scientists in America: Struggles and Strategies to 1940*. Cornell women earned considerably less than men and were paid minimum wages for their summer work, with Comstock making $150 in 1898 and Bailey earning $350. With her lecturer's appointment in 1900, she began to earn $1,200 for the academic year, as did Mary Rogers Miller, whereas Alice McCloskey earned half that amount. See salary discussions in "Proceedings of the Board of Trustees," September 3, 1895, to June 12, 1900, 272 and 394; and "Proceedings of the Board of Trustees," June 20, 1900, to June 13, 1905, 10, 173, 424, and 544, CUA.

65. Liberty Hyde Bailey to Mira Lloyd Dock, November 1, 1899, Dock Papers, Library of Congress. He estimated that he had five or six such women students in the first year. Dock worked around the edges of nature study and by 1907 was hoping to find a position doing editorial work for Bailey; see Bailey's discouraging letter to Dock, January 8, 1907, Bailey Papers, CUA.

66. As a leader in the Chautauqua Horticultural Society, Spencer had been influential in enabling the legislation that appropriated initial funds for the program at Cornell; see Comstock, *Handbook of Nature Study*, xi.

67. Bailey to Messrs. Macmillan and Company, 7 January 1896, Bailey Papers, CUA.

68. Cornell University Agricultural Experiment Station, *Annual Report* (1899): 637-43.

69. Spencer to Bailey, April 21, 1906, Bailey Papers, CUA.

70. L. H. Bailey recalled Spencer as "a very practical man" for applying what he knew to the affairs of life as concretely as he could; typescript of oral interview with by E. L. Palmer (n.d.), Bailey Papers, CUA.

71. Draft letter from John Spencer to Jennie L. Olson, September 15, 1905, Comstock Papers, CUA; this letter is reprinted in Spencer, "Diplomacy of the Good Teacher," *Nature-Study Review* 13 (1917): 22.

72. From the dedication, *Handbook of Nature Study*. Spencer went on half salary in 1909 and gradually retired; Bailey to Spencer, June 22, 1909, Bailey Papers, CUA.

73. Draft of talk to Women's Alliance, Ithaca, April 19, 1905, Cornell/Comstock Press Papers, CUA. The Uncle John's letters were didactic and often repeated sage advice, such as "bending the twig in the direction you would have the tree incline." Memo dated September 19, 1903, Comstock Papers, CUA. Assistants helped with the work; see the obituary, written by editor Anna Comstock, "In Memoriam: Ada E. Georgia," *Nature-Study Review* 17 (February 1921): 94.

74. Copy from lecture notes found in papers of the Bureau of Nature Study and Farmers' Reading Course in Cornell/Comstock Press Papers, CUA.

75. Quoted in Colman, *Education and Agriculture*, 206.

76. Teachers from across the country were involved in this exchange, one with Texas schools and one with sending apple twigs to the New York city schools for nature study programs there; Leigh R. Hunt of Corning, October 30, 1897; Clinton S. Marsh of North Tonawanda, December 6, 1897; Superintendent of Manual Training in New York, February 9, 1898; and others in Spencer Educational Extension Papers, CUA.

77. Ellen E. Doris, "Practice of Nature-Study: What Reformers Imagined and What Teachers Did" (D.Ed. diss., Harvard University, 2002), esp. chap. 5, "'Dear Uncle John:' Teachers, the Junior Naturalists, and Cornell."

78. McCloskey had initiated correspondence with Spencer, including May 6, 1898, and he recognized her potential as a colleague, Spencer Papers, CUA. *Boys and Girls,* printed by Stephens Press in Ithaca, was intended for parents, teachers, and advanced students. Anna Comstock edited its early issues and Martha Van Rensselaer's name is on the masthead of later issues of the journal, which lasted to 1907. First published in 1901, *Boys and Girls* sold for fifty cents for a year's subscription, or five cents per issue, with letters to Uncle John from "nieces" and "nephews," articles, and illustrations; letters reflect distribution well beyond New York State to Iowa and North Carolina. Subscribers were offered discounted subscription rates to other children's magazines. Bailey had not encouraged its continuation but was protective when the Orange Judd Publishing Company challenged it with another children's journal. See Bailey to J. G. Schurman, February 22, 1907, Bailey Papers, CUA.

79. Estimates of enrollment varied, but all agree that at its peak the membership was roughly twenty to thirty thousand and that there were clubs across the country and even in some foreign countries. See Doris, "Practice of Nature-Study," 172. The 4-H movement started in World War I to stimulate the home garden and other war food-conservation efforts. It picked up momentum with the encouragement of the U.S. Department of Agriculture and was given official federal recognition during World War II. See Webb, *Boys' and Girls' 4-H Club Work in the United States;* and Wessel and Wessel, *4-H, An American Idea, 1900-1980: A History of 4-H.*

80. "Cornell University, College of Agriculture," *Nature-Study Review* 2 (April 1906): 129. The sixty credit program concentrated in biological sciences and was intended for "persons who expect to teach nature-study and country life subjects in the public schools."

81. Graduation and alumni details are found in *The Ten Year Book of Cornell University (1898-1908),* 43 and 354, CUA.

82. Rossiter, *Women Scientists in America: Struggles and Strategies to 1940,* 120. The outstanding Van Rensselaer, well known for her leadership in home economics, received her degree from Cornell in 1909 and thus acquired a credential for becoming a more permanent part of the faculty. See Rosa Bouton, Department of Domestic Science, University of Nebraska, to L. H. Bailey, June 13, 1905, Bailey Papers, CUA.

83. Miller left for New Jersey but continued to work on leaflets. Miller to L. H. Bailey, June 20, 1904, Bailey Papers, CUA; and R. H. Halsey to Spencer, December 4, 1897, Spencer Papers, CUA.

84. Bailey, *The Nature-Study Idea,* 68. Julia Ellen Rogers, who had a degree in biology and had taught for a number of years in Iowa, completed an M.A. in agriculture at Cornell in 1902; *Who Was Who in America* (Chicago: Marquis, 1961-68), 4:806.

85. Among those I have found, most were published in very few numbers, with the exception of the ones produced for Hampton Institute, apparently under the guidance of Jean E. Davis; the monthly publications were distributed free to any teacher in a Southern school who requested them until budget cuts in 1904. Other examples are twenty-four Purdue *Nature Study Leaflets,* New Hampshire Agricultural Experiment Station's *Nature Study Leaflets,* and annual *Massachusetts Nature Leaflets.*

86. Myrtilla Avery to L. H. Bailey, May 27, 1905, Bailey Papers, CUA.

87. Bigelow, then editor of *Nature and Science,* who wrote to Bailey on June 12, 1905 [caps in original], Bailey Papers, CUA. The volume proved a useful way to respond to requests for leaflets from outside the state and was reprinted by a private firm, J. B. Lyon Co.; the company declined to publish the *Junior Naturalist Monthly* and the *Home Nature-Study Quarterly.* See Charles M. Winchester to L. H. Bailey, October 17, 1905, Bailey Papers, CUA. Apparently the *Nature Study and Reading Course* compilations were the most frequently requested books by the legislators. See Charles M. Winchester to L. H. Bailey, October 13, 1905, Bailey Papers, CUA; also comments by the clerk of the College of Agriculture to the editor of *Harper's Weekly* (May 22, 1897): 511.

88. Charles A. Weiting to Liberty Hyde Bailey, October 25, 1906, Bailey Papers, CUA.

89. See the report of M. Cook to Spencer, November 15, 1906, Comstock Papers, CUA.

90. L. H. Bailey to W. R. George, January 17 and January 19, 1907, Bailey Papers, CUA.

91. George E. Vincent to L. H. Bailey, March 23, 1905, Bailey Papers, CUA. The Chautauqua had declined in popular attendance near the end of the nineteenth century, but McCloskey noted that her classes nonetheless were well attended. She was so busy teaching five hours each day that she had little time to manage correspondence; she requested a stenographer to help. See letters dated July 7 and 12, 1905, Bailey Papers, CUA.

92. H. L. Drummer, representing the Steuben Nature Study Workers of the Public Schools in planning for their fifth annual Field Day, asked Bailey to deliver an address; letter sent July 13, 1905, Bailey Papers, CUA. Bailey noted Drummer as "one of the effective men in this kind of work" who had accomplished much through his work with local teachers. Bailey to Charles M. Bissell, March 4, 1909, Bailey Papers, CUA.

93. Cited in Colman, *Education and Agriculture*, 131.

94. Rhode Island College instituted a two-week summer course in nature study and issued a monthly pamphlet *Nature Guard*, free to boys and girls. Eschenbacher, *A History of Land Grant Education in Rhode Island*, 79. A nearly complete set of the first fifty issues, addressed to "the boys and girls of New England" with encouragement to form a "merry band," are housed at Ohio State University library. New Hampshire's College of Agriculture issued materials, as did the state of Maine; these are listed in the bibliography of Jewell, *Agricultural Education, Including Nature Study and School Gardens*.

95. L. H. Bailey to W. L. French in Peru, Nebraska, December 7, 1906, Bailey Papers, CUA.

96. Undated memo [December 1906], Bailey Papers, CUA.

97. L. H. Bailey to Elmer Ellsworth Brown, U.S. Commissioner of Education, December 18, 1906; to Stanley Coulter, December 24, 1906; and to Barney Whitney, January 7, 1907, Bailey Papers, CUA.

98. DeAntoni, "Coming-of-Age in the Industrial State," 91.

99. Ellwood P. Cubberley of Stanford University led this movement, as detailed in DeAntoni, "Coming-of-Age in the Industrial State."

100. DeAntoni, "Coming-of-Age in the Industrial State," 91-93.

101. Eva Gordon, "Cornell Nature-Study Leaflets."

102. See unsigned letter to P. S. Aldrich, February 19, 1909, Bailey Papers, CUA.

103. Writing for the Farmers' Institute Board of South Dakota, Bertha Dahl Laws asked if she might get copies of the Cornell Reading Course for Farmers' Wives for her work in Minnesota and South Dakota. Laws to L. H. Bailey, August 26, 1905, Bailey Papers, CUA. Anna W. Blackmer, supervisor of the practice school at Whitewater, Wisconsin, asked if the Cornell leaflets could be distributed beyond the state and if the Junior Naturalist Clubs were open to classes beyond New York. Blackmer to L. H. Bailey, May 7, 1906, Bailey Papers, CUA. The Superintendent of the Farmers Institute of the State of Minnesota made it clear to Martha Van Rensselaer that the institute was willing to purchase bulletins to be distributed in Minnesota if that were possible; letter from O. C. Gregg to Van Rensselaer, April 11, 1906, Bailey Papers, CUA.

104. L. H. Bailey to Julia E. Rogers, January 22, 1909; and to David L. Roberts, June 17, 1910, Bailey Papers, CUA.

105. F. L. Stevens to L. H. Bailey, March 20, 1905, Bailey Papers, CUA; also see Rufus W. Stimson, President of Connecticut Agricultural College, to L. H. Bailey, April 26, 1906, Bailey Papers, CUA.

106. This request came from J. S. Jarman, president of the State Female Normal School at Farmville, Virginia, to Bailey, April 4, 1904, Bailey Papers, CUA.

107. Charles L. Spain to L. H. Bailey, May 5, 1906, Bailey Papers, CUA.

108. See, for example, W. S. Dearmont of Missouri State Normal School in Cape Girardeau, Missouri, November 14, 1906, Bailey Papers, CUA.

109. Helen C. Bennett to L. H. Bailey, November 24, 1906, Bailey Papers, CUA.

110. G. A. Crosthwaite, teaching agronomy in Moscow, Idaho, asked Bailey for help, since his director had made it clear that he would need to teach nature study; Crosthwaite to Bailey, February 20, 1904, Bailey Papers, CUA.

111. Munson had Ph.B. in biology from Yale and then studied at Chicago before taking the job at Washington State Normal School. He suggested that biological study helped people deal with the stresses and strains of relationships both to others and to the physical universe by making behavior more comprehensible. His also wrote a textbook, *Education through Nature Study*.

112. See, for example, Ruth Jackson to Bailey, April 24, 1904, Bailey Papers, CUA.

113. Bailey used the phrase about Anna Comstock in a letter to L. S. Thomas, November 28, 1905, College of Agriculture and Agriculture Experiment Stations Papers, CUA.

114. Bailey attended early Lake Placid conferences to learn more about the direction of the movement; see Rose, Stocks, and Whittier, *Growing College: Home Economics at Cornell University*, 18-44. When Helen Louise Johnson applied for a position, she noted that even the earliest course had "gained an audience for you and interest all over the country." Johnson to L. H. Bailey, May 15, 1906, Bailey Papers, CUA. Also see Rensselaer's *New York Times* obituary (May 27, 1932), 21.

115. Somewhat different views of the Country Life Commission are offered in David B. Danbom, *Resisted Revolution*, and Bowen, *Country Life Movement in America*. Its final report was introduced by Roosevelt in the Commission on Country Life, *Report* (U.S. Senate Document 705, 60th Congress, 2d session, 1909); also published in New York by Sturgis and Walton Co., 1911.

116. See Bailey's own outlook in his *Country Life Movement in the United States*.

117. Harold W. Foght, "Country School," in *Annals of the American Academy of Political and Social Science* 40 (March 1912): 151.

118. Contemporaries identified the problem but had a difficult time knowing how to regenerate a "new rural order"; see Carney, *Country Life and the Country School*, 327.

119. Scott to Spencer, August 11, 1898, Spencer Papers, CUA.

120. On Bessey's career, see Overfield, *Science with Practice*. Bessey occasionally rejoined former colleagues at the Minnesota Seaside School at Vancouver Island near Seattle, engaging in botanical research as well as graduate teaching.

121. A. W. Wood to Bessey, July 7, 1893, Bessey Papers, University of Nebraska Archives hereafter UNA), reported that the Colorado Springs course attracted enthusiastic teachers from eighteen to forty years of age, with three of the dozen being men.

122. Bessey to C. H. Robison (at the State Normal School in Montclair, Illinois), February 27, 1908, Bessey Papers, UNA.

123. A pamphlet simply entitled *Public Schools: Nature Study* (Lincoln, Nebraska, 1893) presents topics organized by season and by grade that teachers might use. It makes reference to both Wilbur Jackman and Arthur C. Boyden; there is no identified author but thanks are given to colleagues Professors Bruner and Henry B. Ward (copy at the University of Illinois library). Robert H. Wolcott, professor of zoology, also suggested that "the scientific work in the University is given from a nature-study point of view" in "Aim and Method in Nature Study," 24-25.

124. Undated memo [1907] to the Chancellor and Board of Trustees, Bessey Papers, UNA.

125. T. L. Lyon, October 28, 1904, Bailey Papers, CUA.

126. See, for examples, letters to Bessey from E. S. Rouse of Plattsmouth City, October 20, 1902; and from W. L. Stephens of Beatrice, April 1, 1903, Bessey Papers, UNA; teachers included Emily Burton from Stoddard, Nebraska, Bessey Papers, UNA. This activity slowed when he became dean and then interim president of the university.

127. Willet M. Hayes to C. E. Bessey, February 24, 1904, Bessey Papers, UNA.

128. C. E. Bessey to M. A. Bigelow, March 19, 1904, Bessey Papers, UNA.

129. Marshall subsequently went to Rockford College in Illinois where she taught nature study, continued her research, and helped establish a nearby nature preserve. See "Ruth Marshall, 1869-1955," *Transactions of the American Microscopical Society* 75 (January 1956): 144.

130. University of Nebraska *Bulletin* (1901-1902): 286; (1909-1911): 200, UNA.

131. L. H. Bailey to C. E. Bessey, May 23, 1901, Bessey Papers, UNA. Bailey accepted the criticisms with good grace and said the test would be in the "free field of public criticism and opinion." See undated letter [1907] to the Chancellor and Board of Regents signed by Bessey, UNA; Bessey's later attitude is also reflected in the assessment of nature study in Overfield, *Science with Practice*, 62.

132. The cordial relationship is documented in intermittent correspondence between the Comstocks and Jordan extending over the course of their professional careers and beyond, with letters in their private papers at both Cornell and Stanford universities.

133. The Comstocks had offered high praise of Jordan when they met with the Stanfords at Cornell just before Jordan was offered the presidency of Stanford University. Kellogg and John Comstock collaborated on a textbook on insects. On Kellogg, see Largent, "These Are Times of Scientific Activism."

134. Announced in the Stanford *Alumnus* 1 (1899): 100, SUA.

135. W. J. V. Osterhaut to Bailey, November 22, 1905, Bailey Papers, CUA.

136. See, for example, his account of "Nature Study" published in the *School Report of Oakland*, 6-10.

137. The Stanford *Annual Registers* for this period provide detail on the summer Hopkins Seaside Laboratory as well as the courses offered. Hopkins Marine Research Station Papers, SUA.

138. Isabel McCracken to Jenkins, December 17, 1899, Jenkins Papers, SUA. McCracken had put together apparatus for physics study, including expansion of liquids and solids; chemistry study on making gases; and electrical apparatus; as well as soils, minerals, and germinating seeds.

139. Jenkins, *Lessons in Nature Study*.

140. For a summary of activities, see B. M. Davis, "Factors and Influences," 257-65.

141. See J. W. Linscott of Santa Cruz to Elmer Ellsworth Brown, December 16, 1897. Brown Papers, University of California at Berkeley, Bancroft Library. These papers are primarily documentation of the projects in the 1890s to engage teachers and write reports on the findings of the committee. Brown's contribution to nature study is indicated in Davis, "Factors and Influences," 262-65.

142. See State Normal School of Los Angeles, *Catalogue*, annual volumes from 1895 to 1919; quotation from 1906-7: 29. In 1915, Miller became head of the department of science and the term nature study disappeared from the curriculum; nature study projects, however, did not.

143. Sherwood, "Children of the Land," 891-99.

144. Greene, "Macdonald Robertson Movement"; and the Canadian program is discussed in Kohlstedt, "Nature Study in North America and Australasia," 439-54. Insight into student work is provided in a small cache of student papers of Eveline Hocking at the University of Guelph, Guelph, Ontario.

145. Brittain, *Teachers' Manual of Nature Lessons*; and Lochhead, "Canadian Department," 17.

146. Greene, "Macdonald Robertson Movement," 123. These special instructors were then given ambitious assignments during an experimental period of teaching nature study and establishing school gardens in their home provinces.

147. Crawford, *Guide to Nature-Study for the Use of Teachers*.

148. Carter, "Ryerson, Hodgins, and Boyle," 127; and MacKay, "Nature Study in the Schools of Nova Scotia," 209-12.

149. The file on teacher Effie Bready, for example, suggests that she came from Ottawa in 1898, where she had taken classes in nature study, to North Battleford, Saskatchewan, Saskatchewan Archives Board, Saskatoon (hereafter SABS). Also see L. H. Bailey's advice to A. H. Gibbard in Grenfall, Saskatchewan, February 3, 1908, Bailey Papers, CUA, urging the agricultural society to locate a program in a provincial school or college. Nature study was part of the curriculum in Alberta in 1922, alongside hygiene and physiology; described in Sheehan, "WCTU and Educational Strategies on the Canadian Prairie," 105.

150. Letter from the Director of the School of Agriculture to W. M. Martin, Minister of Education, July 3, 1917; also see D. P. McCall to W. Scott, November 18, 1917, and February 5, 1918, SABS. The Teachers Training Examination Qualifying Questions for 1912, also in SABS educational files, referenced a number of nature study books and contained questions such as "How does grass differ from other plants?"

151. Campbell, *Reflections of Light*, 24-25. She also noted produce from these school gardens was viewed as contributions to the war effort during World War I.

152. Nature study was featured in the New York City bound volumes put on display in Paris in 1900 and in St. Louis in 1904; these are now held at Teachers College Library, Columbia University (hereafter TC); also see *Journal of the New York Board of Education* (1906): 1236-37. Dora Otis Mitchell, "History of Nature Study," notes that the Bridgewater Normal School model school exhibit under Arthur C. Boyden won the grand prize in St. Louis.

153. Allen, *Naturalist in Britain*, 203; there is no mention of American influence.

154. Geddes, *Introductory Course on Nature Study* and his *City Development: A Study of Parks, Gardens, and Culture-Institutes*.

155. Aberdeen, and indeed all of Scotland, found in nature a source of local and regional identity. See Withers and Finnegan, "Natural History Societies, Fieldwork, and Local Knowledge," 334-53.

156. See Thompson's introduction to Rennie, *Aims and Methods of Nature Study*, xi-xii, 203.

157. A high point may well have been an international meeting recorded in *Official Report of the Nature-Study Exhibition and Conference held in the Royal Botanic Society's Gardens, Regents Park, London*. One of the more popular British texts, owned by Liberty Hyde Bailey, was Ernest Stenhouse's *Introduction to Nature-Study*.

158. A rather unsympathetic interpretation of the British movement, which presented nature study as a conservative, top-down program, appears in Jenkins, "Science, Sentimentalism, or Social Control," 33-43.

159. Frank McMurry and Henry E. Armstrong, "Some Recent Criticism of Nature-Study," 22-26.

160. Stenhouse, *Introduction to Nature-Study*. *Nature Study, or the Naturalists' Journal* was published by the British Field Club of Huddelsfield in 1903 and 1904; and three later issues from the same location were entitled *New Nature Study*, edited by S. L. Moseley and F. O. Moseley. Government support was also evident in T. S. Dymond, *Suggestions for Rural Education*.

161. See the recommendations in [Irish Free State Department of Education], *Teaching of Rural Science and Nature Study in Primary Schools: Regulations and Explanatory Notes for Teachers based on the Syllabuses in Rural Science and Nature Study contained in the Report and Programme presented by the National Programme Conference and adopted by the Department of Education* (Dublin, 1927).

162. Mrs. M. McCrae to L. H. Bailey, March 22, 1905, Bailey Papers, CUA.

163. H. J. Bhabha to L. H. Bailey, December 19, 1908, Bailey Papers, CUA.

164. John A. Leach sent a description of the Australian program to Bailey, who offered to publish parts of it. See Bailey to Leach, January 20, 1908; and reply March 28, 1908, Bailey Papers, CUA. On New Zealand programs, see Ross, "'Two Lucy's'"; and discussion of teachers in Carey, "Departing from Their Sphere"; also see Kohlstedt, "Nature Study in North America and Australasia."

165. Comstock, "Nature-study and Agriculture," 143-44.

166. Dietrich Lange, *Handbook of Nature Study*, v. Lange taught nature study in the St. Paul public schools, and his text included illustrations by Josephine Tilden and Henrietta Fox. Conscious of his own geographic location, Lange pointed out that his material would be most useful for the region from the Atlantic Coast to the Rocky Mountains and from the Canadian provinces to southern Virginia and Kentucky. He also referred to German texts by J. Kiessling and others.

167. Secretary of Agriculture, *Annual Report* (1888), 10.

168. Yale University's loss of the Connecticut Agricultural Experiment Station when it was transferred—only after a Supreme Court fight—to the Connecticut Agricultural College at Storrs was indicative of the fact that funds and status were at stake in this expansion of support for agriculture. Stemmons, *Connecticut Agricultural College*, 67–77.

169. See, for example, the four-page pamphlet by Dick Jay Crosby of the U.S. Department of Agriculture, Office of Experiment Stations, *Circular* 52 (1904), entitled "A Few Good Books and Bulletins on Nature Study, School Gardening, and Elementary Agriculture for the Common Schools." Also see Wayne E. Fullerton, "Making Better Farmers," 154–68.

170. Cited in Colman, *Education and Agriculture*, 149. A number of textbooks picked up this perspective, including Freeman, *Nature Teaching Based upon the General Principles of Agriculture*.

171. A. B. Comstock to L. H. Bailey, August 28, 1905; and September 11, 1904, Bailey Papers, CUA.

172. L. H. Bailey to D. B. Johnson of the Winthrop Normal and Industrial College in Rock Hill, South Carolina, March 16, 1907; and to C. W. Whitney, School Commissioner in Brockton, New York, March 28, 1907, both in Bailey Papers, CUA.

173. L. H. Bailey to M. A. Bigelow, February 1, 1908, Bailey Papers, CUA.

174. Quoted in a special issue on the relationship between agriculture and nature study, edited by Anna Botsford Comstock, in *Home Nature-Study Course* (October/November 1905), 2.

175. James M. Robertson, "Preface," in John Brittain's *Elementary Agriculture and Nature Study*.

176. Schmidt, *Nature Study and Agriculture*, iii. From his point of view, "nature has the same relation to this course that the laboratory has to the study of physics and chemistry." This text balanced attention to informal environmental subjects and more practical ones including field crops, soils, weeds, and insect pests.

177. McCloskey to Bailey, November 25, 1905, Bailey Papers, CUA.

178. In 1908 Bailey indicated that his program published seven thousand regular leaflets and thirty-seven thousand children's supplementary leaflets each month and that the latter were correlated in subject matter to the state nature study syllabus. L. H. Bailey to L. Reber, March 17, 1908, Bailey Papers, CUA. In September 1918, after Bailey had retired, the leaflets provided subject matter "in natural history, agriculture, and home making."

179. On the international movement, see James M. Robertson, "Preface," in John Brittain's texts, *Elementary Agriculture and Nature Study* and *Nature Study and Agriculture*. Also see James Ralph Jewell to L. H. Bailey, October 31, 1905, Bailey Papers, CUA. Bailey's term persisted, as in Kern's *Outline of Course of Instruction in Agricultural Nature Study*.

180. L. H. Bailey to C. R. Davis, March 6, 1908, Bailey Papers, CUA. There was a debate about whether training teachers in agriculture should occur in agricultural colleges or normal schools, which grew especially heated about 1905; a bibliography on agricultural education is attached to a letter from L. A. Kalbach to L. H. Bailey, March 19, 1908, Bailey Papers, CUA.

181. Neely, *Agricultural Fair*, 135, argued that nature study had grown up with and supported the agricultural education movement.

182. Mayberry, *Century of Agricultural Extension*; and Anderson, "Hampton Model of Normal School Industrial Training, 1868–1900," in the edited volume by Walter Feinberg and Henry Rosemont, Jr., *New Perspectives on Black Education*.

183. The "special case" of the South is discussed in Tyack and Hansot, *Managers of Virtue*, esp. 88–89. They compare the South to the North and West and find significant disparities in 1900, with the average expenditure per year per pupil $1.81 versus $4.5; that ratio worsened during the first three decades of the twentieth century for African American students. Southern teachers averaged only $25 a month in salary, and their pupils attended an average of three months each year. A teacher at Calhoun's Colored School reported that some white planters were

opposed to teaching African American children and that having schools run by women seemed the only way to keep the school from getting burned out; see. A. M. Troyer to C. E. Bessey, January 19, 1898; Bessey found this a "strange yet very desirable work" in his response on January 22, 1898, Bessey Papers, UNA. On the political and economic dimensions of the segregated schools, see Park and Ginsberg, *Southern Cities, Southern Schools*.

184. Peabody Educational Fund, *Proceedings of the Board of Trustees for 1894*, 83–84.

185. Russell reported on his trip to Hampton president H. B. Fressell, draft letter, November 12, 1900, Russell Papers, TC.

186. Russell sent Richard E. Dodge, teacher of geography and nature study at Columbia, to a conference on education in the South held at Richmond in April 1903; Russell Papers, TC. By 1900, the Hampton Institute *Catalogue* (1900–1901): 87, mentioned classes in nature study; summer school classes in 1905 were taught by Annie M. Goding, principal of the Normal School in Washington, DC. I used the nearly complete sets at the University of Texas at Austin, Yale University, and the Library of Congress. Also *Nature-Study Review* 1 (June 1905): 43; this account noted that *Hampton Nature-Study Leaflets* could no longer be distributed free.

187. Carver, "Suggestions for Progressive and Correlative Nature Study" (Tuskegee, 1902).

188. T. J. Jackson to Spencer, May 30, 1898, Spencer Extension Education Papers; and R. C. Bedford to L. H. Bailey, Bailey Papers, both CUA. Jackson had earlier written (March 10, 1898) that the "average colored farmer" could not read the leaflets; he offered to have an exchange of letters between children in his school and another.

189. See the inquiry of Samuel H. Bishop of the American Church Institute for Negroes to L. H. Bailey, March 12, 1907; and comments from Bailey to Bishop on March 8 and 16, 1907, Bailey Papers, CUA. Race was a conscious issue. Bailey wrote of N. R. Shields, "a light colored student" who was a graduate of Cornell and engaged at the Agricultural and Normal University at Langston, Oklahoma, and C. C. Poindexter, "a very superior Negro, very light in color," who had a degree from the University of Ohio. Mentors, including George Washington Carver, were engaged as well, and he discussed the progress of John E. Smith at Cornell and the upcoming arrival of a Miss Queen who was preparing to teach at Tuskegee Normal and Industrial Institute; letter to L. H. Bailey, June 15, 1908, Bailey Papers, CUA. On the organizational efforts, see L. H. Bailey to Emma Venable, March 17, 1908, Bailey Papers, CUA.

190. Perkins, "History of Blacks in Teaching," in Donald Warren, ed., *American Teachers: Histories of a Profession at Work*, esp. 350–58. On another summer program for minority children, apparently taught by white teachers affiliated with the Carrol Robbins Training School in Trenton, New Jersey, see Woodward, "Cape May Summer School," 99–102.

191. Miller, *Exercises with Plants and Animals for Southern Rural Schools*.

192. Lazerson, *Origins of the Urban School*, 245.

193. The agricultural education elements of the curriculum of students is suggested in Riney, *Rapid City Indian School*, chap. 3 and epilogue; Lorrawairra, *They Called It Prairie Light*, especially chap. 4; Adams, *Education for Extinction*; and Lindsey, *Indians at Hampton Institute*, 200. Much of the literature on Indian education has concentrated on the boarding schools rather than day schools on reservations, perhaps because boarding schools were mandated to maintain certain records and provide regular reports to Washington; both mission schools and the Bureau of Indian Affairs (hereafter BIA) boarding schools emphasized discipline, a strong set of skills, and a work ethic as the goals of its education program.

194. Quoted in the report by Bengaugh, *Learning How to Do and Learning by Doing*, 80. Goodrich also produced *The First Book of Farming* for use by students.

195. See Brazzel, "Brick without Straw," 41–42.

196. For a discussion of parents' expectations of education at Haskell in Kansas and Flandreau in South Dakota, see Child, *Boarding School Seasons*, chap. 6.

197. Most attention was paid to colleges and universities in the late nineteenth century. The

Commissioner of Education, *Report for 1897-1898*, 2:1578-1601, listed nature study leaflets available for rural schools. This competition is also discussed in Kohlstedt, "'A Better Crop of Boys and Girls,'" 58-93.

198. See, for example, the report on Hampton made after a visit by Bengaugh, *Learning How To Do and Learning by Doing*, 68-80. Nature study seems to have been subordinated to the goal of teaching boys and girls basic agricultural skills.

199. Reel spent six weeks at the Phoenix Indian School introducing teachers to her program, and some elements seemed to flourish under the leadership of the relatively well-liked director Charles Goodman; see Tennert, *Phoenix Indian School*, 86-87, and his "Selling Indian Education at World's Fairs and Expositions." Also see "Nature-study and Gardening for Indian Schools," *Nature-Study Review* 2 (April 1906): 141-43. The most comprehensive survey of schools and practice in the period was completed as a dissertation at Clark University and published by the Bureau of Education, namely Jewell's *Agricultural Education*.

200. *Circular*, BIA, Department of the Interior, March 2, 1902, United States National Archives (hereafter USNA).

201. *Circular*, BIA, Department of the Interior, March 12 and 19, 1904, USNA.

202. For a discussion of the parallel race and class characteristics of urban African American education somewhat earlier, see Nina Lerman, "Uses of Useful Knowledge: Science, Technology, and Social Boundaries in an Industrializing City."

203. Pamphlet entitled *Woonspe Wankantu, Santee Normal Training School, 1897-1898* (Santee Agency, Nebraska [1898]), 18-21.

204. F. W. Parker, "Child," 480. The phrase is underlined for emphasis in Parker's original text, although he credits Froebel with the idea; it recurs throughout the movement, including Dietrich Lange in his "Plea for Nature Study in the Common Schools," typescript address for the Minnesota Educational Association, December 27, 1897, Lange Papers, Minnesota Historical Society.

205. On rural school activism, see Nordin, *Rich Harvest: A History of the Grange*, chap. 3. Women's groups of parents and teachers in rural upstate New York invited speakers and read books on the nature study curriculum. See Martha Van Rensselaer to L. H. Bailey, April 19, 1904, Bailey Papers, CUA.

206. Fuller, *Old Country School*, 222.

207. This ambivalence is discussed in Evelyn Dewey's account of Marie Turner Harvey's leadership of a rural school near Kirksville, Missouri, *New Schools for Old*, esp. chap. 10 titled "Agriculture and the Curriculum."

208. Martha Colson Lippincott to Bailey, April 4, 1904, Bailey Papers, CUA.

209. Mabel Campbell of Cohoes to Spencer, July 15, 1898, Spencer Extension Education Papers, CUA.

210. *Manual of the Public Schools* (Orange County), 1903-4, University of California at Los Angeles Library.

CHAPTER FIVE

1. Blair, "Nature and Other Subjects of Instruction," 259-60.

2. Bigelow, *Spirit of Nature Study*, 99. Whether the story is true, the dilemma is clear for those intent on catching the "spirit" of nature study.

3. See Lucy Wilson, *Nature Study in Elementary Schools: A Manual*. Wilson, who had a Ph.D., was then head of biology at the Philadelphia Normal School for Girls and in charge of nature study in its School of Observation and Practice.

4. Sequel, *Curriculum Field*.

5. Tebbel, *History of Book Publishing*, 600.

6. While Ada Watterson was in charge of "Notes on Recent Pamphlets and Articles," content

included reviews of books and leaflets, as well as articles published in journals. See, for example, Watterson, "Guide to Periodical Literature," *Nature-Study Review* 1 (March 1905): 90-96.

7. L. C. Miall, "Ready-Made Lessons in Nature Study," 101-2. As publisher of *House, Garden and Field* (New York: Longmans, 1904), Miall had provided short essays on nature studies for teachers.

8. Tolley, *Science Education of American Girls*, 20-33.

9. Andrews, *Seven Little Sisters Who Lived on the Round Ball*, was recommended as useful for schools by Holtz, *Nature-Study*, 290; the 1888 edition has an appended "Memorial of Miss Jane Andrews." Her book, *The Stories Mother Nature Told Her Children*, was also popular.

10. Susan Schulten, *Geographical Imagination in America*, uses this phrase to title chap. 5.

11. On geography's integration into nature study, see Phillips, *Modern Methods and the Elementary Curriculum*, 187; also Tolley, *Science Education of American Girls*, 20-33.

12. Scott, *Nature Study and the Child*, 283. He recommended *Heimat-kunde*, which he defined as home geography, as the place to begin with young children.

13. Payne, *Geographical Nature Studies*, 3. Minton, "History of the Nature-Study Movement" (105) credits Payne for introducing the term "nature study" in 1889. Payne used both "nature studies" and "nature study" in his texts and articles throughout the 1890s. Payne's *Lessons in Nature Study around My School* follows a seasonal approach like that of Jackman.

14. Princeton professor Arnold Guyot wrote an *Elementary Geography for Primary Classes*, which started its discussion in the pupil's own backyard as a way of learning basic principles such as distance and direction; this approach was intended to help students to "discover the natural order" of the physical world through their own investigation.

15. The American Book Company (hereafter ABC) failed as a trust of textbook publishers. Tebbel, *History of Book Publishing: The Expansion of the Industry*, 560-65; and Lawler, *Seventy Years of Textbook Publishing*, 209-12, 219-21, and 227-46. In 1891 copyright laws were tightened to curtail pirating of British and European books and that also proved a stimulus to American authors.

16. Appleton's *Scientific Library* series, started in 1871, included sixty volumes based on Edward Youmans's list of important books; it was later sponsored by the Boy Scouts. Tebbel, *History of Book Publishing*, 303. Doubleday had a Nature Library series coordinated by ornithologist Neltje Blanchan. Ginn and Company published Carolyn D. Wood's *Animals: A Nature Study Textbook*; she had been a teacher and nature supervisor in New Bedford, Massachusetts, before moving to the State Normal School in Valley City, North Dakota.

17. Cummings, *Nature Study by Grades*, 3-4. This textbook was unattractive and lacked illustrations. An instructor at Utah Normal School, Cummings claimed that his distribution of outlines at a National Education Association (hereafter NEA) meeting in Los Angeles in 1900 had led to 1,400 inquiries and prompted him to write the book.

18. Carter, *Nature Study with Common Things: An Elementary Laboratory Manual*. Just a year later, ABC published a very similar volume that used outdoor nature items like a mushroom and wasp's nest, written by Frank Overton, assisted by Mary E. Hill, as *Nature Study: A Pupil's Text-book*. Anna Comstock, who had known Hill as an assistant in biology at Cornell, wrote the introduction.

19. Carter, *Nature Study with Common Things*, 7, is quite explicit about this method being appropriate for fourth, fifth, and sixth graders, asserting that it would ultimately be used by the children themselves to examine other objects not in the book.

20. Coulter to Jordan, April 11, 1900, Jordan Papers, Stanford University Archives.

21. *The Course of Study for Elementary Schools for New York City* (May 1903) published by Appleton listed only the books relating to nature study and other subjects which it published; available at Teachers College, Columbia University.

22. Tebbel, *History of Book Publishing*, 560-65.

23. There was reportedly widespread corruption in the marketing of textbooks, with rumors

of bribery, collusion, and even the hiring of alluring women as accomplices to blackmail school officials into selecting certain wares, according to muckrakers Upton Sinclair and Lincoln Steffens, cited in Tyack, *One Best System*, 95.

24. Coulter, Coulter, and Patterson, *Practical Nature Study and Elementary Agriculture*, vii.

25. Advertised in back pages of Wilson, *How to Study Nature in the Schools*.

26. See advertising section at the rear of Wilson, *How to Study Nature in the Schools*.

27. These materials were typically advertised in *Nature-Study Review* and are occasionally found in archival or special collections.

28. Cady to Joseph Grinnell, February 25, 1925, Museum of Vertebrate Zoology, University of California at Berkeley.

29. Anna Comstock's own textbook on insects was *The Ways of the Six-Footed*; she and John Comstock collaborated on *How to Know the Butterflies, A Manual for the Study of Insects*, and *Insect Life: An Introduction to Nature-Study*; these three books all went through multiple editions.

30. Elliot Downing prefaced *Field and Laboratory Guide to Biological Nature Study* with the observation that while he was dealing with plants and animals of most interest to students, his text also used "the scientific method and accumulated knowledge so important in modern life."

31. There has been no systematic compilation of the published and unpublished autobiographies of scientific memoirs of naturalists, but the self-consciousness of many nineteenth-century naturalists is striking. For the Burroughs's inspiration see A. J. McClatchlin to Charles Bessey, January 19, 1903, Bessey Papers, University of Nebraska Archives.

32. Alice Hall Walter to Ira S. Wile, August 8, 1913, Wile Papers, Rush Rhees Library, University of Rochester (hereafter URSC).

33. Davis, *Educational Periodicals during the Nineteenth Century*, 59.

34. Jackman, *Nature Study for the Common Schools*, 2-3.

35. Jackman, *Nature Study for the Common Schools*, iii.

36. A number of the other early texts also followed this plan, including Arthur Boyden's *Nature Study by Months*.

37. Jackman, *Nature Study and Related Subjects*, 11.

38. Ibid., 49. Such statements proved an irritant to some scientists who wanted children to learn what they believed was important about a particular object, but Jackman's goal was to encourage each child's accurate observation of an object's characteristics in ways that could subsequently be used to elaborate scientific principles.

39. Jackman, *Nature Study for the Common Schools*.

40. See the *Bibliography of Science Teaching* (1911): 22.

41. Scott, *One Hundred Lessons in Nature Study Around My School* and *Nature Study and the Child*.

42. Scott, *Nature Study and the Child*, 94.

43. Ibid., 236-64.

44. Bigelow appointed his colleague John Woodhull to be in charge of physical nature study, and there were intermittent articles on physics, astronomy, chemistry, and related subjects, especially in the first five or so years.

45. Tolley, *Science Education of American Girls*.

46. Underhill, *Origin and Development of Elementary-School Science*, 200-207. These so-called select committees were often subcommittees of school boards, sometimes supplemented by parent or administrative members.

47. Woodhull, "Physical Nature Study," 18-19.

48. John W. Cook to the editors of *Inter-State School Review*, Danville, Illinois; note in Charles A. McMurry Papers, Illinois State University at Normal (hereafter ISUN).

49. Although Rein thought of himself as a follower of Herbart, historians of education have argued that the earlier educator's emphasis on ideas made him very different from educational

psychologists who concentrated on behavior at the end of the century. In *Herbart and Herbartianism*, Dunkel notes the short-lived visibility of the movement but points out that key texts by DeGarmo and others were kept in print until the 1930s (278).

50. Frank McMurry received a Ph.D. in Germany, subsequently teaching at New York State Normal School at Buffalo before going to Teachers College, Columbia University. See Ohles, ed., *Biographical Dictionary*, 2:844–45.

51. Charles A. McMurry Papers, ISUN. The single best contextual account of the philosophy is in Cruikshank, "Rise and Fall of American Herbartianism." Illinois Normal had a strong science faculty and close ties to Bridgewater Normal through Richard Edwards, who served as the Illinois Normal's school president during its formative years from 1861 to 1873; Kuslan, "Rensselaer and Bridgewater," 64–68.

52. Lida McMurry, *Nature Study Lessons for Primary Grades*, xi.

53. Charles DeGarmo's comment is in a letter written for the fiftieth anniversary of McMurry's years of educational service, February 1, 1927. Another writer, Joseph B. Richey, February 12, 1927, commented that the debate between the McMurrys and Harris at a NEA meeting in Cleveland in 1896 was "the most spirited discussion I have ever heard." McMurry Papers, Vanderbilt University Archives (hereafter VUA).

54. DeGarmo, "Ethical Training in the Public Schools." Jackman alluded to the moral and civic values that might be enhanced by nature study, but DeGarmo elaborated this expectation for other aspects of the curriculum.

55. These examples are given by Charles A. Allen, February 15, 1927, McMurry Papers, VUA. Other letters attest to McMurry's influence in the South, especially Mississippi and Alabama in the 1910s and 1920s.

56. Undated and unsigned short presentation about McMurry in McMurry Papers, VUA. Also see Robert A. Lovely, "Mastering Nature's Harmony: Stephen Forbes and the Roots of American Ecology."

57. There is a considerable literature on the formation of the field of ecology at the University of Chicago, which includes Joel Hagen, *An Entangled Bank*; Mitman, *The State of Nature*; and Cittadino, "A 'Marvelous Cosmopolitan Preserve.'"

58. The National Herbart Society's *Second Annual Yearbook* (1896) lists the Executive Council, which had considerable overlap with early nature study advocates: Charles DeGarmo (Swarthmore), President; Charles H. McMurry (at University of Chicago for the year), Secretary; as well as Nicholas Murray Butler (Columbia), John Dewey (Chicago), Wilbur Jackman (Cook County Normal), Elmer E. Brown (University of California, Berkeley), Frank M. McMurry (State Normal at Buffalo), Levi Seeley (State Normal School, Trenton), and C. C. Van Liew (Illinois State Normal). Butler and Dewey do not seem to have stayed involved as the movement faded. The Society later became the National Society for the Study of Education.

59. Both quotations are from Harper, *Development of the Teachers College*, 204–5. McMurry could, in fact, be quite critical of Herbart, finding the German educator overly abstract, metaphysical, and even obscure.

60. This was the summary of a faculty meeting in 1894, cited in Harper, *Development of the Teachers College*, 219.

61. Cited in Harper, *Development of the Teachers College*, 233.

62. A long, undated autobiographical statement from Frank McMurry was submitted on the occasion of his brother's fiftieth anniversary in education. McMurry Papers, VUA.

63. The Herbartian influence is evident in Lida Brown McMurry's *Correlation of Studies with the Interests of the Child*, a reprint from the second edition of the *First Yearbook of the National Herbart Society*.

64. Lida Brown McMurry had a degree from Illinois Normal School at Bloomington in 1874 and eventually became a training teacher there; she joined the migration of faculty members who were enticed by President John Cook to move to Northern Illinois State Teachers Univer-

sity in 1900. She had two sons, but apparently had separated from her husband by the time she moved to DeKalb. See Lida Brown McMurry Papers, NISU, and *Who Was Who in America*, 1943–1950 (Marquis, 1950), 2:364.

65. McMurry, *Nature Study Lessons for Primary Grades*, 167–71. The text provides many questions that were to help the teachers direct the attention of pupils and engage them in conversation about the particular animals and plants that they introduced to their classes.

66. Ibid., ix-x. Charles McMurry went on to explain that "children should gain a positive enrichment of knowledge and observation and should give expression to a definite fund of ideas and experience. In other words, the lessons should be fruitful in ideas and in the power to express them."

67. Wilson, who had taught school and then completed a Ph.D. at the University of Pennsylvania, was active in promoting nature study and home economics before turning her attention to anthropology. Some data on Wilson's later career is sketched in Mathien, "Lucy L. W. Wilson. Ph.D.," 135–45.

68. Cited in Wilson, *Handbook of Domestic Science*; Ellen H. Richards, noted for her leadership in the early home economic movement, wrote the preface.

69. This approach was common, as in Warren, *From September to June with Nature*; and Wright, *Gray Lady and the Birds*.

70. Wilson, *Nature Study in Elementary Schools*, 3–4. Francis Parker wrote an endorsing preface. Wilson did not disparage the importance of detail and observed, "the better her [the teacher's] stack of facts, the better will be her perspective; the less imperative her desire to make every one of her facts a part of the mental equipment of the child (4)."

71. Ibid., 8. As a result, Wilson listed potential verse at the heading of every chapter by such authors as William Wordsworth, Lucy Larcom, and Helen Hunt Jackson, as well as stories from Greek mythology.

72. Ibid., 11–13.

73. Wilson wrote numerous textbooks, including the frequently reprinted *Handbook of Domestic Science*.

74. The child study advocates reacted to the assumption that children were simply miniature adults; on the formulation of programs in child study, see Cravens, *Before Head Start: The Iowa Station*.

75. Cruikshank, "The Rise and Fall of American Herbartianism," 167–77.

76. Hall had studied what children knew at various ages and published his results in "The Contents of Children's Minds," 249–73.

77. Kliebard, *Forging the American Curriculum*.

78. These comments are found in McMurry's brief undated notes on outstanding personalities and movements in contemporary education, McMurry Papers, NISU.

79. Schmucker, *The Study of Nature*, 9.

80. Dewey was apparently not highly effective as a classroom teacher, but his theory about using the "natural interests" of children was disseminated through his graduate students. They were also bemused by stories of his theory in action, as when visitors to his home noted water running down a stair and asked if the plumbing was leaking. Dewey investigated and casually reported, "Oh, they just left the bathtub water running over. They're testing their boats." Quoted in Corinne A. Seeds, "Uses of the History of Creative Elementary School" (typescript oral history), Corinne A. Seeds Papers, University of California at Los Angeles (hereafter UCLA), 94. Seeds moved to Los Angeles city schools where she ran an Americanization Center and Socialized Evening School from 1914 to 1920 before receiving an M.A. in supervision from Teachers College, Columbia, and becoming head of the campus elementary school at UCLA (until 1927 the State Normal School at Los Angeles).

81. Hodge, *Nature Study and Life*, xiv.

82. Hodge made it clear that he cared little for prerogatives and titles and wanted to influence

teachers to concentrate on teaching pupils that they could "control the forces of nature for the highest human happiness and the best human good." See series of letters to John Spencer, March 30, May 4, and May 28, 1898, Spencer Papers, Cornell University Archives (hereafter CUA).

83. Hodge, *Nature Study and Life*, 1.

84. On the complex tension among intellectuals wrestling with ideas of modernity and their own ethical concerns about society and ethics, see Lears, *No Place of Grace*.

85. Pets played a significant role in nineteenth-century households, and Jennifer Mason argues in *Civilized Creatures* that such encounters help explain a new affinity for animals in this period. Hodge, it seems, recognized the pervasiveness of pet ownership and encouraged teachers to build on children's familiarity and interest to advance other kinds of systematic study.

86. Hodge, *Nature Study and Life*, 478.

87. Others applied his example in various ways, including campaigns to eliminate an infestation of bagworms in Washington, DC, and other cities; see Barton, "The Bagworm Drive," 109–12.

88. Hodge, *Nature-Study and Life*, xiv.

89. Hodge, "Civic Biology: A High School Science Devoted to Community Welfare and the National Life," reprint from the *Pennsylvania School Journal* (February 1909): 5, found in the Ira S. Wiley Papers, URSP.

90. Bigelow, "Best Books for Nature Study," *Nature-Study Review* 2 (April 1906): 168–77.

91. Bailey, *Outlook to Nature*. Bailey's final chapter details his personal point of view on evolution. The lack of discussion about evolution or natural selection in the nature study textbooks themselves was perhaps, as Edward Larson argues, because proponents believed that the theory was too difficult for younger children; see his *Trial and Error: The American Controversy over Creation and Evolution*.

92. *Cornell Rural School Leaflet* 4 (September 1910): 4. Leaflets often had photographs and sketches, and a few had colored plates, sometimes by bird illustrator Louis Agassiz Fuertes.

93. Bailey wrote *Wind and Weather*, a collection of his poems, and *The Holy Earth*, a treatise on the debt humans owed to the earth that sustained them.

94. *Cornell Rural School Leaflet* 4 (September 1910): 1. Bailey's poetry is scattered throughout the leaflet series, along with frequent quotations from poets Alfred Tennyson, Walt Whitman, and Henry David Thoreau.

95. Bailey, *Outlook to Nature*, 37.

96. Bigelow, *How Nature Study Should be Taught*. Bailey's book is based on the lectures that he gave to teachers, where he consistently argued, "nature study is emotional, as science is intellectual" (45). The marginalia are in the author's copy of Bigelow's copy of *The Nature-Study Idea*.

97. Liberty Hyde Bailey's letter to C. A. Wieting, Commissioner of Agriculture, opens volume one of the *Cornell Nature-Study Leaflets* (Albany: J. B. Lyon, 1904), n.p. Stephanie L. Sarver calls Bailey an "agrarian environmentalist" in *Uneven Land*.

98. See, for example, the materials developed by Alice M. Macomber and Delia I. Griffin for teachers in the expanding western suburbs of Boston in *Outline of Nature Study for Primary and Grammar Grades*.

99. Holtz, *Nature-Study: A Manual for Teachers and for Students*, 65. Holtz had directed the program at Mankato State Normal School in Minnesota before moving to the Brooklyn Training School for Teachers.

100. Schmucker, *Study of Nature*.

101. Gehrs was on the faculty at the Southeast Missouri State Teachers College when he wrote *Agricultural Nature Study*.

102. Overton and Hill in *Nature Study: A Pupil's Text-book* presented a series of specific lessons each calling for multiple pupil tasks. Jean Broadhurst reviewed it as one of the "best yet published" of those intended for pupils but pointed out that teachers would still need to provide explanations; "Notes on New Books and Pamphlets," *Nature-Study Review* 1 (September 1905): 229.

103. This rare lesson book, published by ABC in 1904, opened from the back to show three flowers that bloom in May; a copy is held by the Special Collections Library at the University of Toronto. A similar student book was Marion Mille's *An Out-of-Door Diary for Boys and Girls*.

104. Overton and Hill, *Nature Study: A Pupil's Text-book*. The emphasis was on clarity of expression, with only selective use of the "ogre of school work in English, the blue pencil" (8); the comment on grading was made in the introduction by Anna Botsford Comstock.

105. Carter, *Nature Study with Common Things*, 8.

106. The term "evolution" was not in evidence in most of the nature study textbooks, perhaps because in the pre–high school years there was less attention to theory, although themes of development were often woven into accounts of particular species. This could be a topic for more extensive investigation, as recent work on high schools has suggested; see Shapiro, "Civic Biology and the Origin of the School Antievolution Movement," 409–33.

107. The historians of education David Tyack and Elizabeth Hansot argue that when educational leaders of this period admired efficient organization, scientific facts, or methodological planning, they were not turning away from religious values but selectively emphasizing certain parts of their heritage; see *Managers of Virtue: Public School Leadership*, 116.

108. "Nature Study and Religious Training," *Educational Review* 30 (June 1905): 12–30. Marion E. Cady wrote *Bible Nature Studies: A Manual for Home and School*.

109. Scott, *Nature Study and the Child*, 116–25.

110. Hodge, *Nature Study and Life*, 2–4.

111. Bailey, *Outlook to Nature*, 62.

112. Tallmadge, *First Book of Nature*. He quoted the familiar lines, "He prayeth best who loveth best / All things both great and small; / For the dear God who loveth us, / He made and loveth all." This outlook was also reflected in Arthur F. Newcomb, "Nature Study as Material for Religious Education" (M.A. thesis, University of Chicago, 1910).

113. See Sheldon, *Blind Child in the World of Nature*.

114. Osborn and Sherwood, *Museum and Nature Study in the Public Schools*, 10.

115. Houghton Publishing Company had been heavily invested in natural histories, local color sketches, pastoral memoirs, nature poetry, and travel writing, but in the 1880s refined their list of outdoor books as they began to concentrate on a distinctly American tradition of essays on nature observations with authors like John Burroughs and Olive Thorne Miller, as well as John Muir and Dallas Lore Sharp. Eric Lupfer, "Reading Nature Writing." I thank Jennifer Gunn for bringing this essay to my attention. Carl Kaestle comments on broad patterns in "Literacy and Diversity."

116. *Secret Gardens*; also see Eddy, *Bookwomen*.

117. Scott, *Nature Study and the Child*, 272.

118. Holtz, *Nature-Study*, 65.

119. Holt, *Sixty Years as a Publisher*; he also noted that the shelf life of a textbook was typically twenty years or less.

120. Lutts, *Wild Animal Story*, ix.

121. Ritvo, "Learning from Animals," 72–93.

122. Gates, *Kindred Nature: Victorian and Edwardian Women*, 51–64. Buckley had worked for years with geologist Charles Lyell and thus knew well the scientific outlook of contemporaries.

123. Buckley, *By Pond and River*, 38; her *Life and Her Children: Glimpses of Animal Life from Amoeba to the Insects* was frequently reprinted in Britain and the United States. These books helped shape popular science literature at the end of the century. See Lightman, "Marketing Knowledge for the General Readers," esp. 100–106.

124. Potter, *Tale of Peter Rabbit*.

125. Now classic historical studies of nature writing in this period include Nash, *Wilderness and the American Mind*; Worster, *Nature's Economy*; and Lawrence Buell, *Environmental Imagination*.

126. Tebbel, *History of Book Publishing*, 574. The juvenile literature tended to have less violence, but it was not totally eliminated with stories like *Black Beauty* remaining top sellers through this period.

127. Burroughs, *Wake Robin*, xvi.

128. Burt, who had taught at Cook County Normal School, introduced his work in her *Little Nature Studies for Little People from Essays by John Burroughs*. She argued that, after reading Burroughs' *Birds and Bees*, her pupils "read better . . . [and] came rapidly to a better appreciation of finer bits of literature in their regular readers" (iii), as quoted in Lupfer, "Reading Nature Writing," 37–58.

129. The data are provided in Lupfer, "Reading Nature Writing," 56.

130. Ibid.

131. Besides *Birds that Every Child Should Know*, Neltje Blanchan (her full name was Neltje Blanchan DeGraff Doubleday) wrote one of several books with titles like *Bird Neighbors, How to Attract the Birds*, and *Birds that Hunt and Are Hunted*. Photography required complicated planning with fixed equipment, and so nesting birds were often pictured rather than the sketched birds in flight found earlier.

132. Blanchan, *Birds that Every Child Should Know*, 11. Her style is engaging and familiar, rich in metaphor. On ornithology and growing popular attention, see Barrow, *A Passion for Birds*; and Paul Farber, *Discovering Birds*.

133. On the breadth and intensity of this debate about anthropomorphism, see Crist, *Image of Animals*.

134. Olive Thorne Miller [actually Harriet Mann Miller] knew the Cornell faculty and sought to leave them her natural history slide collection; see A. B. Comstock to L. H. Bailey, January 11, 1907, Bailey Papers, CUA. Among her popular juvenile titles were *Little Brothers of the Air, The First Book of Birds*, and *True Bird Stories from My Note-books*.

135. Royalty statement January 12, 1898, Cooke Papers, Chicago Historical Society. Cooke's manuscript collection is an excellent resource on education during the Progressive Era.

136. Rogers had received her M.A. from Cornell in 1902 and moved to New Jersey where she lectured and wrote books. Perhaps the best known were *The Tree Book* and *The Shell Book*, which were easily used by teachers and advanced students. J. E. Rogers to L. H. Bailey January 21, 1908; and Bailey to Rogers, August 13, 1907 and January 22, 1908, Bailey Papers, CUA. See *Biographical Dictionary of Women in Science*, 1120.

137. Rogers, *A Key to the Nature Library*; the seventeen-volume series was published between 1905 and 1908.

138. See J. E. Rogers to L. H. Bailey, January 21, 1908. Bailey wrote a letter of recommendation for her to P. G. Holden at Iowa State College, March 23, 1908, noting she was originally from Iowa and was "a very superior teacher." Both letters are in Bailey Papers, CUA.

139. The introduction by his son emphasizes that there is "no fiction, no 'nature faking'" anywhere in this rambling, autobiographical account (v). Verrill, *True Nature Stories*.

140. Kellogg's *Insect Stories* was part of the *American Nature Series* edited by Rogers; also see Kellogg, *Nuova: The New Bee*, which included songs written by his wife.

141. Largent, "These are Times of Scientific Ideals."

142. Kellogg, *American Insects*, v–vi. I thank Mark Largent for bringing this observation to my attention.

143. Jenkins, *Interesting Neighbors*, with 81 illustrations by W. S. Atkinson.

144. Hornaday, *American Natural History*, v–vi.

145. Abbott, *Young Folks' Cyclopedia of Natural History*, 6.

146. Patch, *A Little Gateway to Science: Hexapod Stories*.

147. *Tami, The Story of a Chipmunk* has engaging illustrations done by Cady and her daughter Carol.

148. Kate Whiting Patch, "Mother Butterfly's Baby," in a compilation of stories edited by Sullivan, *Friends of the Fields*, 24.

149. Warren, *From September to June with Nature*, 157; and L. A. F., "Being a Frog," in Sullivan, *Friends of the Fields*, 146–47.

150. Lutts, *Wild Animal Story*, 281, argues that the decline was real, and nature stories were not revived until Rachel Carson and others began to publish after World War II.

151. Lillie, *Course in Nature-Study for Elementary Grades*.

152. Lange, *Handbook of Nature Study*.

153. Anna B. Comstock, "What Nature-Study Does for the Child and for the Teacher," 134–35.

154. Ephemeral pamphlet literature includes the Society for the Prevention of Cruelty to Animals (SPCA) leaflet entitled *Canary, Parrot, and Pigeon* (n.p., n.d.) "prepared for the use of the teacher in instructing children in the intelligent and humane treatment of caged birds." The New York Humane Education Association produced *A List of Books* that highlighted those effective in teaching humane nature study.

155. See, for example, *Bulletin 675* of the State University of New York, *Syllabus for Nature Study: Humaneness, Elementary Agriculture, and Homemaking*. See Coleman, *Humane Society Leaders in America*. There is an extensive literature on the SPCA in Britain and considerable attention to this issue because of a case that captured international attention; see Lansbury, *Old Brown Dog*. I thank Jan Zita Grover for this reference.

156. Historians of science have connected nature study to conservation, including Barrow, *A Passion for Birds*, 130–31.

157. See Long, *Gene Stratton Porter*, 24.

158. These were often listed on state and city course outlines; see, for example, *Nature Study, Grades 2–6* (Pittsburgh, Kansas, 1930–1931), Teachers College Library, Columbia University (hereafter TC).

159. *National Geographic* hired Jessie L. Burrell in 1919 for its fledgling School Service Division to provide lantern slides and cardboard-backed photographs for classrooms that would supplement use of the magazine. She described her work in "Sight-Seeing in School: Taking Twenty Million Children on a Picture Tour of the World"; see Bryan, *The National Geographic Society*, 299–300.

160. Page, "Use of the Axe," 41–64; and Armitage, "Bird Day for Kids," 528–51.

161. Hodge, "Nature Study and the Bobolink," 52–64, and "Passenger Pigeon Investigation," *Nature-Study Review*, 110–11; 248–51. The last known passenger pigeon died in a Cincinnati zoo in 1914; a compilation of data of its original population density and the story of its extinction is in Schorger, *Passenger Pigeon*.

162. Lange, "Nature-Study in the Public Schools," 407–8.

163. Andrei, "The Accidental Conservationist," 109–13.

164. In some instances, this was the explicit purpose of a textbook or manual as in Kraus, *Manual of Moral and Humane Education*.

165. Coulter, "Nature-Study in Indiana," *Nature-Study Review* 5 (May 1909): 33–35. Also see Kohler, *All Creatures*, 206.

166. "The Place of Forestry in General Education," was the topic of a special issue of *Nature-Study Review*, September 1911.

167. Holmes, *Organization of Co-Operative Forest-Fire Protective Areas in North Carolina*.

168. *Nebraska Alumnus* (March 1951), 11, University of Nebraska Archives, Lincoln.

169. Hodge, *Nature Study and Life*, 1–15.

170. Reese, "The School Health Movement," 209–37. Reese points out that the movement to provide nutritious school meals was closely linked to the medical inspection movement to insure healthy bodies, a movement applauded by some but challenged by those who resisted compulsory vaccination.

171. Walker, *Anatomy, Physiology, and Hygiene*, vii.

172. Merryman, *The Indiana Story: Pennsylvania's First State University*, 166-70.

173. Monroe, *A Cyclopedia of Education*, 353-54. The encyclopedia also included an entry on teaching hygiene, but concentrated on specific sciences rather than nature study.

174. Public schools were part of the campaign for healthy buildings. Edward Hyatt, a nature study teacher and superintendent of public instruction, promoted a movement for "open air schools" based on successful experiments with such buildings in Pasadena and Fresno in his pamphlet "California School House for $500." Hyatt, a graduate of Ohio State University with an interest in entomology, conducted institutes on nature study and on hygiene while supervisor of schools in Riverside. Hyatt Papers, UCLA.

175. Parker, *Purifying America*; and Mattingly, *Well-Tempered Women*. Most accounts emphasize state and national political campaigns rather than the extensive educational activities of local chapters of the Women's Christian Temperance Union (hereafter WCTU).

176. Blaisdell, *Our Bodies and How We Live*; and Hutchinson *Laws of Health: Physiology, Hygiene, Stimulants, Narcotics*; despite the title of the latter book, relatively little space is devoted to the effects of alcohol and tobacco.

177. See this orientation toward "useful biology" and the influence of organizations such as the WCTU on schools in New York in Pauly, "Development of High School Biology," 662-88; and Tomes, *Gospel of Germs*, 121-23.

178. Trafton, "Outline of Nature-Study," 92-167. Trafton (1874-1943) taught at Mankato from 1911 to 1943, and the science building on campus honors his name.

179. *Nature-Study Review* 6 (February 1910): 33-39. Goals were explicit in Alice Jean Patterson's *Nature Study and Health Education*, 16; the message was that children should have a full bath once a week, brush their teeth daily, and have four glasses of water but no tea or coffee.

180. Gregg, "Hygiene as Nature Study," 225-29. The same issue reviewed Lydston's *Sex Hygiene for the Male*.

181. Bigelow, *Sex-Instruction as Phase of Social Education*. He was active in the American Federation for Sex Hygiene and coauthored a report with Baillet and Morrow, *Report of the Special Committee on the Matter and Methods of Sex Education*. Considerable impetus came from the Rockefeller Bureau of Social Hygiene; also see the Social Welfare Archives at the University of Minnesota.

182. Cady and Cady, *The Way Life Begins*. Bertha subsequently received a Ph.D. in education at Stanford in 1923, and Vernon received his Ph.D. there in 1925.

183. A short brochure on Bigelow's work at Columbia and in New York City is in the papers of James Earl Russell's son and his successor at Teachers College, the William Fletcher Russell Papers, TC. Bigelow's book, *Sex-Education*, seems to mark the transition. He spent much of his later career lecturing for the American Social Hygiene Association and advocating a conservative view that even "the scientific word 'contraception' and the Sears, Roebuck term 'feminine hygiene'" were unacceptable. Bigelow to Louis Hayden Meek, April 19, 1932, also in Russell Papers, TC.

184. Royston, *Unity of Life*, 255.

185. Cady and Cady, *The Way Life Begins*, 77, passim. The account began with the lily and advanced from the moth, through fish, to the rabbit, and to the child. Chapter one, entitled "The Deeper Meaning of Nature Study," argued that nature study must include the complex and interrelated "web of life" from birth to death. The book was positively reviewed in *Nature-Study Review* 13 (March 1917): 125-26, for its demonstration that "sex reproduction is [not] debased or alien to our humanity."

186. Bigelow's *Sex-Education* was originally published in New York by the American Social Hygiene Association in 1917; it was reprinted in 1921 and subsequent years as well.

187. See Raftery, *Land of Fair Promise*. Rafferty points out that school nurses were controversial because their public health roles impinged on family prerogatives.

188. Moran, *Teaching Sex*, 81.

189. Bigelow, *Sex-Education*, 129-30. Some of this work was done with his wife Anna N. Bigelow, who taught nature study at Miss Chapin's School for Girls.

190. Nietz, *Evolution of American Secondary Textbooks*, 104-9.

191. Bailey, *Outlook to Nature*, 174.

192. Hodge and Dawson, *Civic Biology*. This widely used textbook was one under contention at the Scopes Trial in 1926. Hodge continued to link nature study teaching to community issues in "Preparation of Teachers for National, State and Civic Biology," 294-307; his move is noted in "University and Education News," *Science* 38 (October 17, 1913): 545. Hodge taught at Oregon until 1919 and then moved to Florida where he worked in extension programs until his retirement. See *American Men and Women of Science* (New York: Science Press, 1944), 823.

193. F. L. Holtz, "Standardizing Nature-Study," 52-54, quotation on 54.

194. See the Louisiana Department of Education's *Course of Study for the Elementary Schools for 1913*, 45. Given its multiple aspirations, it is perhaps not surprising that the course follows Munson's *Education Through Nature Study* and also references the quite different textbooks by Scott, Holtz, and Hodge. This admixture is found in many of the courses of study, changing surprisingly little over the years but with less attention to W. T. Harris's ideas about returning to the topics in two-year cycles that had been advocated in *Course of Study in the Public Schools of East Orange, New Jersey* (August 1896), 62.

195. The phrase is that of Boyden in "American Nature-Study Society," 184.

CHAPTER SIX

1. Patterson, "Present Trends in the Teaching of Nature Study," 1-5.

2. The term was given currency in Rossiter, "Women's Work in Science," 381-98.

3. Contemporaries and historians have discussed the feminization of the teaching force, noting that by 1900 nearly 75 percent were women in the United States, with the proportion highest in the northeast and lowest in the South. This reflected the rapidly rising demand for teachers, the lengthening of the school term, the professional alternatives for men in some regions, and the salary differential between men and women. School board assumptions that women were nurturing and more compliant may also have played a role. Discussion of the expanding women's participation in teaching is in Rury, "Who Became Teachers?" in Donald Warren, ed., *American Teachers: Histories of a Profession at Work*, 23-28; Biklen, *School Work*, 50; and Albisetti, "Feminization of Teaching," 253-63.

4. Ogren, "Where Coeds were Coeducated," 1-26.

5. Early normal schools, such as Bridgewater, were all-female, and women in school classrooms were replacing young men who often had a rather different preparation in male academies or colleges. This led, Paul Mattingly argues, to an increasing distinction between normal school graduates and college graduates; see Mattingly, *Classless Profession*, 164.

6. By the Civil War in educationally conscious Massachusetts, for example, one in five adult women had taught school at some point, according to Bernard and Vinovskis, "Female Teacher in Ante-Bellum Massachusetts," 332-45.

7. Rossiter, *Women Scientists in America: Struggles and Strategies to 1940*, esp. chaps. 1 and 8; and Tolley, "Science Education of American Girls, 1784-1932," 208-9.

8. Stage and Vincenti, *Rethinking Home Economics*.

9. Ogren in *American Normal School* provides a comprehensive list of state normal schools with their founding dates.

10. Tyack, *One Best System*, 45.

11. Herbst, "Teacher Preparation in the Nineteenth Century," 217-20.

12. Mulligan, "Common Cares," esp. 113 and 116.

13. Some sense of the necessary organization and discipline required is found in the pocket-sized *Outline of State Institute Work for Minnesota Together with a Graded Course of Study for the District Schools* (n.p., 1885), found in Minnesota Historical Society, St. Paul (hereafter MHS).

14. Because various levels of government administered these annual examinations, from a local school administrator or board to formal statewide assessments, variations were considerable. Training for teaching in normal or high schools could provide a substitute credential as well as greater standardization. See Sedlak, "'Let Us Go and Buy a School Master,'" 262–67.

15. For a useful discussion of institutes as "accessible, acceptable, affordable, and adaptable" to the needs of teachers and regions, see Cordier, *Schoolwomen of the Prairies and Plains*, 53–63. The institutes, begun in the 1830s in Connecticut, reached California by the late 1860s and were required of teachers in some counties. See Weiler, *Country Schoolwomen*, 99. Multiweek summer training schools became especially common in the 1890s in an effort to acquaint local communities with changes in education, as well as to enable practicing teachers to meet higher standards for permanent positions. See, for example, "The Summer School—for What?" in *School Education* 14 (December 1895): 5–6, accompanied by a report of teachers' training schools around the state of Minnesota. Institutes are also discussed in Mattingly, *The Classless Profession*.

16. Ogren, *American Normal School*; and Beatty, "Child Gardening," 23.

17. Frank McMurry's typescript "Some Recollections of the Past Forty Years of Education," in McMurry Papers, Vanderbilt University Archives, Nashville (hereafter VUA).

18. The enthusiastic Mary Bradford, who had studied science at the University of Wisconsin, emphasized nature study in her summer institute, discussed in her *Memoirs of Mary D. Bradford*, 267.

19. She particularly remembered Marian Lounsbury and "the pleasant communion we had together." Originally from Ohio, Price had moved steadily westward, teaching in Indiana, Iowa, and finally reaching the plains states in the 1870s and 1880s. Her biography and selections from her diary are in Cordier, *Schoolwomen of the Plains and Prairies*, 175–207, quotation on 186.

20. Special programs across the country were taught by normal school faculty, field station scientists, and experienced teachers at campuses, county seats, and naturalist locations like those at the Biological Station at Flathead Lake; see "Lectures at Flathead Lake," University of Montana *Bulletin* 5 (1902).

21. Letters home from Lottie M. Howard, who attended Moorhead Minnesota State Normal School in 1895–96, reveal the discipline and long hours of study, as well as the camaraderie among young women who left their homes, often for the first time, to attend normal schools. Howard was a high school graduate and had taught eighteen months before attending Moorhead. Lottie M. Howard and Family Papers, 1891–96, MHS.

22. Only a few letters written by Lottie Howard survive, although she carefully saved those from her mother, Lavinia Howard. Lottie to Lavinia, May 13, 1996, Lottie M. Howard and Family Papers, MHS.

23. Bailey diary, April 28, 1899, Louise Bailey Papers, Wisconsin Historical Society, Madison (hereafter WHS).

24. Quoted in Harper, *Development of the Teachers College*, 104–5. Brown went on to note that "I cooked our meals and my brother [took responsibility for] running the errands. On Friday evenings he helped me do the washing, and on Saturday I ironed . . . Saturday was baking day also—a very full day."

25. See, for example, the discussion of John R. Kirk of the State Normal School in Kirksville, Missouri, to L. H. Bailey, April 26, 1904, Bailey Papers, Cornell University Archives (hereafter CUA).

26. Undated lecture materials in the Charles A. McMurry Papers, Northern Illinois State University, DeKalb (hereafter NISU).

27. Joseph T. Marsh, "History of Teacher Education at Northern Illinois University" (D.Ed. diss., Indiana University, 1969), 85–87.

28. Ibid., 85.

29. Hayter, *Education in Transition*, 107.

30. Handwritten examples of her advice can be found in the Lida Brown McMurry File at NISU.

31. *Annual Catalogue and Course of Study of Northern Illinois State Normal School at DeKalb, Illinois* (1900): 75, NISU.

32. *Northern* [Northern Illinois University yearbook] 2 (1901), n.p., NISU.

33. Charles's program for the first four grades is in the Northern Illinois State *Nature-Study Bulletin* 6 (May 1909), at NISU.

34. Marshall, *Grandest of Enterprises: Illinois State Normal University*, 373-84. Colton's *An Elementary Course in Practical Zoology* was widely used in the Midwest in part because it used land animals and plants. It was a deliberate counterpart to the other widely used text by Alpheus Packard, *Zoology for Students and General Readers*, which had been written at seaside and used marine forms.

35. See Patterson's faculty File at Illinois State University at Normal (hereafter ISUN), including Patterson's obituary in *The Vidette*, 41 (June 17, 1929).

36. Kohlstedt, "'A Better Crop of Boys and Girls,'" 58-93.

37. Coulter, Coulter, and Patterson, *Practical Nature Study*. John G. Coulter only taught at Illinois Normal for about four years; he was then asked to resign over financial issues. He later published a School Science series in Bloomington. John G. Coulter File, ISUN.

38. Handwritten copies of some of his lectures and biographical information come from the Edward Hyatt Papers, University of California at Los Angeles, Archives and Special Collections (hereafter UCLA); Charles Wilson kindly made these unprocessed files available to me.

39. Edwards, "The Los Angeles Nature-Study Exhibition," 263-70. On the Mt. Wilson trip, see Edwards's "Nature Study Supervision in the Los Angeles City Schools," 32. Edwards had also contributed to *A Preliminary Plan for the Los Angeles Zoological Park and Aquarium* with the goal of making such facilities useful to the schools in their nature study teaching.

40. Hughes on "Nature Study," in Rosa V. Winterburn, *Methodology in Teaching*, 174.

41. On illiteracy, poverty, and political problems in Southern schools, see Charles R. Wilson and William Ferris, eds., *Encyclopedia of Southern Culture* (Chapel Hill: University of North Carolina Press, 1989), 264-65.

42. See Franklin, *George Peabody*, 160-67. In 1911 Peabody trustees gave a final grant of over $2,000,000 to the college as a permanent endowment and ceased its philanthropy there. Also see Conklin, *Peabody College*. Frances Kohler kindly pointed out this volume to me.

43. Kegley, "The Peabody Scholarships, 1877-1899."

44. See Peabody Educational Fund, *Proceedings of the Trustees for 1894* (Cambridge: John Wilson and Sons, 1900): 21-22, 55-62.

45. Peabody Educational Fund, *Proceedings of the Trustees for 1894*, 84-86.

46. *Peabody Normal College Bulletin* (1903-1904): 54. Galloway had a master's degree from Harvard and a Ph.D. from the University of Cumberland. During this decade, Peabody faculty distinguished between normal school students who were going to be teachers and "teachers college" students who, like many graduating from Chicago and Columbia universities, were planning on careers in administration, where they would "direct educational affairs" (19). This became even more explicit in 1912, when the Peabody foundation made its final contribution and the *Bulletin* for 1912-13 announced that it would be a "college of higher education of teachers for all the South" (5).

47. *Peabody Normal College Summer Session Bulletin* (1904).

48. Peabody College for Teacher, *Announcement* (1914-15): 15 and 33; together with various related materials in the Bruce R. Payne Papers, VUA.

49. Nashville also had the widely recognized Seaman A. Knapp School of Country Life, which was a farm-based education program; see Kohlstedt, "'A Better Crop of Boys and Girls,'" 73.

50. "News and Notes," *Nature-Study Review* 2 (December 1906): 319–20.

51. Elliot Downing compiled the "List of Instructors in Nature-Study" for *Nature-Study Review* 12 (February 1916): 69–78.

52. This remained the case through the 1920s, even in Nashville; see Bachman, *Public Schools of Nashville*. He did note, however, that subjects like art, music, and nature study received increasing, although still inadequate, attention (32).

53. Isbell, *History of Eastern Michigan University*, 257–58.

54. Munson, *Education through Nature Study*. Munson had a Ph.B. in biology from Yale and then studied at Chicago before taking the job at Washington State Normal School. His text argued that biological study helped people deal with the stresses and strains of both relationships to others and to the physical universe by making behavior more comprehensible.

55. *Wisconsin Alumni Directory* (Madison: University of Wisconsin, 1921), 212.

56. See Ruth Marshall's proposal to University of Nebraska Regents, October 1, 1907, signed by Bessey, in Bessey Papers, and her letter to F. H. Abbott, June 4, 1908, Abbott Papers, both in the University of Nebraska Archives (hereafter UNA).

57. Wolcott, "What Nature Study Should Be," 7–8.

58. Marshall had received her Ph.D. under Henry B. Ward; her collection of water mites is now housed at the Field Museum; notice in *Nebraska Alumnus* (March 1951): 11, UNA. She also ran, unsuccessfully, as progressive candidate for the county superintendent of schools in 1914. She subsequently taught at Northern Illinois Normal School; see *Nature-Study Review* 12 (February 1916): 71.

59. American Academy of Medicine, *Studies of the Present Teaching of Hygiene through "Domestic Science" and through "Nature Study,"* 35.

60. At Rockford she continued work on nature study started in Nebraska; see Marshall, "Course in Nature-Study for Teachers," *Nature-Study Review* 15 (October 1909): 184–89. Marshall coordinated the October volume on Nebraska nature study, including an opening essay by Charles Bessey in which he claimed "the fairy-tale type of nature has now practically disappeared from Nebraska."

61. Nelson, "Nature-Study in Our City Parks," *Nature-Study Review* 10 (September, 1914): 210–13.

62. See Evelyn I. Fernald to Barbara Campbell, December 10, 1950, Rockford College Archives, courtesy of Joan B. Surrey. Vera Norwood, in *Made from this Earth*, refuses to use the term as a pejorative and argues persuasively that "sentiment" was a useful meld of intellect and emotion and that nature was a site where these were in balance for many women.

63. Alice Rich Northrop, who started the New York School Nature League in New York City, left within two years, in 1917, to move to Great Barrington, Massachusetts, where she began to establish a camp before her death in 1922. Northrop Papers, Schlesinger Library, Radcliffe Institute. Her posthumously published book, *Through Field and Woodland*, was detailed and well illustrated with photographs; a short biographical sketch opens the volume. The League continued to be active through the 1930s and 1940s, with publications and records in the American Museum of Natural History (hereafter AMNH).

64. See letter regarding a position at the *Syracuse Herald*, L. H. Bailey to F. E. Dawley, February 6, 1907, Bailey Papers, CUA.

65. Burns, *Gentle Hunter*; and Shearer and Shearer, eds., *Notable Women in the Life Sciences*, 119–21. Margaret Rossiter kindly brought this to my attention.

66. For a discussion of the progressive period, see Linn, Songer, and Eylon, "Shifts and Convergences in Science," 445–47.

67. The struggle for teachers to gain some self-determination in educational policy is discussed in Ruediger, *Agencies for the Improvement of Teachers in Service*, 151–53.

68. See *Harper's Weekly* (December 1, 1900), 1130.

69. For a thoughtful account, see Ellen E. Doris, "The Practice of Nature-Study."

70. Women with children also found school vacations useful for family life. Gordon, *Gender and Higher Education*, 197-98.

71. Richardson, "Picturing Nature," provides a detailed picture of Stanwood's interest in nature and photography; also see Norwood, *Made from This Earth: American Women and Nature*, 74-76.

72. Flora J. Cooke's "Science in Primary Grades" was a talk given at the dedication of Emmons Blaine Hall at the University of Chicago in 1904; typescript in the McCormick Papers, WHS.

73. Teachers' unions offered a venue where the discussion about autonomy was actively engaged. Wilbur Jackman often championed teachers in *Elementary Science Teacher* where he argued that common school teachers deserved more trust and should be given more latitude in their classrooms; also see Jackman to Mr. White, June 22, 1905; copy in McCormick Papers, WHS.

74. Lupfer, "Reading Nature Writing," 37-58.

75. Liberty Hyde Bailey encouraged pupils to be comrades of their teachers in investigating nature, a stance highlighted in Doris, "Practice of Nature-Study," 32-34 and 46-51.

76. In the intensely scrutinized schools for American Indians under the auspices of the Bureau of Indian Affairs, they were required. See Kohlstedt, "'Better Crop of Boys and Girls,'" 70-71.

77. This effect of undermining teacher confidence is well documented in Cruikshank, "Rise and Fall of American Herbartianism," chap. 2.

78. Michigan Association for the Childhood Education Research Committee, *Nature Study Units and Suggestions for the Early Elementary Grades* (The Association, [1930]).

79. H. C. Bryant to Cheney, Grinnell Papers, University of California at Berkeley, Museum of Vertebrate Zoology (hereafter UCB-MVZ). Insight into the women, including the founding patron, Annie Alexander, who was closely affiliated with the Museum of Vertebrate Zoology, is found in Stein, *On Her Own Terms*.

80. Violet D. Holgersen to Joseph Grinnell, November 5, 1930, Grinnell Papers, UCB-MVZ.

81. Corinne Aldine Seeds, "Uses of the History of Creative Elementary School" (typescript oral history), 9; UCLA.

82. M. A. Bigelow to C. B. Davenport, May 8, 1919, and Davenport to Bigelow, May 13, 1919, Davenport Papers, American Philosophical Society (hereafter APS). Mark Largent kindly brought this letter to my attention. The laboratory conducted a summer nature study course from 1901 through 1903; see http://www.cshl.org/CSHLlib/archives/courses.htm (last accessed June 19, 2009) for a list of courses taught between 1898 and 1923.

83. Bertha Chapman Cady to Joseph Grinnell and his reply [July 1919], Grinnell Papers, UCB-MVZ.

84. By the 1910s, these were courses in "elementary agriculture, nature study, manual and industrial work, and domestic art and science," according to Ruediger, *Agencies for the Improvement of Teachers in Service*, 50.

85. Gilbert H. Trafton was then supervisor in the Passaic, New Jersey, system. See his letter to L. H. Bailey, November 6, 1905, Bailey Papers, CUA.

86. Graham, "Expansion and Exclusion," 759-773. Margaret Rossiter, "Sexual Segregation in the Sciences," also points out that, in a general way, women found it easiest to move into new and unestablished areas (or into declining) fields of science (146-51).

87. Alice Hall Walter to Ira B. Wile, August 1, 1913, Wile Papers, Rush Rhees Library, Special collections, University of Rochester. Walter, educational editor for *Bird-Lore*, noted the variety of ways that school systems in Chicago and California worked to offset the problems of lack of training and enthusiasm among teachers or the paucity of materials. She argued that New York City had considerable resources in its parks and other institutions and thus ought to be a leader in nature study.

88. Massachusetts Board of Education, *Annual Report for 1890-1891* (1892): 80; and Arthur

Boyden, *Nature Study in the Public Schools of Massachusetts*, 5–6. Arnold subsequently became first Dean of Women at the newly founded Simmons College.

89. Dartt to John Spencer, November 16, 1898, Spencer Papers, CUA.

90. Tolley, "Science Education of American Girls, 1784–1932," 246–48. Elizabeth Peeples was supervisor in Washington, DC (identified on title page of *Elementary Science by Grades* written by both Peeples and Ellis Persing [New York: D. Appleton, 1928]). Louise Miller was identified as "for several years in charge of Nature Study work for the schools of Detroit"; in Miller, *Course in Nature Study for Use in the Public Schools*, Department of Agriculture, Commonwealth of Pennsylvania, *Bulletin* 63 (1900): 3. Lenore Conover was Supervising Instructor of Nature Study for the Detroit Teachers College; Conover to E. Laurence Palmer, January 22, 1926, American Nature-Study Society (hereafter ANSS) Papers, CUA. Hattie Rainwater served as supervisor of public schools in Atlanta, Georgia, according to the undated manuscript membership roster of ANSS (about 1928) in ANSS Papers, CUA.

91. Separately published report by Jenkins, *Nature Study for the Oakland Public Schools Report Supplement* (1896–1897): 15. He indicated that the program was at an experimental stage and he encouraged latitude because "teachers have on their own judgment selected those subjects that they considered their children capable of handling."

92. The general program with its emphasis on teacher initiative is discussed in Winterburn, *Methodology in Teaching*, including a practical chapter on nature study by Edward Hughes. Hughes was principal of a grammar school and supervisor of nature study for the county and offered "methods of practice rather than theory."

93. Comstock to David Starr Jordan, April 13, 1900, Jordan Papers, Stanford University Archives (hereafter SUA). On McFadden, see *Stanford University Alumni Directory* for 1904 and 1931, SUA.

94. Helen Swett to Charles Schwartz, December 15, 1900, Swett Papers, University of California at Berkeley, Bancroft Library (hereafter UCB-BL).

95. Bertha Chapman Cady's career is traced in the *Stanford University Alumni Directory* for 1904, 1910, 1920, and 1931. After her work in Oakland, she had gone to Chicago as instructor of nature study before marrying Vernon Mosher Cady. See *Nature-Study Review* 4 (March 1909): 99; also see *American Men and Women of Science* (1944): 263.

96. *Junior Science Leaflet* 1 (October 1901).

97. B. M. Davis, "Factors and Influences Now Contributing to the Progress of Nature Study in California," 257–65.

98. Chapman, "School Gardens in the Refugee Camps of San Francisco," *Nature-Study Review* 2 (1907): 225–29.

99. Pruitt, "Otis William Caldwell," 285–86. Sometimes identified as the "father of general science education," Caldwell taught science education at Chicago from 1907 to 1917 when he went to Teachers College, Columbia, as professor of education and director of its Lincoln School.

100. In "The Normal Schools and Colleges and the Problem of Sex Education," Cady called for more teacher preparation and specifically help from deans of women students to introduce these topics to young women.

101. She held the standard two-year term, 1928–29, but the death of Alice Jean Patterson, elected for the 1930–31 term, led Cady to continue through an extra term.

102. Also see Swett's self-reports in the *Stanford Alumnus* 4 (1902): 37–38; and the *Stanford University Alumni Directory* for 1904.

103. *Program for the July Summer Session on Nature Study, June 18 to July 16, 1900*; leaflet for the San Francisco State Normal School.

104. Full-time high school teachers earned $100 per month. Helen Swett to Charles Schwartz, June 5, 1901, Swett Papers, UCB-BL.

105. Helen Swett to Charles Schwartz, August 26, October 3, and November 10, 1900, Swett

Papers, UCB-BL. Swett mentions correspondence with Vernon Kellogg and with Cecil Stebbins, then affiliated with the Agricultural Experiment Station at Berkeley. The couple actively worked on the *Stanford Alumnus*.

106. Helen Swett to Charles Schwartz, September 12, 1900, Swett Papers, UCB-BL.

107. Helen Swett to Charles Schwartz, August 26, 1900, Swett Papers, UCB-BL.

108. Helen Swett to Charles Schwartz, January 13, 1901, and February 24, 1901, Swett Papers, UCB-BL.

109. Helen Swett to Charles Schwartz, May 1, 1901, and August 1, 1901, Swett Papers, UCB-BL.

110. Schwartz's death was reported in the *Stanford Alumnus* (December 1901): 37–38.

111. Northrop, then President of the School Nature League, discussed her work with Straubenmueller in a memo to Osborn, July 11, 1922, Osborn Papers, Central Files, AMNH.

112. As a teacher, Straubenmueller had written to Anna Botsford Comstock about building classroom aquariums; letter, dated May 28, 1898, Spencer Extension Education Papers, CUA. In 1912 he was Associate City Superintendent and chaired the Committee on Studies and Textbooks; *City Directory of Teachers in the Public Schools* (New York City: Board of Education, 1912). Straubenmueller's obituary is in the *New York Times*, May 14, 1934, p. 17. He was born in Baltimore in 1860 but moved to New York where he attended public schools and graduated from the normal school that ultimately became the College of the City of New York in 1880. John A. Hollinger of Pittsburgh was also a supervisor; see *Course of Study in Nature Study and Elementary Science* (Pittsburgh: Board of Public Education, 1928), CUA.

113. Lange's normal school degree was from Minnesota State Normal School at Mankato and his B.A. (1900) from the University of Minnesota. See Lange Papers, MHS; and *Who Was Who in America*, 2:312. His report in the *Annual Report for 1903-1904* of the St. Paul Board of Education recommended L. H. Bailey's *Nature-Study Idea* and John Dewey's *My Education Creed* to teachers in the system; the study of nature was, he argued, an alternative to "too much dead mechanical teaching in our schools" (66). He also produced *Course of Study in Nature Study, Grades One to Eight*.

114. Tolley claims that this group was concerned with women's issues. No official records seem to have survived, but references are found in *Nature-Study Review* and a list of officers is in *Science Education* 31 (March 1947): 78–80. Also see Palmer to George R. Green, October 6, 1926, ANSS Papers, CUA.

115. Palmer to George R. Green, October 6, 1926, ANSS Papers, CUA.

116. In 1925, for example, they met with the National Education Association in Cincinnati and Alice Jean Patterson was reelected president of the group; see *Nature Magazine* 3 (April 1925): 253. The new journal took over the existing *General Science Quarterly*, founded in 1916. E. L. Palmer to George R. Green, April 26, 1927; and Theodosia Hadley to E. L. Palmer, May 29, 1928, both in ANSS Papers, CUA.

117. Hadley had an M.A. (1904) in zoology from the University of Chicago and taught biology at Northern Michigan State Normal School, where she published, with Eliot R. Downing, *Physiography of the Marquette Region with the Hypotheses of Its Origin*, before she took the normal school position in Kalamazoo. Hadley was an active suffragist, worked as a nurse's aid in Europe during World War I, and helped establish the Kleinstruck Reserve that eventually became a wildlife sanctuary near Kalamazoo. *Western Michigan College News Magazine* 5 (1947): 8–9. I thank Sharon Carlson for this reference.

118. MacPhail, *Kate Sessions*. Sessions was well-known for her nursery of plants for the California climate and became titled as the "Mother of Balboa Park" for her landscaping work there.

119. Hanson, *Animal Attractions*.

120. Kuslan, "Elementary Science in Connecticut, 1850-1900," 288.

121. Porter, *Girl of the Limberlost*. Karen Reeds kindly pointed out this connection. Also see Amy Green, "'She Touched Fifty Million Lives.'"

122. Not only did the school board president suggest that this new position would be "delightful and much easier than the routine grind of other teachers" but it would pay $200 toward expenses as well as a salary of $750. Porter, *Girl of the Limberlost*, 263.

123. Some women teachers were more visibly activist than others, and at least one author suggests that this should not lead historians to minimize the challenges that they faced as women; see Martusewicz, "The Will to Reason," 65. Also see Kohlstedt, "In from the Periphery," 81–96.

124. Adam Rome, "'Political Hermaphrodites,'" 442–43, 446.

125. See Geraldine Joncich Clifford, "Man/Woman/Teacher," 305–6; and Hoffman, *Women's True Profession*.

126. The directory of members comprised the entire issue of *Nature-Study Review* 5 (March 1909): 68–88. Downing's "List of Instructors in Nature-Study" reinforces the fact that women remained a perhaps surprising quarter of such instructors in 1916 (107 of 435 of those identified by name rather than simply initials) in *Nature-Study Review* 12 (1916): 69–78.

127. This division of labor coincides with the findings in Tyack and Hansot, *Managers of Virtue*; and Tyack, *One Best System*, 60. Positions with the highest status and authority, as well as the best salaries, typically went to men, creating a multitiered and gendered system.

128. Edward Lee Thorndike, "Feminization of American Education," 48–62. Also see Blount, *Destined to Rule the Schools: Women and the Superintendency, 1873–1995*, esp. chap. 3.

129. Tolley, "Science Education of American Girls," 277–324; women were admitted in 1974. The gender tension resulted in a male Principles Association and an Association of Women Teachers of the City of New York; see New York City's *Journal of the Board of Education* (1908): 804 and (1907): 265.

130. Bailey to Charles E. Rand, May 3, 1908, Bailey Papers, CUA. C. H. Robison believed that inadequate salaries led to a "lack of available men" to teach agriculture; see Robison to L. H. Bailey, March 26, 1908; and Bailey's reply, March 28, 1908, Bailey Papers, CUA.

131. Ida A. Reveley to L. H. Bailey, May 19, 1908; and his reply May 21, 1908, Bailey Papers, CUA. Bailey suggested that she might apply for a position at Morrisville Normal School, which would open in a year or two.

132. J. F. Millspaugh's request was written as he left the State Normal School at Winona, Minnesota, to become president of the Los Angeles Normal School; Millspaugh to L. H. Bailey, May 14, 1904, Bailey Papers, CUA. A similar letter came from the University of Virginia, seeking a man for summer school teaching in agriculture and nature study or zoology; Bruce R. Payne to L. H. Bailey, Bailey Papers, CUA.

133. Johnson to L. H. Bailey, July 26, Bailey Papers, CUA. Johnson was concerned that this person be of good character and a model for students. Women's colleges often hired women in the late nineteenth century, sometimes even creating protégé chains in certain fields, according to Rossiter, *Women Scientists in America: Struggles and Strategies*.

134. Endorsements of Wilson's manual by faculty at Wake Forest College, North Carolina, and the State Normal School at New Paltz, New York, are at the back of her *Nature Study in Elementary Schools: First Reader*. Reviewers suggested that a good guide could make such teaching "a delight rather than an added burden." See W. L. Eikenbery, *Teaching of General Science*, 66–67.

135. See the Jenkins's "Nature Study" report in *School Report of Oakland, 1896–1897*.

136. A litany and review of these concerns is in Wagner, "On the Training of Teachers of Nature-Study," 47–55. He blamed training rather than the teachers: "The materials of Nature-Study are drawn from a very large field—from biology, astronomy, geology, chemistry and physics—and without very broad training the immature teacher, attempting to follow an outline, is speedily and hopelessly lost" (48).

137. Stanley M. Coulter, "Educational Values and the Aims of Nature-Study," *Nature-Study Review* 1 (1905): 50-53.

138. The anecdote may well be apocryphal, but it reflected an attitude expressed among some scientific commentators on the entrepreneurial activity found around nature study. See *Nature-Study* 2 (February 1902): 152.

139. The emphasis on sex difference was common in medical literature and psychology in this period, but many sexist arguments were contested by women entering the social sciences. See Rosenberg, *Beyond Separate Spheres*.

140. Clifford, "Man/Woman/Teacher," 305 and 306.

141. Maxine Sellers, "G. Stanley Hall and Edward Thorndike on the Education of Women: Theory and Policy in the Progressive Era" (unpublished paper). On the identification of science as a masculine endeavor at the turn of the century, see Hevly, "Heroic Science of Glacier Motion," 66-86; Herzig, "In the Name of Truth"; and Bederman, *Manliness and Civilization*.

142. There is no evidence about the taste of either girls or boys regarding nature study texts. Hall's comment is in Hodge, *Nature Study and Life*, xv.

143. This undated talk was probably delivered in the 1910s by Jesse F. Millspaugh while head of the Los Angeles Normal School from 1904 to 1917. Millspaugh Papers, UCLA.

144. *Science* 15 (1902): 793-794. On Bessey, see Tobey, *Saving the Prairies*, 38.

145. Bessey several times commented about preferences for male over female graduates in letters to school superintendents, as to one in Lead, Nebraska: "If for any reason you conclude not to employ a man for this position, let me know at once as I have an excellent young woman." Bessey to A. H. Bigelow, April 15, 1907, Bessey Papers, UNA.

146. C. B. Wilson, "How Can Advanced Science in the College and University and Nature Study Be Rendered Mutually Beneficial?" 593.

147. Interestingly, these are the criticisms that have been echoed in historical literature that was dismissive of nature study, including David Allen who denigrated it as science "watered down for juvenile capacities" in his otherwise fine account of *The Naturalist in Britain*, 202. Historians of education sometimes mention key figures like Jackman and Bailey in passing but spend little time on curriculum, as in Cremin, *Traditions of American Education*, 134. Louis I. Kuslan suggested it was a "craze of the 1890s" foisted by administrators on reluctant teachers in "Elementary Science in Connecticut, 1850-1900," 286-89. A British revisionist historiography suggests nature study was an effort to further the interests of rural industry, as in Selleck, *New Education: The English Background*.

148. George A. Works, *Rural School Survey of New York State: A Report to Patrons, by the Joint Committee on Rural Schools* (Ithaca: n.p., 1922), 220.

149. Comstock, "Editorial," *Nature-Study Review* 10 (October 1914): 287-89; see next chapter for a discussion.

150. Ibid.

151. Cruikshank, "Rise and Fall of American Herbartianism," chap. 2. Kim Tolley also follows the thread of gender issues facing women teachers of science in her book *Science Education of American Girls*, esp. 98-125.

152. Rury, "Who Became Teachers?" 23, 31-33.

153. There are several accounts of naturalists and conservationists who credit early nature study teachers with awakening their concerns and teaching them skills. See, for example, "Loye Holmes Miller: An Interpretive Naturalist," oral interview, UCB.

154. Letter from unidentified teacher at the end of a school term, May 10, 1906, Cornell/Comstock Press Papers, CUA.

155. See, for example, Wagner, "On the Training of Teachers of Nature-Study," 47-55. His essay presents nature study teachers as female and pupils as male, a common practice in educational journals of the period.

156. Kohlstedt, "Single-sex Education and Leadership," 93–112.

157. D. C. Heath (president of the corporation) to David Starr Jordan, October 10, 1898, Jordan Papers, SUA.

158. Bailey to Sarah Payne, December 21, 1907, Bailey Papers, CUA. His *Nature-Study Idea* was dedicated to Julia Field King, the teacher who encouraged his early interest in nature.

159. Kate Van Buskirk's introduction to the *Course of Education for Elementary Schools of Arizona*, Teachers College Library, Columbia University. A school is named in her honor in Tucson.

CHAPTER SEVEN

1. Hodge, "Established Principles of Nature Study," 7.

2. C. Maynard had published eleven issues of *Nature Study in the Schools* in 1899 but apparently was not able to develop an adequate subscriber list. The Manchester, New Hampshire, Institute of Arts and Sciences called its periodical *Nature Study* and published it monthly from 1900 to 1904, but a fire in 1902 destroyed the building and collections that had been the mainstay of the institute. *Boys and Girls: A Nature Study Magazine*, intended for a wide audience including pupils, only survived from 1903 to 1907, and in its last volume included advertising for the *Review*.

3. Maurice Bigelow, who had completed his Ph.D. in 1901, established his credentials by publishing "Outlines of Nature-Study in the Horace Mann School."

4. The U.S. Bureau of Education issued an annual *Bulletin* with a "Bibliography of Education." These annual lists typically had a dozen or so new educational journals each year at the turn of the century and often suggest the diversity of approaches within particular fields. *An Index to the Reports of the Committee on Education, 1867–1907*, was published as its *Bulletin* 7 (1909).

5. "Nature Study and Moral Culture," 149. In the 1890s the magazine had affiliated with the American Association for the Advancement of Science (hereafter AAAS) and was rapidly becoming the most widely distributed among scientists and those who followed science. The major independent British science magazine reported on the nature study movement in the United Kingdom; see "Nature-Study in Rural Schools," *Nature Magazine: A Weekly Illustrated Journal of Science* 61 (1899–1900): 553.

6. Coulter, "Nature Study and Intellectual Culture," 740–44.

7. Findling, ed., *Historical Dictionary of World's Fairs and Exposition*; and Tennert, "Selling Indian Education at World's Fairs and Expositions," 203–22.

8. M. J. Holmes to L. H. Bailey, May 7, 1904, Bailey Papers, Cornell University Archives (hereafter CUA).

9. Bigelow, "Introduction," *Nature-Study Review* 1 (January 1905): 1. The language of "faddism" was common in this period because so many new ideas were, in fact, in play. See Tyack, *One Best System*.

10. He included the caustic comment of Henry E. Armstrong of London who wrote, "Nature-Study lessons as I witnessed, when not specifically botanical or zoological and scientific in character, were eminently superficial and worthless," in Armstrong, "Some Recent Criticisms of Nature Study," 22–26.

11. Bailey, *Nature-Study Idea*, 3. The book's subtitle was "Being an Interpretation of the New School Movement to put the Child in Sympathy with Nature."

12. Bailey, *Nature-Study Idea*, 3.

13. Roberts, *Nature-Study Quarterly*, 220.

14. J. M. Coulter, "Nature Study and Intellectual Culture," 744.

15. Perez, "Fancy and Imagination." On the ambiguity among scientists themselves and the larger public about their political and moral authority in the twentieth century, see Shapin, *Scientific Life*.

16. Wilson "How Can Advanced Science in the College and University and Nature Study Be Rendered Mutually Beneficial?" 593.

17. Olmsted's comments in National Educational Association (hereafter NEA), *Journal of Proceedings and Addresses* (1903), 418.

18. Caldwell, NEA, *Journal of Proceedings and Addresses* (1904), 895.

19. Quoted in Joncich, *Sane Positivist*, 165.

20. Thorndike, "Sentimentality in Science Teaching," 60-61.

21. Thorndike, "Sentimentality in Science Teaching," 56-64; and his "Reading as Nature Study," 386-91. Also see Seller, "G. Stanley Hall and Edward Thorndike on the Education of Women," 95-107.

22. W. J. Beal, "What is Nature Study?" 991-92.

23. These comments come from W. J. Beal, A. S. Packard, John M. Coulter, C. P. Gillette, W. M. Davis, A. E. Verrill, David Starr Jordan, and Thomas H. McBride, "What is Nature Study?" 910-13.

24. Coulter, Coulter, and Patterson, *Practical Nature Study*, 37. J. M. Coulter to C. E. Bessey, April 21, 1899, Bessey Papers, University of Nebraska Archives (hereafter UNA). This was part of a much larger debate between Bessey and Coulter, as noted in Tobey, *Saving the Prairies*, 38-39.

25. For a discussion of this disciplinary professionalization, see Geiger, *To Advance Knowledge*.

26. Many nature writers, women and men, casually mixed imagination, poetics, and careful observation. For examples, see Bonta, *Women in the Field*; Gates, *Kindred Nature*; and Gates and Shteir, *Natural Eloquence*.

27. McMurry, *Special Method in Natural Science for the First Four Grades*, 64. Elsewhere he observed that Herbart identified three "fountains of interest and inspiration from nature," namely empirical (with observations of change, variety, and attractiveness), speculative (noting causal relationships), and aesthetic (personal responsiveness), in McMurry, *Special Method in Elementary Science for the Common School*, 54.

28. Troeger's *Harold's Rambles* is second in the series of *Nature-Study Readers*, esp. 31-33 and 448-51.

29. Meyers, "Evolution of Aim and Method in the Teaching of Nature Study," 205-13.

30. Lupfer, "Reading Nature Writing," 177-204. Lupfer uses the term "literary naturism" and makes the claim that "nature writing" is an anachronism and describes only a narrow genre.

31. See, for examples, Philippon, *Conserving Words*; Buell, *Environmental Imagination*; and Lutts, *Nature Fakers*, 31.

32. British women writers on nature also had a popular audience in North America; see Gates, *Kindred Nature*.

33. Ritvo, "Learning from Animals," 72-93.

34. For a thoughtful and balanced account, see Crist, *Images of Animals*.

35. McMurry and Spofford, *Songs of the Treetop and Meadow*.

36. Lupfer, "Reading Nature Writing," 50.

37. Van Slyck, *Carnegie Libraries and American Culture*.

38. Roger Burlingame, *Of Making Many Books: A Hundred Years of Reading, Writing and Publishing* (New York: Charles Scribner's Sons, 1946), 132.

39. J. Burroughs, "Real and Sham in Natural History," *Atlantic Monthly* 91 (1903): 299; he reviewed Ernest Thompson Seton's *Wild Animals I Have Known* and argued that the "line between fact and fiction is repeatedly crossed."

40. The best overview of the debates is Lutts, *Nature Fakers*; attention to gendered aspects of the controversy about animal portraits is in chap. 4 of Donaldson, "Picturesque Scenes, Sentimental Creatures."

41. Quoted in Frank M. Chapman's defense of Long in "The Case of William Long" *Science* 4 (1904): 387–89.

42. Schmucker, "Science and Nature Study," 48–52.

43. Eccles, "Nature Study Revisited," 18.

44. E. R. Whitney, "Nature Study as an Aid to Advanced Work in Science," NEA, *Addresses and Proceedings* (1904): 889–96, quotation on 892.

45. "Inaccurate Nature Books," 240–42.

46. Some discussions of the perceived problems are reviewed in Mitchell, "History of Nature Study," *Nature-Study Review* 19 (September 1923): 295–321.

47. Bigelow's letter to college men of science like Bailey and Bessey asked their opinions of a list of potential board members, acknowledging significant differences over the question of whether nature study should be "science made elementary." M. A. Bigelow to L. H. Bailey, March 1 and 4, 1904, Bailey Papers, CUA. M. A. Bigelow to C. Bessey, February 25, 1904; also Bessey to Bigelow, September 8 and May 13, 1904, Bessey Papers, UNA. Bigelow coauthored *The Teaching of Biology in Secondary Schools* with Teachers College colleague Francis Lloyd.

48. *National Cyclopaedia of American Biography*, 44:65–66; *Journal of Social Hygiene* (March 1946): 151–56; and obituary in the *New York Times* (January 7, 1955), 21. Bigelow was active in a number of scientific and professional societies, gradually shifting his interests to social hygiene, practical biology, and eugenics; he was president of the American Eugenics Society for six years and kept it running during World War I.

49. The editorial committee was listed on the inside cover of *Nature-Study Review* 1 (July 1905). H. W. Fairbanks, primarily a geologist, had published his *Home Geography for the Primary Grades* in 1902.

50. *Nature-Study Review*, 1–2. He interpreted nature study in its "literal and widest sense" as "all natural science" studies of the lower school; "the natural history of plants and animals (nature-study in its common and most limited sense), school gardening, and the closely allied elementary agriculture, elementary physical science, the physical science of geography, and physiology and hygiene with special reference to the human body."

51. Bessey's invitation from Bigelow remarked, "It is important that the nature studies of the lower schools have the sympathetic support of college men." He also included a one-page prospectus for the journal in Bigelow to Bessey, February 29, 1904, Bessey Papers, UNA.

52. Muldrew, "Nature-Study and Elementary Agriculture in Canada," 20–22.

53. Bailey objected to the term "elementary science" first proposed by Bigelow. M. A. Bigelow to L.H. Bailey, March 1 and 4, 1904, Bailey Papers, CUA.

54. M. A. Bigelow to L. H. Bailey, October 19, 26, and November 2, 1905, Bailey Papers, CUA. The *Review* cost one dollar for an annual subscription, with individual issues available for twenty cents each. The first issue appeared in January 1905. Volume two and subsequent volumes were published nine times a year coincident with the school calendar.

55. Bigelow was listed as the Managing Editor and Publisher on the inside cover of *Nature-Study Review* 1 (January 1905).

56. Issues after the first were published by W. F. Humphrey of Geneva, New York, until it was taken over by the Comstock Publishing Company in 1910.

57. See Officers Records, John Francis Woodhull Papers, TC. Woodhull had spent some summers in Woods Hole while completing his Ph.D. at Columbia in 1899; he then joined the faculty at Teachers College.

58. M. A. Bigelow requested the advice of Bailey about her participation, April 12, 1904, Bailey Papers, CUA.

59. "News Notes," *Nature-Study Review* 3 (September 1907): 190. The editor, clearly bemused, offered a description: "The agent signs himself F. W. Cooley, is a Quaker about 70 years old, gray hair, blue eyes, slender form, not well dressed (perhaps he can afford better clothing now) and minus two fingers on his right hand."

60. Fairbanks in "Nature Study and Its Relation to Natural Science: A Symposium by H. W. Fairbanks, C. F. Hodge, T.H. McBride, F. L. Stevens, and M. A. Bigelow," *Nature-Study Review* 1 (January 1905): 4-6.

61. L. H. Bailey to John Dearness, teacher at the Normal School in London, Ontario, and one of the leaders in the Canadian nature study movement, March 5, 1907, Bailey Papers, CUA.

62. Hodge, *Nature Study and Life*, 14. He was also aware that parents needed to be convinced of the value of "original research in hand-to-hand contact with nature."

63. Hodge wrote a series of essays on the "Foundations of Nature Study" for Clark University's educational journal that argued that the fundamental aspect of nature study should be in the "reaction to human interests" of the topics engaged. He also took on, without naming proponents, the "aesthetic method" that deteriorated into fancy work and the "museum method" with its magpie instinct. Given his caustic style, it is a tribute to tolerance that he was later elected president of the Nature-Study Society. The official name was the American Nature-Study Society, but contemporaries almost never added the term American; this book acknowledges the official name in the abbreviation but commonly uses the shorter version in the text, as did contemporaries. For Hodge's opinions, see his "Foundations of Nature Study," esp. 214 and 222.

64. Geography was undergoing dramatic shifts during this period as textbooks began to explain human civilization "not strictly as a function of climate and race, but in terms of climate and commerce." See the discussion of school geography in Schulten, *Geographical Imagination in America*, esp. chaps. 5 and 6.

65. Coulter, "Principles of Nature-Study," 58.

66. Bigelow in "Nature Study and Its Relation to Natural Science: A Symposium by H. W. Fairbanks, C. F. Hodge, T.H. McBride, F. L. Stevens, and M. A. Bigelow," *Nature-Study Review* 1 (January 1905): 14-19; also Bigelow, "Scope and Method of Nature Study," 35-36.

67. In general, university science faculty contributed to early issues of the *Review*, and educational faculty with backgrounds in philosophy and psychology published their research elsewhere. Although psychologists wondered whether children were "born naturalists," the scientists tended to assume that they were, needing only to have their natural curiosity channeled. Normal school instructors wrote primarily about the best ways to prepare teachers to teach effectively, concentrating on content and method.

68. Bigelow, "Editorial," *Nature-Study Review* (September 1905): 233-34. Noting that he had received letters from superintendents and teachers who wanted very specific materials each month, Bigelow indicated that the journal could not be a substitute for training school classes. However, occasionally the journal did publish accounts that offered detailed project instruction, such as Adele M. Fielde, "Communal Life of Ants," *Nature-Study Review* 1 (November 1905): 239-51.

69. Bigelow, "Introduction," *Nature-Study Review* 1 (January 1905): 2.

70. This vocabulary described the *Review* in advertisements that appeared throughout the 1907 volume of *Boys and Girls* magazine.

71. Bigelow, "Articles by Those Who Think and Do," *Nature-Study Review* 1 (March 1905) 80-81. His optimistic goal was ten thousand subscribers.

72. Unfortunately there are no Bigelow manuscripts, although a few letters from him remain in other collections. See letters from M. A. Bigelow to L. H. Bailey, April 5 and November 2, 1905, Bailey Papers, CUA. Bigelow apparently sent five thousand subscription circulars to normal schools, as well as to all teachers, supervisors, and school officials known to be interested in nature study.

73. M. A. Bigelow to L. H. Bailey, October 19 and 26 and November 2, 1905, Bailey Papers, CUA.

74. Bigelow, "Training Teachers of Nature-Study," 121-35.

75. Wilson devoted an hour-long class to evolution in the junior year to offset the students'

perception that "evolution is a religious theory, or rather an antireligious one, or else that its cardinal tenet is that man has descended from monkeys." Reported in Bigelow, "Training Teachers of Nature-Study," *Nature-Study Review* 2 (April 1906): 123.

76. William Hailmann's "German Views of American Education," was translated and published by the Department of the Interior, Bureau of Education, *Bulletin* 2, no. 361 (1907): 18. Hailmann concluded after his tour in 1904 that there was more nature study and industrial education in the United States, and less attention to language, history, geography, and arithmetic than in Prussian schools.

77. Bigelow, "Introduction," *Nature-Study Review* 1 (January 1905): 1-2.

78. "The American Nature-Study Society," 159-61; Bigelow also created cards to advertise the new organization; copy found in the Davenport Papers, American Philosophical Society, Philadelphia (hereafter APS). I thank Mark Largent for information from this collection.

79. "Constitution," *Nature-Study Review* 4 (1908): 5-7.

80. The fact that this was part of the AAAS 1907 annual meeting (held in January 1908) led even some contemporaries to be confused about whether ANSS was founded in 1907 or 1908 since ANSS would hold its second joint meeting on December 29-30, 1908, with the AAAS in Baltimore.

81. Bigelow, "First Meeting of the American Nature-Study Society," *Nature-Study Review* 4 (January 1908): 4. Bailey cautioned in his report that the statements of approval that he heard from "scientific men" concerning nature study were largely approval for "conservative and scientific study of nature-study problems." He reported that Professor McMurrich of Toronto had made a plea for nature study in the schools in his presidential address to the American Society of Naturalists.

82. J. G. Coulter to L. H. Bailey, March 20, 1908, Bailey Papers, CUA. Bailey's reply, March 23, 1908, expressed admiration for Hodge but acknowledged that "he cannot separate his work from personalities and he handicaps his effort thereby."

83. This was clearly meant to honor Bailey's role as articulate spokesperson and to take advantage of his high visibility. M. A. Bigelow to L. H. Bailey, January 27, 1908, and reply also dated January 27, 1908, Bailey Papers, CUA.

84. L. H. Bailey to M. A. Bigelow, February 1, 1908, Bailey Papers, CUA.

85. M. A. Bigelow to L. H. Bailey, June 30, 1908, Bailey Papers, CUA.

86. The directory comprised the entire issue of *Nature-Study Review* 5 (March 1909): 68-88. A supplement appeared in *Nature-Study Review* 5 (December 1909): 239-43.

87. The New York section met at Teachers College and its rapid growth is described in "New York Section of the American Nature-Study Society," *Nature-Study Review* 5 (November 1909): 107-8. See announcement for meeting on March 2, 1911, in Alice Rich Northrop Papers, Schlesinger Library, Radcliffe Institute, Harvard University. Once the New York section numbered over 125, Northrop was elected to serve on the Nature-Study Society Council.

88. The section in St. Louis flourished while led by Alfred F. Satterthwaite, who eventually became ANSS president (1931-32). The Rockford group in northern Illinois probably was formed through the initiative of Ruth Marshall, who was by that time teaching at Rockford College. Other sections mentioned in the *Review* in California, Indiana, Chicago, and other cities and states were apparently short-lived. See the list of "Local Sections," *Nature-Study Review* 9 (January 1913): 32.

89. There was no society meeting with the AAAS in 1913 in Atlanta because officers felt too few members would attend the southern meeting; "Annual Meeting and Election of Officers," *Nature-Study Review* 9 (November 1913): 235.

90. This question of nature study in southern schools deserves further attention. *Harpers' Weekly* (October 10, 1903), 1619 and 1642, for example, in an article on educational improvements in Georgia records that "nature study and manual training have been established in many

of the counties, and are being introduced in others." Nature study is also key to the section on "Methods in Teaching Primary Science," in *Outline of Methods for Georgia Teachers for the Summer Institutes for 1896* (Atlanta: Franklin Printing and Publishing, 1896), 113. Further research may well locate nature study practices beyond those mentioned throughout this book, perhaps taught using the terms elementary or primary science. There are occasional clues in correspondence such as C. A. Ives of Franklin, Louisiana, who wrote, "Of course nature study has formed a part of school work for some years." Ives to L. H. Bailey, August 30, 1904, Bailey Papers, CUA. Indications that nature study and agricultural education were part of a school crusade, undermined either by lack of resources or back lash against northern origins and state authority, are found in Link, "The School That Built a Town: Public Education and the Southern Landscape"; and Heffron, "Nation Building for a Venerable South: Moral and Practical Uplift in the New Agricultural Education, 1900–1920."

91. There was some dissent, as when Manley Townsend wrote "What Do I Expect That Nature-Study Should Do for My Child?" *Nature-Study Review* 19 (February 1923): 61; he felt compelled to comment: "I want her (both of my boys are girls) to 'feel the sweet influence of the Pleiades'" and used the female pronoun for pupils in his essay.

92. *National Cyclopedia of American Biography* (1945), 33:315.

93. Hornaday, "The Weakness in Teaching Nature-Study," *Nature-Study Review* 2 (October 1906): 241–43. Also see W. T. Hornaday to Charles Bessey, November 13 and 15, 1906; Bessey to Hornaday, November 20, 1906, Bessey Papers, UNA. M. A. Bigelow urged Bailey and his staff to respond; Bigelow to Bailey, November 22, 1906, Bailey Papers, CUA.

94. Hornaday, "Weakness in Teaching Nature-Study," *Nature Study Review* 2 (October 1906): 243. Hornaday anticipated a strong reaction, as he indicated in a letter to C. E. Bessey, November 13 and 16, 1906, Bessey Papers, UNA. His opening statement was clearly a challenge and resulted in "frank and friendly discussion" in letters to the editor about his article "full of ideas radically opposed to those commonly accepted by science teachers," *Nature-Study Review* 3 (January 1907): 24–31. Somewhat later Layton S. Hawkins in his *Methods of Nature Study* suggested that Hornaday's criticism was directed at upper schools but that his own guide for teachers was intended only through grades five or six.

95. See Kellogg's comments in "Discussion," *Nature-Study Review* 3 (January 1907): 30–32. Kellogg actually remained a quiet supporter over the years, serving as a featured speaker for a banquet at the annual meeting in 1923 while head of the National Research Council. See, for example, Boyden, "American Nature-Study Society," *Science* 57 (February 9, 1923): 184. Further dissent against Hornaday's position came in the next issue in Charles W. Eliot's comments on "Dr. Hornaday's 'Weakness of Nature Study,'" *Nature-Study Review* 3 (February 1907): 52; additional letters followed on 52–59.

96. Bigelow, "Are Children Naturally Naturalists?" *Nature-Study Review* 3 (November 1907): 236–39.

97. Bigelow, "Protective Color of Animals: Review of an Article by John Burroughs," *Nature-Study Review* 1 (July 1905): 159–64.

98. Examples are in Luther A. Hatch, "Why Many Fail in Teaching Nature-Study," *Nature-Study Review* 1 (May 1907): 97–100; and S. B. Sinclair, "Time Required for Nature-Study," *Nature-Study Review* 1 (May 1905): 122–24.

99. "Committees of the American Nature-Study Society," *Nature-Study Review* 5 (April 1909): 106–7. Other committees under consideration were on elementary school physiology and hygiene, principles of nature study, nature study in relation to geography, nature study literature, and training teachers in normal schools.

100. Bergen, "Economic Biology in the Schools," *Nature-Study Review* 5 (April 1909): 108–9; Schulten, *Geographical Imagination*, 105–6.

101. John Dewey, "Science as Subject Matter and as Method," 123. Frederick L. Holtz

responded to the criticism in "Course of Nature-Study in the Elementary School," *Nature-Study Review* 6 (October 1910): 191. Ironically, Dewey's own contemporaries perceived him to lack system and attention to scientific detail, as in Mayman, *Teaching Elementary Science in Elementary Schools*, 10–11; Mayman sided with Thorndike in emphasizing facts and laws of science through "habits of accuracy in observing and recording data."

102. Mayman, "An Experimental Investigation of the Book Method, Lecture Method, and Experimental Method," 9–10. He pointed out that there were significant variations in the course of study, the materials presented, the caliber of teachers, and the usefulness and efficiency of the methods.

103. Hunter, "Correlation between Nature-Study and High School Biology," 127–31. Also see Pauly, "Development of High School Biology," 662–88.

104. He gave notice that the subsidy would end in December of 1909 when he stepped down as editor. See "American Nature-Study Society," *Nature-Study Review* 5 (May 1909): 139.

105. Charles, "Editorial Notes," *Nature-Study Review* 6 (February 1910): 47. Frederick Charles was a particularly active editor, pressing directors like Delia Griffin to recruit new members and contribute materials related to well-planned topical issues. Charles to Griffin, January 28, 1910, Directors' Records, St. Johnsbury Museum and Planetarium, St. Johnsbury, Vermont (hereafter StJM).

106. See, for example, Hyman Cohen, "Dental Hygiene," *Nature-Study Review* 8 (March 1912): 113–21; Edith Prindeville, "Children's Pets as Disease Carriers," *Nature-Study Review* 7 (December 1911): 260–63.

107. On Bigelow's opinions, see "Washington Meeting," *Nature-Study Review* 8 (January 1912): 2–7. Bigelow held other offices, but, after his defeat for a vice-presidential post in 1916, he was not involved in the association and thus never became president of the organization he had helped to found. Caldwell's comments are in "The Problem that Faces Us," *Nature-Study Review* 6 (January 1910): 21–23.

108. Charles earned a B.S. from Northwestern and M.S. from the University of Chicago and spent two summers doing research at Woods Hole. He taught in Chicago high schools before joining Northern Illinois Normal at DeKalb in 1899 and moving on to the University of Illinois, where he died, an apparent suicide, in May 1911. See "Fred L. Charles," *Nature-Study Review* 7 (November 1911): 196–97, and other notes throughout that volume. Davis had a B.A. from Butler University and a Ph.D. from the University of California at Berkeley (1906), teaching at California normal schools in Los Angeles and Chico before going to Miami in 1907.

109. "Dr. Elliott Downing," *Nature-Study Review* 13 (February 1917): 44.

110. Bigelow, "Washington Meeting," *Nature-Study Review* 8 (January 1912): 2.

111. Downing, "Editorial," *Nature-Study Review* 9 (September 1914): 192. Downing argued that nature study was method rather than content, its aim "not to impart a mass of information but to give to the pupil a mental attitude, habituate him to think clearly upon a wide range of personally acquired facts, to marshal them wisely, to relate them cautiously, [and] to draw conclusions discretely." The problem, as he saw it, was that "we shelter our pupils, crush their expansive minds with the weigh of ready-made, second hand opinions we force upon them. Then we bemoan their lack of virility, acumen, and stamina." Also see B. M. Davis to Delia Griffin, October 21, 1911, Directors' Records, StJM.

112. See, for example, "Nature Articles in Recent Magazines," *Nature-Study Review* 10: (May 1914): 202.

113. Downing, "Editorial," 9:191–92.

114. Downing, "The Present Status of Nature-Study," 741–42; and more data in Downing, "Preparation of Teachers for Nature Study and Elementary Agriculture," 609–21. He wrote to 245 public and 40 private schools training teachers and had 165 replies; of those, 157 taught nature study in the elementary grades of the practice school.

115. B. M. Davis to Delia I. Griffin, October 21, 1911, Directors' Records, StJM.

116. A quarter-page advertisement for three monthly issues cost $6.00; M. A. Bigelow to C. A. Davenport, January 27, 1908, Davenport Papers, APS.

117. Comstock, "Nature-Study as Servant," *Nature-Study Review* 8 (April 1912): 131–32.

118. Elsa G. Allen, "Winter Birds," *Nature-Study Review* 19 (May 1923): 204–22.

119. "A Report of the 1920 Annual Meeting," *Nature-Study Review* 17 (February 1921): 47.

120. "A Report of the 1921 Annual Meeting," *Nature-Study Review* 18 (January/February 1922): 62–63.

121. Downing, *A Source Book of Biological Nature Study*, xii.

CHAPTER EIGHT

1. Mitchell, "History of Nature-Study," 315.

2. Historians continue to debate the term progressive and its chronological parameters in education; for an assessment see Kloppenberg, *Uncertain Victory*.

3. Fear, "Nature and the Americanization as Allies," 1–4.

4. Burnham, in *How Superstition Won and Science Lost*, shows that war propaganda taught scientists how to influence public opinion and led to a new, influential Science Service for the media. Other useful discussions of modernity, science, and public culture are provided in Tichi, *Shifting Gears*; and Deacon, *Elsie Clews Parsons: Inventing Modern Life*, esp. 112–13 and 132.

5. Elwood P. Cubberley's introduction to Trafton, *Teaching of Science in the Elementary School*, v–vi. As evidence, Cubberley pointed to the role of science in the war and to educational surveys in the period, including Abraham Flexner's *A Modern School*.

6. James G. Needham found supervisors "sadly lacking in knowledge of nature" and argued that "administration tends to conformity" in his "Agreement as to the Nature-Study Program," *Nature-Study Review* 13 (June 1917): 2–3. Rural residents were not fully persuaded that school consolidation into larger and more complex systems really improved education. See Edward W. Updike to L. H. Bailey, January 23, 1906, Bailey Papers, Cornell University Archives (hereafter CUA).

7. See, for example, the explicit denunciation of evaluation standards in Holtz, "Standardizing Nature-Study," 52–54. The momentum was difficult to resist; see Giordano, *How Testing Came to Dominate American Schools*. Those who held on to nature study vocabulary in their systems tried to adapt these techniques, as in Irwin, "Measurement of Nature Study in the Primary Grades," 23–32.

8. In *Every Farm a Factory: The Industrial Ideal in American Agriculture*, Deborah Fitzgerald documents a fundamental tension between those who researched to understand agricultural problems and those who sought to help individual farm families.

9. Counts, *School and Society in Chicago*, 212–13.

10. This led the anonymous author of a Louisiana nature study text to justify his recommendation to use Spanish moss by noting it was "so prolific that there is little danger of destroying the base of supply" by student collecting; see *Louisiana Nature Guardian Handbook for the Schools of the State*.

11. Joseph Grinnell to Bertha Chapman Cady, December 14, 1918, Grinnell Papers, University of California at Berkeley, Museum of Vertebrate Zoology (hereafter UCB-MVZ).

12. Joseph Grinnell to Susie L. Dyer of Alameda High School, n.d., Grinnell Papers, UCB-MVZ.

13. See some of the recommendations, for example, in the manuscript typescript Des Moines Public Schools, *General Science Course of Study* (June 1933), Teachers College Library, Columbia University (hereafter TC).

14. Minton, "History of the Nature-Study Movement," 122.

15. Boston School Committee, *School Document* no. 7 (1917): 3. Although the term nature study did not remain much in evidence after about 1906 in Boston, the practices of using materials at hand, training observation, and organizing field trips stayed in place. Ellor Carlisle Ripley's *Provisional Course of Study in Elementary Science for the Elementary Schools* (Boston School Committee, *School Document* no. 5 [1911]) suggested that the "science lesson is the children's hour" and noted that the Park Department and Public Grounds Department of Boston offered specimens to teachers (5-6).

16. Trafton, *Teaching of Science in the Elementary School*, vii.

17. Trafton, *Nature Study and Science for Intermediate Grades*.

18. Copies of these outlines from grades one to four are in the Teachers College curriculum collection; the Long Beach school system had earlier printed nature study bulletins for teachers from 1920 to 1925.

19. Van De Voort, *Teaching of Science in Normal Schools and Teachers Colleges*, 51-52. The courses ranged from those on methods to those on supervising such work in the schools. Thirty-three reported that nature study was taught in their training school. Also see Meier, "Economic Biology in the Elementary Schools."

20. Downing, *Yearbook of the American Nature-Study Society*, 234-36.

21. Downing, *Field and Laboratory Guide in Biological Nature Study*. Downing also wanted to reintroduce more physical science though his *Field and Laboratory Guide in Physical Nature Study* and *Our Physical World: A Source Book of Physical Nature Study*. At the University of Chicago, the College of Education, which had trained teachers since its founding in 1901, was closed in 1931, although the academic School of Education remained.

22. Edith M. Patch's *First Lessons in Nature Study*, with drawings by Robert J. Sim, was reprinted five times and then revised and reprinted seven times by 1941. Patch had a Ph.D. in entomology from Cornell in 1911, later heading Maine's Agricultural Experiment Station and teaching at the University of Maine in Orono.

23. Bryan, *National Geographic Society*, 299-300.

24. Author's personal copy.

25. Mitchell, "History of Nature-Study."

26. Competition and collaboration were part of a civic strategy in this period; see Bender, *New York Intellect*.

27. Fisher and Langham, *Required Nature Study in the New York City Public Schools*.

28. Paroni, *Course of Study Monographs: Nature Study*, 6; she reported optimistically that Bailey, Hodge, Comstock, and Downing had "helped stabilize this work [of teaching through nature] in the modern school systems of our country."

29. Esther L. Guthrie, Sacramento Department of Science, to Joseph Grinnell, October 30, 1930, Grinnell Papers, UCB-MVZ. San Jose State College produced six issues of *Western Nature Study* between 1930 and 1933, but it proved too expensive for most rural teachers, who had learned to expect the small free or low-cost leaflets; see Hadsall, "Extension Activities of Certain Publicly Supported Institutions," 8.

30. This point was made by long-term advocate Edward Bigelow, *Spirit of Nature Study*, 99.

31. "Course of Study for the Minneapolis Public Schools" (1913-14), 73-77; Minneapolis Public Schools, *Annual Report for 1923-1924*, 91-92.

32. Ruggles, *Semi-Centennial Historical Sketch and Notes: Winona State Normal School*; and Selle, ed., *Winona State Teachers College Historical Notes*.

33. Evidence is in the Peabody College for Teachers, *Bulletin*, as, for example, in 1924, p. 24, Vanderbilt University Archives (hereafter VUA).

34. Typescript of "A Teacher Laboratory at Peabody College," Lucy Gage Papers, VUA.

35. Handwritten notes entitled "Ebb and Flow" in folder with lecture notes, undated but probably mid- to late 1920s. Charles McMurry Papers, Illinois State University at Normal, Illinois; also see his autobiographical statement in the archive files.

36. Content became repetitive under Downing, and course outlines were common. Evidence of Downing's lack of editorial oversight includes the increasing typographical errors in 1915 and 1916. There were a few dedicated contributors like Comstock, Robert W. Schufeldt, and Alice Patterson who wrote new material on familiar topics. Downing himself often published in other educational journals.

37. For examples, see Maude A. Doolittle, "Rambler at Summer Camp," *Nature-Study Review* 19 (April 1923): 156-68; Ruby M. Joliffe, "Palisades Interstate Park," *Nature-Study Review 19* (April 1923): 148-51; and Elliot R. Downing, "Summer Outing," *Nature-Study Review* 12 (May 1916): 229-44. Joliffe was head of the park commission's group camp from 1920 to 1948 and a friend of Eleanor and Franklin Roosevelt. Downing's essay follows topics similar to those in Girl Scout and Boy Scout handbooks about how to set up camp, organize tents, and prepare outdoor meals.

38. She is pictured with the Nature-Lore staff in *Nature-Study Review* 19 (April 1923): 138. William Gould Vinal was professor of nature study at Rhode Island College of Education in Providence before establishing his nature training school at Western Reserve University; he was also President of the National Association of Girls' Camps. "Nature Lore School," *Nature-Study Review* 19 (August 1923): 227-29.

39. Anna Botsford Comstock to David Starr Jordan, February 10, 1920, Jordan Papers, Stanford University Archives. She also wrote to ask if Jordan could send her something for a promised issue on Pacific Coast nature study in May, noting, "I am up this tree and am asking some of my friends to put up a ladder or so to help me down."

40. In honor of her contributions more than fifty people attended a celebratory dinner at the December meeting with the American Association for the Advancement of Science in 1922; "American Nature-Study Society," *Science* 57 (February 9, 1923): 183-85.

41. Champagne and Klopfer, "Pioneers of Nature Study: III. Ephraim Laurence Palmer," 557-90; and Bellisario, "E. Laurence Palmer."

42. Palmer apparently tried to negotiate a higher salary. J. E. Russell wrote directly to Livingston Farrand, President of Cornell, July 9, 1923, and also sent a letter to Palmer stating bluntly, "your attachment to Cornell seems to me to jeopardize your success in Teachers College. Under the present circumstances I should be sorry to have you come here." Russell Papers, TC.

43. Comstock, "A New Venture in Nature Lore," *Nature-Study Review* 19 (March 1923): 136 and "The Future of the Nature-Study Review," *Nature-Study Review* 19 (November 1923): 368-69. The role of Comstock as "Editor for the Nature Study Department" was announced in the front matter of the December 1923 issue of *Nature Magazine*.

44. Comstock, *Comstocks of Cornell*, 264-65.

45. Pack attended the 1926 meeting of the society where "he asked for and received criticism of the contents and mechanics of his publication." Minutes of the meeting, held December 27-28, 1926, American Nature-Study Society (hereafter ANSS) Papers, CUA.

46. *The Nature Almanac: A Handbook of Nature Education* (Washington, DC: American Nature Association, 1930), 34-35. Bertha Chapman Cady, as president, expressed her frustration about the new journal and its "commercial" basis to secretary E. Laurence Palmer; Cady to Palmer, August 8, 1928, and Palmer to Cady, August 8, 1928, both in ANSS Papers, CUA.

47. Palmer's *Fieldbook of Nature Study* had later editions entitled *Fieldbook of Natural History* that were produced by other publishers. A three-ring binder of complementary materials for teachers, "Field Book of Nature-Study, with Classroom Outline and Helps in Five Parts" (c. 1928), is held in the library at Cortland State University, Cortland, New York.

48. This *Yearbook of the American Nature-Study Society* was published in the name of the Society in Toledo by its Board of Education in 1925; Martha R. Van Cleve was society president that year and also Director of Nature-Study and General Science for the Toledo schools.

49. *Yearbook of the American Nature-Study Society*, 2-7. The turn to statistics in education is discussed in Callahan, *Education and the Cult of Efficiency*, 12.

50. *Yearbook of the American Nature-Study Society*, 8-17.

51. Ellen Eddy Shaw summarized these efforts in the late 1920s in "The American Nature-Study Society," *Nature Almanac: A Handbook of Nature Education*, 40-42. *Nature Almanac* appeared in 1927 and again in 1930, but then combined with *Nature Magazine*.

52. Outgoing president Van Cleve and board member Mrs. John Sherman particularly reacted to the nomination of Van Evrie Kilpatrick from various comments in the miscellaneous and incomplete records of the ANSS at Cornell. Palmer's role was unclear since he was active in both; see George R. Green to E. L. Palmer, October 4, 1926, ANSS Papers, CUA.

53. A partial list of its officers is summarized in "National Council of Elementary Science," *Science Education* 31 (March 1947): 78-80.

54. Palmer to G. R. Green, February 22, 1926; and Green to Palmer, August 14, 1925, and June 3, 1929, ANSS Papers, CUA. Green claimed that about a thousand teachers were taking nature study extension classes in Pennsylvania.

55. The meetings were reported briefly in *Science* during this period, although it is not clear if the group actually met as planned in Nashville in 1927. For Kansas City, see *Science* 63 (January 29, 1926): 134-35; Philadelphia, *Science* 65 (January 28, 1927); for Nashville's plans, *Science* 66 (December 2, 1927): 532; for New York City, *Science* 68 (September 21, 1928): 614; for Des Moines, *Science* 71 (February 7, 1930): 168; and Cleveland, see *Science* 73 (February 6, 1931): 168.

56. "Dinner for Writers on Nature," *New York Times* (December 23, 1928), 22; "Scientists Open Conclave Today," *New York Times* (December 16, 1928), 14; and *Science* 68 (September 21, 1928): 614.

57. The list provided names and addresses for most people, with the majority clustered in New York and the upper Midwest, alongside pockets in major cities from Portland, Oregon, to Atlanta, Georgia; ANSS Papers, CUA.

58. See B. C. Cady to E. L. Palmer, October 22, 1928, ANSS Papers, CUA.

59. Two self-published anniversary histories, *50th Anniversary of the Webster Grove Nature Study Society, 1920-1970* and *The Webster Grove Nature Study Society, 1920-1995: A Tribute to the 75th Anniversary of WGNSS* (the former with a committee attribution for authorship and the latter without any), were kindly provided to me by Randy Korotev of Washington University. J. A. Drushel, teacher at the Harris Teachers College in St. Louis and president of the ANSS in 1920-21, helped coordinate the local group in 1920. By 1921 there were 129 active members and 116 junior members. The society at one point had property with a lodge and published monthly *Nature Notes*. In the early years it had topical subgroups on herpetology, geology, and ichthyology, and later on botany and ornithology. It also hosted a "nature melody group."

60. A form letter from A. F. Satterthwait to members, November 1, 1931, outlined projects. He described the ANSS as a liaison for "the naturalist, nature lover, field worker, and nature teacher" and a forum for "discussion of methods of presentation of science subject material through lectures, museums, zoological and botanical gardens, visual means, organizations, travel and hiking trips, and schools." ANSS Papers, CUA.

61. When Jennie Hall resigned as secretary after serving for four years, she made the worrisome observation that the local chapters were draining away funds and commitment to the national society. "Report of the Secretary . . . for 1934," ANSS Papers, CUA.

62. The discussion was prompted in part by the fact that the National Council of Supervisors of Elementary Science had changed its name from "nature study and school gardening." An ANSS committee studying the relationship between the two groups recommended vaguely that the society promote school programs "through research, bulletins of subject matter and teaching devices, lectures, museum aids, and other agencies" but avoid duplicating activities of other organizations. The council at the Boston meeting unanimously favored a name change, but a ballot to

the eighty-seven paid members was strongly divided. See memos from Hall to the president and members of the council for 1933 and 1934. There were new but apparently short-lived chapters in Union County, New Jersey, under J. Andrew Drushel, and Chicago under O. D. Frank. ANSS Papers, CUA.

63. Examples from the NEA's annual *Journal of Addresses and Proceedings* include Emma C. Dans, " Nature Study in the Hawaiian Islands" (1926): 631-33; Theodosia Hadley, "Larger Aspect of Nature Study" (1928): 571-75; Loye Holmes Miller, "Nature Study: A Fundamental in Education" (1931): 568; Carroll D. Scott, "Esthetic Value of Nature Study" (1931): 569; John Adams Hollinger, "Most Important Phase of Nature Study in the Schools of Pittsburgh" (1932): 473-74; Ellen Eddy Shaw, "Nature Education at the Brooklyn Botanical Garden" (1932): 474-75.

64. The membership had grown from eight thousand members in 1917 to two hundred thousand in 1930, but while the membership was largely women, male administrators remained in firm control. See Urban and Wagoner, Jr., *American Education: A History*, 241-42.

65. Nature study was one of several prominent themes in youth organizations in Canada; see McKee, "Voluntary Youth Organizations in Toronto, 1880-1930," esp. chap. 6.

66. Nice, *Research is a Passion with Me*, 38; she held nature study classes for her daughter and friends. Margaret Rossiter brought this account to my attention.

67. See his description in Bigelow to Joseph Grinnell, October 15, 1927, Grinnell Papers, UCB-MVZ; and *Nature Almanac*, 63-64. Bigelow remains a somewhat shadowing figure who persisted in nature study for more than forty years, writing and lecturing, but without clear influence and no steady income. Some basic information is in *Who Was Who in America*, 1:93 and *National Cyclopaedia*, 29:104-6. A plaintive letter suggested that his limited education and lack of permanent position made his life as lecturer unstable, as he candidly admitted to John Spencer, November 26, 1898, Spencer Papers, CUA.

68. The early positive anticipations of both men were clear, as were the growing frustrations on both sides when the project eventually used some $15,000 of Tod's funds. Bigelow was worn down from dealing with recalcitrant builders and unsympathetic local businessmen, and the physical labor of trying to landscape the grounds alone. These are detailed in a hundred-page privately printed anonymous booklet *Correspondence between J. Tod Kennedy and Edward F. Bigelow, Innis Arden House, Sound Beach (Old Greenwich) Connecticut*. This effort was made between 1909 and 1911; copy found at Stanford University Library.

69. Wessel and Wessel, *4-H: An American Idea*, 31, 41-44. Warren apparently was the first to use the term 4-H in a government publication in 1918 and remained with the organization for the rest of her career.

70. These rural clubs started just after the turn of the century in Illinois, Ohio, Texas, Iowa, and some southern states, originally primarily for boys but gradually including girls; by 1915, according to one account, over 250,000 girls and boys were enrolled in cooperative extension clubs. Alfred Satterthwait, "A Survey of Twenty Years' Progress," *Nature-Study Review* 17 (February 1921): 71-80. On clubs more generally, see Hadley, "The Relation of Nature-Study to Boys' and Girls' Club Work," *Nature-Study Review* 16 (February 1920): 54-61. The 4-H work was supported in various ways by the U.S. Department of Agriculture and the General Education Fund of the Rockefeller Foundation.

71. Wessel and Wessel, *4-H: An American Idea*, 15.

72. T. Gilbert Pearson, a founder of the National Association of Audubon Societies, discussed its program to distribute bird study materials to schools (for a fee) in "Twenty-Five Thousand Dollars to Aid Teachers in Bird-Study," *Nature-Study Review* 11 (1915): 33-34.

73. The turn to regional study and semi-investigated landscapes in this period is discussed by Kohler, *All Creatures: Naturalists, Collectors and Biodiversity*.

74. Kohlstedt, "Collections, Cabinets and Summer Camp," 10-23.

75. On summer camps, see the text and extensive bibliography in Van Slyck, *Manufactured Wilderness*; and Susan Miller, "Girls in Nature/the Nature of Girls."

76. The summer camp movement generated its own literature as, for example, *Dan Beard's Animal and Camp Fire Stories*.

77. For a discussion of the ways the Camp Fire Girls' outdoor work created sometimes inconsistent themes of individual development and a maternal outlook in its early decades, see Helgren, "Inventing American Girlhood."

78. See http://www.extramile.us/honorees/gulick.cfm (accessed June 18, 2009).

79. J. E. Russell to Francis P. Dodge, August 16, 1917; and other materials that show his involvement with Juliette Low and Mrs. Herbert Hoover (Russell Papers, TC). Margaret Jennings Rogers suggests that the Girl Scouts continued social feminism and, encouraged by Sarah Louise Arnold, emphasized scientific housekeeping in "From True to New Womanhood: The Rise of the Girl Scouts." The scouts turned from the Agassiz Association's emphasis on collecting toward observing nature and recording it in illustrations and photographs.

80. Boy Scout camps typically featured nature museums, which Abigail Van Slyck suggests were more important symbolically than instructionally; see *Manufactured Wilderness*, 77-78.

81. Westell, *Nature Stalking for Boys through Field Glasses, Stereoscopes and Camera*; the volume was introduced by R. S. S. Baden-Powell, British founder of the scouting movement.

82. Bertha Chapman Cady and Lou Hoover had been student colleagues at Stanford and remained friends; see letters from Cady to Hoover in the L. H. Hoover Papers, Hoover Presidential Library. Hoover, who had led a troop in Washington, was elected President of the National Council of Girl Scouts in 1922 while her husband Herbert Hoover was Secretary of Commerce. Her troop turned her front yard into a war garden in 1917, and she accompanied girls on hikes, attended regional conferences, and visited summer scout camps; see Pryon, *Lou Henry Hoover, Gallant First Lady*.

83. The short-lived Coordinating Council on Nature Activities attempted to create collaboration among Boy Scouts of America, Camp Fire Girls, Girl Pioneers, and Girl Scouts; amateur groups like the Horticultural Society of New York, the American Forestry Association, the Torrey Botanical Club, and the Wild Flower Preservation Society; the American Social Hygiene Association and the National Board of the Young Women's Christian Association; and the Nature-Study Society, the Playground and Recreation Association of America, the School Nature League, and the National Council of Supervisors of Nature Study and Gardening. See letterhead, B. C. Chapman to E. L. Palmer, August 6, 1928; and a letter from E. L. Palmer to G. R. Green, April 26, 1927, both in ANSS Papers, CUA. The American Nature Association, with Sarah Louise Arnold on its board, similarly tried to coordinate youth and young adult programs.

84. Comstock Publishing Company produced this undated book.

85. These *Guides* were published in Ithaca by the Slingerland-Comstock Press in 1930.

86. Cady to Grinnell, February 18, 1925, and Grinnell's reply, February 26, 1925, Grinnell Papers, UCB-MVZ.

87. The mixed response is reported in Stanley B. Mulaik, "Nature Counselor Preparation in Relation to the Status of Nature Study in the Camps;" also see Emmy Lou Cooke, "A Study of the Materials and Methods of Nature Study in Camps as Shown by Literature on the Subject."

88. Vinal outlined some distinctions between nature study and nature lore in "Summer Camp and Nature-Study," *Nature-Study Review* 18 (April 1922): 145-46; "The First Nature Lore School," *Nature-Study Review* 17 (April 1921): 145-46; and his [Cap'n. Bill], *Nature Recreation: Group Guidance for the Out-of-Doors*. Written with humor, this book emphasized a nature study outlook and reminded prospective nature leaders, "Nature trips should be pleasurable activity, not endurance tests" (101). See also *Nature Almanac* (1930), 58.

89. Vinal, *Nature Guiding*.

90. Palmer, "The American Nature-Study Society," 183.

91. See the entire *Nature-Study Review* issue on camping, 17 (April 1921), passim; and E. Laurence Palmer, "Nature-study and the Scouting Movement or Vice Versa," *Nature-Study Review*, 168-76; and "Editorial" 14 (May 1919): 223. Gender emphasis is at the core of Miller, "Girls in Nature/the Nature of Girls," and on the Comstock publications, see p. 170.

92. Albright and Taylor, *Oh, Ranger: A Book about the National Parks*, 87.

93. Pauline Sauer Papers, Special Collections, University of Northern Iowa, Cedar Falls, Iowa.

94. Dann, *Across the Great Border Divide*, 67-81.

95. Enos A. Mills, "The Long's Peak Trail School and Nature Guiding," 95-98. Mills created an exhibit at a local inn where flowers, trees, and rock minerals were clearly identified, and he trained the first woman guide licensed to take parties through the park. John Muir had already made his reputation taking visitors such as Teddy Roosevelt to see natural wonders, and it was not uncommon for other serious naturalists to accompany groups who could afford their services. Parks were now making similar service to middle-class visitors.

96. Various accounts of this encounter are found in the Loye Holmes Miller Papers, University of California at Berkeley, Bancroft Library (hereafter UCB-BL). See the entire issue devoted to the origins of "The Nature Guide Service" in the July 1960 issue of *Yosemite*. See Bryant to Joseph Grinnell, January 22, 1920; and Grinnell to Loye Miller, September 12, 1921, Grinnell Papers, UCB-MVZ.

97. Miller worked with C. M. Goethe to form the California Nature Study League whose goal was "conservation through education." See Miller, "Memo of September 22, 1959," Miller Papers, UCB-BL. Also see Goethe, "The California Nature Study League," *Nature-Study Review* 16 (May 1920): 204-7; among its projects were nature study libraries in selected state and national parks.

98. H. C. Bryant to Joseph Grinnell, March 2, 1925, Grinnell Papers, UCB-MVZ.

99. Pappas, "Forest Scholars," 91-97.

100. Polly Welts Kaufman's *National Parks and the Woman's Voice* provides a very useful introduction to the development of the park service as well as women's roles in it.

101. Kaufman, *National Parks and the Woman's Voice*, 32-36. Also see the General Federation of Women's Clubs' support discussed in educational terms in Comstock, "Editorial," *Nature-Study Review* 14 (February 1918): 83.

102. Cited in Pappas, "Forest Scholars," 187; also see Glenda Riley, *Women and Nature* and Kaufman, *National Parks and Woman's Voice*.

103. Kaufman, *National Parks and the Woman's Voice*, 67-68.

104. Ibid., 139.

105. McMurry, *Special Method in Elementary Science for the Common School*.

106. There are numerous examples from across the country, but see Kansas City Missouri School District Board of Education, *Tentative Course of Study in Nature Study and Elementary Science for Grades I-VI* (Kansas City: Kansas City Public Schools, 1930); Long Beach Board of Education, *Nature Study and Elementary Science, Course of Study for Grades One, Two . . . and Six* (Long Beach: Long Beach City Schools, 1931); and Pittsburgh Board of Public Education, *Course of Study in Nature Study and Elementary Science* (Pittsburgh: Board of Public Education, 1928); all these are in the TC files.

107. Persing and Peeples, *Elementary Science by Grades: A Nature Study and Science Reader*.

108. Kim Tolley makes the claim that biological rather than physical sciences dominated because of the preponderance of women teachers with expertise in such natural sciences (*Science Education for Girls*, 218-19). Interestingly, under the editorship of Anna Comstock, *Nature-Study Review* rather consistently published articles on astronomical topics; see *Nature-Study Review* 17 (January 1921): 1-34, which included essays on building a sundial and had accounts of the sun, moon, and constellations.

109. Craig, for example, presided over the opening session of the ANSS annual meeting in

New York City in 1928 and also led discussion on state programs and supervision. See pamphlet copy of the *Program of the Thirty-Second Annual Meeting, held at the American Museum of Natural History* (n.p., 1928).

110. Champagne and Klopfer, "Pioneers of Elementary-School Science: IV. Gerald Spellman Craig," 7–24.

111. Craig, "Certain Techniques in Developing a Course of Study in Science." Also see Arnold, "Nature Study: An Analytical History of Its Demise."

112. A copy is in the Gerald Spellman Craig Papers, Wisconsin Historical Society, Madison. Most of the collection is from the period after 1932, but there are some notes from the earlier period at Columbia as well as many of his publications.

113. During the 1910s, elementary education could be overshadowed as high schools expanded and their standards pushed lower grades toward particular curricula. See Krug, *Shaping of the American High School*.

114. Champagne and Klopfer, "Pioneers of Elementary School Science IV: Gerald Spellman Craig," 9. This statement rings hollow, since even his supportive colleagues found Jackman's writing somewhat dense and he was rarely quoted directly.

115. Billig, *Technique for Developing Content*. On her career, see Champagne and Klopfer, "Pioneers of Elementary School Science V: Florence Grace Billig," 149–68.

116. The Association attracted men in university positions in science education who sought a stronger professional image. Elliot Downing was a founding member and president from 1930 to 1932; Otis Caldwell was also active. See discussion of the group in Tolley, *Science Education of American Girls*, 188–89; and biographical sketches of Downing in Ohles, ed., *Biographical Dictionary of American Education*, 1:393–97; and of Caldwell in Ohles, ed., 1:227–28.

117. The debates must have been lively among the committee members, although no records have been found to document them; the report was patronizing and nature study was effectively displaced. *Elementary School Science*, 3–4, TC.

118. Palmer initially seemed to believe that compromise would be possible with educators like Craig, but, by 1936, he resented the fact that leaders who proposed "science education" had deliberately "overlooked" the offerings of Jackman, Comstock, Bailey, Bigelow, and others and thus had "failed to profit by what they did." Palmer to J. A. Drushel, October 14, 1936, ANSS Papers, CUA.

119. E. L. Palmer to Elizabeth Downhour, April 15, 1930, Palmer Papers, CUA. Palmer noted, "Calling a good course one thing or another will not make it better any more than calling a bad course something else will make it good."

120. Craig's ideas are outlined in a thirty-page publication entitled *A New Science Program for Elementary Schools* (the phrase was copyrighted by Craig), 24.

121. Craig, *New Science Program*, 24.

122. Champagne and Klopfer, "Pioneers of Elementary School Science: IV. Gerald Spellman Craig," 16.

123. The calculation is in Tolley, *Science Education of American Girls*, 270, n. 69.

124. Tolley, "Science Education of American Girls," 312–15.

125. Gerald S. Craig, *Science for the Elementary School Teacher*, 27, 17. By the 1958 edition of his text, nature study was simply ignored.

126. The 1920s were a "time of great accomplishment for public education" in terms of teacher salaries, school facilities, and curricular diversification, but the outcomes also meant a streamlined educational ladder as "pedagogical progressives lost out to the administrative progressives." See Urban and Wagoner, Jr., *American Education: A History*, chaps. 7–8, quote on 226.

127. Lane, *Teacher in Modern Elementary Science*, 190–91.

128. Powers, "Preface," in Underhill, *Origin and Development of Elementary School Science*, vii. The text carries similar themes.

129. Harrington Wills of the Santa Barbara Field School of Nature Study pointed out the multiple techniques in his *Teaching of Nature Study and the Biological Sciences*, 7.

130. These and other programs are discussed in Hadsall, "The Extension Activities of Certain Publicly Supported Institutions," 7–11. On the growth of science on the radio, see Lafollette, "A Survey of Science Content in U.S. Radio Broadcasting," 4–33; and her *Science on Air*.

131. Wayne Urban in "Historical Studies of Teacher Education" points out that in university education programs the disciplinary faculty tended to follow their own interests and typically denigrated the schools even as they called on teachers to be more scientific in their teaching.

132. Emerson, *Nature Study Teachers Manual, Elementary Schools, State of Oregon*.

133. Open letter to ANSS members by Jennie Hall, November 14, 1932. The next summer a more dramatic call-to-arms was issued by Theodosia Hadley, chair of the membership committee, suggesting that members had been "poor salesmen" for nature study and needed to advertise its value to parents, education departments, and teacher-training institutions: "We must wake up. Organize! Pool our interests! Be a vital force in our country. Sell our value to the public." Letter dated June 3, 1933. Both in ANSS Papers, CUA.

134. E. L. Palmer to G. R. Green, April 26, 1927, ANSS Papers, CUA.

CONCLUSION

1. This quotation from Rachel Carson's *The Sense of Wonder*.

2. Environmental historians have long noted this tension, from the early work of Samuel Hays in his *Conservation and the Gospel of Efficiency* to accounts like that of Thomas Dunlap's *Saving America's Wildlife*. Mark Barrow makes an explicit connection between the thinking of nature study advocates and the early environmental movement in his "Naturalists as Conservationists."

3. To be fair, recent authors have found few institutional accounts of nature study and relied on commentaries from the 1940s, often critical, to describe it in passing. See notes in the introduction and Pauly, *Biologists and the Promise of American Life*; and Richard A. Overfield, *Science with Practice*.

4. Kohlstedt, "Nature Not Books," 324–52.

5. On the withdrawal of scientists from public engagement and education, see Burnham, *How Superstition Won and Science Lost*. On the importance of science education, see Klug, *Shaping of the American High School*; and Coulter, "Mission of Science in Education," 281–93; this frequently cited commencement address discussed science education as a significant factor in contemporary culture.

6. Beatty, "Psychologizing the Third R." Beatty also makes the point that all these educational psychologists spent time teaching and did some of their best work while they were actively involved with teachers.

7. Anna Botsford Comstock referenced the relatively small numbers of teachers who joined the Nature-Study Society and challenged the use of numbers as a measure of influence in an editorial comment in *Nature-Study Review* 19 (December 1923): 413.

8. Patterson, "Survey of Twenty Years Progress," 55–62.

9. Opportunities faded in the late 1920s and 1930s for women in science generally. See Rossiter, *Women Scientists in America*; and Kaufman, *National Parks and the Woman's Voice*, 70–73.

10. David Tyack, *One Best System*, noted that the more centralized the system, the more the "employment of women as teachers thus augmented the authority of largely male leadership," 60.

11. See Bigelow's chapter on "Sissies and Tomboys" in *Spirit of Nature-Study*, 14–26.

12. Tolley, *Science Education of American Girls*, 228.

13. Herbst, "Teacher Preparation in the Nineteenth Century," 231–33. New York State's

example is instructive, as its eleven normal schools established between 1844 and 1897 had their curriculum and certification programs brought under centralized control by 1917; many were subsequently renamed teachers colleges. These were consolidated into the State University of New York system in 1948; see Corby, "Centralization of Educational Administration."

14. Tyack and Hansot, *Managers of Virtue*, 108.

15. Urban and Wagoner, Jr., *American Education: A History*, chaps. 7 and 8.

16. While introducing an expanded curriculum to include more history, geography, and natural science was a significant change in the late nineteenth century, it does seem that there was more "tinkering" with both content and method in the twentieth century; see Tyack and Cuban, *Tinkering toward Utopia*.

17. Munson, *Education through Nature Study*, 25.

18. This insight came from reading Seymour Papert's review of *Tinkering toward Utopia* by Tyack and Cuban in *The Journal of the Learning Sciences* 6 (1997): 417–27.

19. It is beyond the scope of this volume to detail the long-term traces of nature study. In fact, the Webster Grove Nature Study Society still exists and so do other centers that use the phrase, sometimes making a conscious connection to the early movement. Today nature study is viewed as essentially synonymous with outdoor education and conservation education, according to the description of "Environmental and Ecological Education," in Husin and Postlethwaite, *International Encyclopedia of Education*, 1993.

20. Much of this childhood background is documented in Linda Lear's authoritative *Rachel Carson: Witness for Nature*, 7–21.

21. Rachel Carson's best remembered books are *The Sea Around Us*, *The Edge of the Sea*, and the powerful insecticide indictment *Silent Spring*, which was initially serialized in *The New Yorker* magazine. Her final book, *The Sense of Wonder*, was based on an earlier essay, "How to Help Your Child Wonder," which was published posthumously.

22. Richard Louv's *Last Child in the Woods* (2005) strikes many of the editorial chords played by the founding advocates of nature study. As this book is in its final stages of completion, there are local programs and an effort to gain federal support using such titles as "no child left indoors" and "no child left inside" that emphasize the importance of enhancing the lives and education of children through nature.

Bibliography

MANUSCRIPT COLLECTIONS

The following manuscript list provides abbreviations for those libraries and repositories that appear more than twice in the notes.

AAS   *American Antiquarian Society, Worcester*: Worcester Natural History Society Papers

AMNH   *American Museum of Natural History, Special Collections, New York*: Agassiz Association Papers; Henry Fairfield Osborn Papers

APS   *American Philosophical Society, Philadelphia*: Charles Davenport Papers

BSNH   *Boston Society of Natural History, now in the Boston Museum of Science Archives, Boston*: Samuel Scudder Papers

BPL   *Boston Public Library, Boston*

CHS   *Chicago Historical Society (now the Chicago History Museum), Chicago*: Flora Juliette Cooke Papers

CSL   *Connecticut State Library, Hartford*

CUA   *Cornell University Library, Division of Rare and Manuscript Collections, Ithaca*: American Nature-Study Society Papers; Clara Keopka Trump Papers; College of Agriculture and Extension Papers; Cornell University Press/Comstock Publishing Company Papers; E. Lawrence Palmer Papers; James George Needham Papers; John and Anna Botsford Comstock Papers; Liberty Hyde Bailey Papers; Spencer Extension Education Papers

ClUA   *Clark University, Worcester*: G. Stanley Hall Papers

HHPL   *Herbert Hoover Presidential Library, West Branch*: Lou Henry Hoover Papers

ISUN   *Illinois State University at Normal, Normal*: Charles A. McMurry Papers; John G. Coulter Papers

LC   *Library of Congress, Washington, DC*: Mira Dock Papers; William Hornaday Papers

MHS   *Minnesota Historical Society, St. Paul*: Dietrich Lange Papers; Hiram Slack Papers; Lottie M. Howard and Family Papers

NISU   *Northern Illinois State University, DeKalb*: Charles A. McMurry Papers; Lida Brown McMurry Papers

NYPL   *New York Public Library, New York*

PUL   *Princeton University Firestone Library, Princeton*: Alpheus Hyatt Papers

RCA	*Rockford College Archives, Rockford*: Ruth Marshall File
SCA	*Simmons College Archives, Boston*: Sarah Louise Arnold Papers
SIA	*Smithsonian Institution Archives, Washington, DC*: Spencer F. Baird Papers
SLR	*Schlesinger Library, Radcliffe Institute, Harvard University, Cambridge*: Alice Rich Northrup Papers; Elizabeth Agassiz Papers
SPL	*Syracuse Public Library, Syracuse*: Syracuse City School District Papers
SABS	*Saskatchewan Archives Board, Saskatoon, Saskatchewan*
SCHS	*Suffolk County Historical Society, Rivershead*: Grace Miller Papers
StJM	*St. Johnsbury Museum and Planetarium, St. Johnsbury*: Director's Papers (Delia Griffin)
SUA	*Stanford University Archives, Stanford*: David Starr Jordan Papers; Hopkins Marine Station Papers; Oliver P. Jenkins
TC	*Teachers College Library, Columbia University, New York*: James Earl Russell Papers; John Francis Woodhull Papers; William Fletcher Russell Papers
UCB-BLB	*University of California at Berkeley, Bancroft Library, Berkeley*: Division of Agricultural Education Papers; Elmer Ellsworth Brown Papers; Helen Swett Papers
UCB-MVZ	*University of California at Berkeley, Museum of Vertebrate Zoology, Berkeley*: Joseph Grinnell Papers
UCA	*University of Chicago Archives, Regenstein Library, Chicago*: Chicago Institute Papers; College of Education Papers; Francis Wayland Parker Papers; School of Education Papers
UCLA	*University of California at Los Angeles, Archives and Special Collections, Los Angeles*: Corrine A. Seeds Papers; Edward Hyatt Papers; Jesse Fonda Millspaugh Papers
UG	*University of Guelph, Guelph, Ontario*: Eveline Hocking Papers
UIA	*University of Illinois Archives, Urbana-Champaign*: Natural History Survey, Chief's Office (Stephen A. Forbes)
UMN	*University of Minnesota, Minneapolis*: Social Welfare Archives, Anderson Library
UNA	*University of Nebraska Archives, Lincoln*: Charles Bessey Papers
URSC	*Rush Rhees Library, Special Ccollections, University of Rochester*: Ira S. Wile Papers
USNA	*United States National Archives, Washington, DC*: Records of the Bureau of Indian Affairs; Records of the Office of the Commissioner of Education
VUA	*Vanderbilt University Archives, Nashville*: Bruce R. Payne Papers; Charles C. McMurry Papers; Lucy Gage Papers
WHS	*Wisconsin Historical Society, Madison*: Gerald Spellman Craig Papers; Louise Bailey Papers; McCormick Family Papers (Anita McCormick Blaine)
WSUC	*Washington State University at Cheney Archives, Cheney*

CHILDREN'S POPULAR BOOK AND TEXTBOOK COLLECTIONS

Blackwell History of Education Museum and Research Collection, Northern Illinois State University

Children's Literature Research Library, University of Minnesota

Collection of Syllabi, Milbank Memorial Library, Teachers College, Columbia University

Emmanuel Rudolph Textbook Collection, Ohio State University

Monroe C. Gutman Library Special Collections, Harvard University

School Collection of Children's Literature, University of Illinois, Urbana-Champaign

GOVERNMENT SERIALS

Boston School Committee, *Annual Report*
Cornell University Agricultural Experiment Station, *Annual Report*
Cornell University College of Agriculture, *Annual Report*
Massachusetts Board of Education, *Bulletin*
——, *Annual Report*
Minneapolis Board of Education, *Annual Report*
New York City Board of Education, *Annual Report*
New York City Superintendent of [sometimes indicated as Public] Schools, *Annual Report*
New York State Education Department, *Annual Report*
——, *Journal*
St. Paul Board of Education, *Annual Report*
State Normal School of Los Angeles, *Catalogue*
United States Bureau of Education, *Bulletin*
United States Commissioner of Education, *Annual Report*
United States Secretary of Agriculture, *Annual Report*

FREQUENTLY CITED
BIOGRAPHICAL DICTIONARIES

American Men and Women of Science. New York: Science Press, 1944.
Biographical Dictionary of American Educators, ed. John F. Ohles. 3 vols. Westport: Greenwood
　　Press, 1978.
*Biographical Dictionary of Women in Science: Pioneering Lives from Ancient Times to the Mid-
　　Twentieth Century*, ed. Marilyn Bailey Ogilvie and Joy Dorothy Harvey. 2 vols. New York:
　　Routledge, 2000.
A Cyclopedia of American Education, ed. Paul Monroe. 6 vols. New York: Macmillan, 1914.
National Cyclopedia of American Biography. Clifton: T. H. White, 1893–
Notable American Women, 1607–1950, ed. Edward T. James and Janet James. 3 vols. Cambridge:
　　Harvard University Press, 1971.
Notable American Women: The Modern Period, ed. Barbara Sicherman and Carol Hurd. Cambridge:
　　Belknap Press, 1980.
Notable American Women: A Biographical Dictionary Completing the Twentieth Century, ed. Susan
　　Ware and Carol Hurd Green. Cambridge: Belknap Press, 2004.
Notable Women in the Life Sciences: A Biographical Dictionary, ed. Benjamin F. Shearer and Bar-
　　bara S. Shearer. Westport: Greenwood Press, 1980.
Who Was Who in America. Chicago: Marquis, 1899–.

OTHER PUBLISHED SOURCES

Abbott, Charles C. *Young Folks' Cyclopedia of Natural History.* New York: A. L. Burt, 1895.
Abbott, Jacob. *The Teacher; or Moral Influences Employed in the Instruction and Government of the
　　Young; Intended Chiefly to Assist Young Teachers in Organizing and Conducting their Schools.*
　　Boston: William Peirce, 1836.
Abbott, Jacob. *Rollo's Museum.* New York: Crowell and Co., [1854] 1855.
Adams, David Wallace. *Education for Extinction: American Indians and the Boarding School Experi-
　　ence, 1875–1928.* Lawrence: University of Kansas Press, 1995.

Agassiz Association, The. Stamford, n.d.

Agassiz Association. *The First General Convention of the Agassiz Association.* Philadelphia: Times Printing House, 1884.

Agassiz, E[lizabeth] C. *A First Lesson in Natural History.* Boston: Ginn and Heath, 1879.

Agassiz, Louis. "Methods of Study in Natural History." *Atlantic Monthly* 10 (September 1862): 336.

Ahearne, Margaret. "Nature Study in the Gary Schools." *Nature-Study Review* 11 (February 1915): 59–63.

Albright, Horace M., and Frank J. Taylor. *Oh, Ranger: A Book about the National Parks.* Palo Alto: Stanford University Press, 1928.

Albisetti, James C. "The Feminization of Teaching in the Nineteenth Century: A Comparative Perspective." *History of Education* 22 (1993): 253–63.

Alexander, Edward P. *The Museum in America: Innovators and Pioneers.* Walnut Creek, CA: Altamira Press, 1997.

Allen, David E. *The Naturalist in Britain.* London: Urwin Publishing Co., 1976.

Allen, Elsa G. "Winter Birds." *Nature-Study Review* 19 (May 1923): 204–22.

American Academy of Medicine. *Studies of the Present Teaching of Hygiene through "Domestic Science" and through "Nature Study": Being the Second Section of the Report of the Committee to Investigate the Teaching of Hygiene in Public Schools.* Pamphlet dated April 1906.

"The American Nature-Study Society." *Science* 57 (February 9, 1923): 183–85.

Anderson, James D. "The Hampton Model of Normal School Industrial Training, 1868–1900." In *New Perspectives on Black Education.* Edited by Walter Feinberg and Henry Rosemont, Jr. Urbana: University of Illinois Press, 1975.

Andrews, Jane. *Seven Little Sisters Who Lived on the Round Ball that Floats in the Air.* Boston: Lee and Shepard, [1861] 1888.

———. *The Stories Mother Nature Told Her Children.* Boston: Ginn and Co., 1888.

Andrei, Mary Anne. "The Accidental Conservationist: William T. Hornaday, the Smithsonian Bison Expeditions and the US National Zoo." *Endeavor* 29 (September 2005): 109–13.

"Annual Meeting and Election of Officers." *Nature-Study Review* 9 (November 1913): 235.

Applegate, Cecelia. *A Nation of Provincials: The German Idea of Heimat.* Berkeley: University of California Press, 1990.

Armitage, Kevin Connor. "Bird Day for Kids: Progressive Conservation in Theory and Practice." *Environmental History* 12 (July 2007): 528–51.

———. "Knowing Nature: Nature Study and American Life, 1873–1923." Ph.D. diss., University of Kansas, 2004.

Armstrong, Henry E. "Some Recent Criticisms of Nature Study." *Nature-Study Review* 1 (January 1905): 22–26.

Arnold, Charles. "Nature Study: An Analytical History of Its Demise." M.S. thesis, Cornell University, 1976.

Arnold, Sarah Louise. *Waymarks for Teachers Showing Aims, Principles, and Plans of Everyday Teaching with Illustrative Lessons.* New York: Silver, Burdett, and Co., 1894.

Azelvandre, John P. "Forging the Bonds of Sympathy: Spirituality, Individualism and Empiricism in the Ecological Thought of Liberty Hyde Bailey and Its Implications for Environmental Education." Ph.D. diss., New York University, 2001.

Bachman, Frank P. *Public Schools of Nashville, Tennessee: A Survey Report.* Nashville: Peabody College for Teachers, 1931.

Bailey, Liberty Hyde. *The Country Life Movement in the United States.* New York: Macmillan, 1911.

———. *The Holy Earth.* New York: Scribner's Sons, 1915.

———. "How the Squash Gets Out of the Seed." *Nature-Study Leaflet* 1 (1896).

———. *Nature-Study for the Grammar Grades: A Manual for Teachers and Pupils below the High School in the Study of Nature.* New York: Macmillan, 1898.

———. *The Nature-Study Idea: Being an Interpretation of the New School Movement to Put the Child in Sympathy with Nature.* New York: Doubleday and Co., 1903.

———. "Nature Study Movement." National Education Association. *Journal of Addresses and Proceedings* (1903): 109–16.

———. *The Outlook to Nature.* New York: Macmillan, 1905.

———. *Talks Afield about Plants and the Science of Plants.* Boston: Houghton, 1885.

———. *Wind and Weather.* New York: Scribner's Sons, 1916.

Bailey, Margaret Emerson. *Good-bye, Proud World.* New York: Charles Scribner's Sons, 1945.

Ballard, Harlan H. *Hand-book of the St. Nicholas Agassiz Association.* Boston: Lothrop, Lee, and Shepard, 1884.

———. "History of the Agassiz Association." *Science* 9 (January 28, 1887): 93–96.

———. *Three Kingdoms: A Handbook of the Agassiz Association.* New York: Writers Publishing Co., 1888.

Barringer, Mark Daniel. *Selling Yellowstone: Capitalism and the Construction of Nature.* Lawrence: University Press of Kansas, 2002.

Barrow, Mark V., Jr. "Naturalists as Conservationists: American Scientists, Social Responsibility and Political Activism before the Bomb." In *Science, History, and Social Activism: A Tribute to Everett Mendelsohn.* Edited by Garland Allen and Roy M. MacLeod. Dordrecht: Kluwer Academic Publishers, 2001.

———. *A Passion for Birds: American Ornithology after Audubon.* Princeton: Princeton University Press, 1998.

Barbauld, Anna Letitia. *Lessons for Children from Four to Five Years Old.* Philadelphia: B. F. Bache, [1778] 1788.

Barton, Anna Parfitt. "The Bagworm Drive." *Nature-Study Review* 16 (February 1920): 109–12.

Baym, Nina. *American Women of Letters and the Nineteenth-Century Sciences: Styles of Affiliation.* New Brunswick: Rutgers University Press, 2001.

Beadie, Nancy, and Kim Tolley, eds. *Chartered Schools: Higher Schooling and American Social Life, 1740–1940.* New York: Routledge, 2002.

Beal, W. J. "What is Nature Study?" *Science* 15 (June 20, 1902): 991–92.

Beal, W. J., A. S. Packard, John M. Coulter, C. P. Gillette, W. M. Davis, A. E. Verrill, David Starr Jordan, and Thomas H. McBride. "What is Nature Study?" *Science* 16 (December 5, 1902): 910–13.

Beard, Dan. *Dan Beard's Animal and Camp Fire Stories.* New York: Moffat and Co., 1907.

Beatty, Barbara. "Child Gardening: The Teaching of Young Children in American Schools." In *American Teachers: Histories of a Profession at Work.* Edited by Donald Warren. New York: Macmillan, 1989.

——— "Psychologizing the Third R: Hall, Dewey, Thorndike, and Progressive Era Ideas of the Learning and Teaching of Arithmetic." In *When Science Encounters the Child: Education, Parenting, and Child Welfare in 20th-century America.* Edited by Barbara Beatty, Emily D. Cahan, and Juliet Grant. New York: Teachers College Press, 2006.

Beaver, Donald deB. "Altruism, Patriotism, and Science: Scientific Journals in the Early Republic." *American Studies* 12 (1971): 5–19.

Bederman, Gail. *Manliness and Civilization: A Cultural History of Gender and Race in the United States, 1880–1917.* Chicago: University of Chicago Press, 1995.

Bellisario, Joseph. "E. Laurence Palmer: His Contribution to Nature, Conservation, and Science." Ph.D. diss., Pennsylvania State University, 1969.

Bender, Thomas. *New York Intellect: A History of Intellectual Life in New York City, from 1750 to the Beginnings of Our Own Time.* New York: Knopf, 1987.

Bengaugh, Thomas. *Learning How to Do and Learning by Doing.* Toronto: Cameron, 1902.

Benson, Keith R. "From Museum Research to Laboratory Research." In *The American Development of Biology.* Edited by Ronald Rainger, Keith Benson, and Jane Maienschein. Philadelphia: University of Pennsylvania Press, 1988.

Bergen, J. Y. "Economic Biology in the Schools." *Nature-Study Review* 5 (April 1909): 108–9.

Bernard, Richard M., and Maris A. Vinovskis. "The Female Teacher in Ante-Bellum Massachusetts." *Journal of Social History* 10 (1977): 332–45.

Bibliography of Science Teaching. United States Department of Education. *Bulletin* 446 (1911).

Biennial Report of the [Nebraska] State Superintendent of Public Instruction for 1901–1903. Lincoln, 1903.

Bigelow, Edward F. *How Nature Study Should be Taught: Inspiring Talks to Teachers.* New York: Hinds, Noble, Eldredge, 1904.

———. *The Spirit of Nature Study.* New York: A. S. Barnes, 1907.

Bigelow, Maurice A. "The American Nature-Study Society." *Nature-Study Review* 3 (September 1907): 159–61.

———. "Are Children Naturally Naturalists?" *Nature-Study Review* 3 (November 1907): 236–39.

———. "Articles by Those Who Think and Do." *Nature-Study Review* 1 (March 1905) 80–81.

———. "Best Books for Nature Study." *Nature-Study Review* 2 (April 1906): 168–77.

———. "Editorial." *Nature-Study Review* 1 (May 1905): 140.

———. "Editorial." *Nature-Study Review* 1 (September 1905): 233–34.

———. "First Meeting of the American Nature-Study Society." *Nature-Study Review* 4 (January 1908): 4.

———. "Introduction." *Nature-Study Review* 1 (January 1905): 1–2.

———. "Nature Study and Its Relation to Natural Science: A Symposium by H. W. Fairbanks, C. F. Hodge, T. H. McBride, F. L. Stevens, and M. A. Bigelow." *Nature-Study Review* 1 (January 1905): 14–19.

———. "Outlines of Nature-Study in the Horace Mann School." *Teachers College Record* 5 (March 1904): 35 ff.

———. "Protective Color of Animals: Review of an Article by John Burroughs." *Nature-Study Review* 1 (July 1905): 159–64.

———. "The Scope and Method of Nature Study." *Nature-Study Review* 2 (January 1906): 35–36.

———. *Sex Education: A Series of Lectures Concerning Knowledge of Sex in Its Relation to Human Life.* New York: Macmillan, 1918.

———. *Sex-Instruction as a Phase of Social Education.* New York: American Social Hygiene Association, 1913.

———. "Training Teachers of Nature-Study." *Nature-Study Review* 2 (April 1906): 121–35.

———. "The Washington Meeting." *Nature-Study Review* 8 (January 1912): 1–7.

———. "Wilbur S. Jackman." *Nature-Study Review* 3 (March 1907): 65–67.

Bigelow, Maurice A., Thomas M. Baillet, and Prince Albert Morrow. *Report of the Special Committee on the Matter and Methods of Sex Education.* New York: American Federation for Sex Hygiene, 1913.

Bigelow, Maurice A., and Francis Lloyd. *The Teaching of Biology in Secondary Schools.* New York: Longman's Green, 1904.

Biklen, Sari Knopp. *School Work: Gender and the Cultural Construction of Teaching.* New York: Teachers College Press, 1995.

Billig, Florence Grace. *A Technique for Developing Content for a Professional Course in Science for Teachers in Elementary Schools.* New York: Teachers College, Columbia University, 1930.

Blair, Francis G. "Nature and Other Subjects of Instruction." *Nature-Study Review* 4 (December 1908): 259–60.

Blaisdell, Alfred. *Our Bodies and How We Live: An Elementary Textbook of Physiology and Hygiene for Use at Schools with Special Reference to the Effects of Alcohol Drinks, Tobacco, and Other Narcotics on the Body.* Boston: Ginn and Co., 1896.

Blanchan, Neltje. *Birds that Every Child Should Know.* New York: Doubleday, Page, and Co., 1914.

Blount, Jackie M. *Destined to Rule the Schools: Women and the Superintendency, 1873–1995.* Albany: State University of New York, 1998.

Blum, Ann Shelby. *Picturing Nature: American Nineteenth-Century Zoological Illustration*. Princeton: Princeton University Press, 1993.

Bonta, Marcia Myers. *Women in the Field: America's Pioneering Women Naturalists*. College Station: Texas A&M University Press, 1991.

Boston School Committee. *Annual Reports*. 1890–1905.

Bowen, William. *The Country Life Movement in America*. Port Washington: Kennikat Press, 1974.

Boyden, Arthur C. "American Nature-Study Society." *Science* 57 (February 9, 1923): 184.

———. *A History of Bridgewater Normal School*. Bridgewater: Bridgewater Normal Alumni Association, 1933.

———. *Nature Study by Months: Part 1, For Elementary Grades*. Boston: New England Publishing Co., 1898.

———. *Nature Study in the Public Schools of Massachusetts*. Boston: Nathan Sawyer and Sons, 1893.

———. "Nature Study: Then and Now." *Nature-Study Review* 19 (March 1923): 93–96.

Bradford, Mary D. *Memoirs of Mary D. Bradford: Autobiographical and Historical Reminiscences of Education in Wisconsin, through Progressive Service from Rural School Teaching to City Superintendent*. Evansville: Antes Press, 1932.

Branch, E. Douglas. *The Sentimental Years, 1835–1860*. New York: Appleton, Century, and Co., 1934.

Brazzel, Johnetta Cross. "Brick without Straw: Missionary Sponsored Higher Education in the Post-Emancipation Era." *Journal of Higher Education* 63 (January–February 1992): 41–42.

Brittain, John. *Elementary Agriculture and Nature Study*. Toronto: Educational Book Co., 1909.

———. *Nature Study and Agriculture*. Toronto: Educational Book Co., 1911.

———. *Teachers' Manual of Nature Lessons for the Common Schools*. Saint Johns: J. and A. McMillan, 1896.

Broadhurst, Jean. "Notes on New Books and Pamphlets." *Nature-Study Review* 1 (September 1905): 229.

Brumberg, Joan Jacobs, and Nancy Tomes. "Women in the Professions: A Research Agenda for American Historians." *Reviews in American History* 10 (1982): 275–95.

Bryan, C. D. B. *The National Geographic Society: 100 Years of Adventure and Discovery*. New York: Harry N. Abrams, Ind., 1987.

Buckley, Arabella B. *By Pond and River*. New York: Funk and Wagnalls Co., 1901.

———. *Life and Her Children: Glimpses of Animal Life from Amoeba to the Insects*. New York: Appleton, [1873] 1898.

Buell, Lawrence. *The Environmental Imagination: Thoreau, Nature Writing, and the Formation of American Culture*. Cambridge: Harvard University Press, 1995.

Bullough, William A. *Cities and Schools in the Gilded Age: The Evolution of an Urban Institution*. Port Washington: Kennicutt Press, 1974.

Burlingame, Roger. *Of Making Many Books: A Hundred Years of Reading, Writing, and Publishing*. New York: Charles Scribner's Sons, 1946.

Burnham, John. *How Superstition Won and Science Lost: Popularizing Science and Health in the United States*. New Brunswick: Rutgers University Press, 1987.

Burns, Virginia Law. *Gentle Hunter: A Biography of Alice C. Evans, Bacteriologist*. Laingsburg, MI: Enterprise Press, 1993.

Burroughs, John. "Real and Sham in Natural History." *Atlantic Monthly* 91 (1903): 299.

———. *Wake Robin*. Boston: Houghton Mifflin, 1895.

Burstyn, Joan. "Early Women in Education: The Role of the Anderson School of Natural History." *Journal of Education* 159 (1977): 50–64.

Burt, Mary. *Little Nature Studies for Little People from Essays by John Burroughs*. New York: Ginn and Co., 1897.

Cady, Bertha Chapman. "The Normal Schools and Colleges and the Problem of Sex Education." *Social Hygiene* 3 (July 1917): 367–77

———. *Tami: The Study of a Chipmunk*. Ithaca: Comstock Publishing Co., 1927.

Cady, Bertha Chapman, and Vernon Mosher Cady. *The Way Life Begins: An Introduction to Sex Education*. New York: American Social Hygiene Association, 1917.

Cady, Marion E. *Bible Nature Studies: A Manual for Home and School*. Oakland: n.p., 1902.

Caldwell, Otis W. "The Problem that Faces Us." *Nature-Study Review* 6 (January 1910): 21–23.

Callahan, Raymond. *Education and the Cult of Efficiency*. Chicago: University of Chicago Press, 1962.

Campbell, Eleanor. *Reflections of Light: A History of the Saskatoon Normal School, 1919–1953, and the Saskatoon Teachers' College, 1953–1964*. Saskatoon: University of Saskatchewan Press, 1996.

Carey, Jane. "Departing from Their Sphere: Australian Woman and Science, 1880–1960." Ph.D. diss., University of Melbourne, 2003.

Carney, Mabel. *Country Life and the Country School: A Study of the Agencies of Rural Progress and of the Social Relationship of the School to the Country Community*. Chicago: Row, Peterson, and Co., 1912.

Carss, Elizabeth. "Outline of a Course on Nature Study in the Horace Mann School." *Teachers College Record* 1 (March 1900):1–64.

Carson, Rachel. *The Edge of the Sea*. New York: Houghton Mifflin, 1955.

———. *The Sea Around Us*. New York: Oxford University Press, 1951.

———. *The Sense of Wonder*. New York: Harper Collins [1956] 1998.

———. *Silent Spring*. New York: Houghton Mifflin, 1962.

Carter, John C. "Ryerson, Hodgins, and Boyle: Early Innovators in Ontario School Museums." *Ontario History* 86 (June 1994): 21–27.

Carter, Marion Hamilton. *Nature Study with Common Things: An Elementary Laboratory Manual*. New York: American Book Co., 1904.

Carver, George Washington. *Suggestions for Progressive and Correlative Nature Study*. Tuskegee, AL, 1902.

Catalogue of the Educational Exhibition of the Commonwealth of Massachusetts of the World's Columbian Exposition. Chicago, 1893.

Champagne, Audrey B., and Leopold E. Klopfer. "Pioneers of Elementary School Science: I. Wilbur Samuel Jackman." *Science Education* 63 (1979): 145–75.

———. "Pioneers of Elementary-School Science: II. Anna Botsford Comstock." *Science Education* 63 (1979): 299–322.

———. "Pioneers of Elementary-School Science: III. Ephraim Laurence Palmer." *Science Education* 63 (1979): 557–90.

———. "Pioneers of Elementary-School Science: IV. Gerald Spellman Craig." *Science Education* 64 (1980): 7–24.

———. Pioneers of Elementary-School Science: V. Florence Grace Billig." *Science Education* 64 (1980): 149–67.

Channing, William. "On the Moral Uses of the Study of Natural History." In *Lectures*. American Institute of Instruction. Boston: Carter, Hindee, and Co., 1835.

Chapman, Bertha. "School Gardens in the Refugee Camps of San Francisco." *Nature-Study Review* 2 (1907): 225–29.

Chapman, Frank M. "The Case of William Long." *Science* 4 (1904): 387–89.

Charles, Fred L. "Editorial Notes." *Nature-Study Review* 6 (February 1910): 46–48.

Cheney, Ednah D. *Memoirs of Lucretia Crocker and Abby W. May*. N.p., 1903.

Chew, Samuel C. *Fruits among the Leaves*. New York: Appleton, Century, Crofts, 1950.

Chicago Institute. *Catalogue of the Chicago Institute: Academic and Pedagogic, 1900–1901*. Chicago: Chicago Institute, 1900.

———. *The Course of Study: A Monthly Publication for Teachers and Parents* 1 (July 1900).

Child, Brenda J. *Boarding School Seasons: American Indian Families, 1900–1940*. Lincoln: University of Nebraska Press, 1998.

Cittadino, Eugene. "A 'Marvelous Cosmopolitan Preserve': The Dunes, Chicago, and the Dynamic Ecology of Henry Cowles." *Perspectives on Science* 1 (1993): 520–59.

City Directory of Teachers in the Public Schools. New York City: Board of Education, 1912.

Clapp, Henry L. "The Nature and Purpose of Nature Study and Intellectual Culture." *Education: A Monthly Magazine Devoted to the Science, Art, Philosophy and Literature of Education* 15 (1894–95): 597 ff.

Clement, Priscilla Ferguson. *Growing Pains: Children in the Industrial Age, 1850–1890*. New York: Twayne, 1997.

Clifford, Geraldine Joncich. "Man/Woman/Teacher." In *American Teachers; Histories of a Profession at Work*. Edited by Donald Warren. New York: Macmillan, 1989.

Cohen, Hyman. "Dental Hygiene." *Nature-Study Review* 8 (February 1912): 113–21.

Cohen, Ronald D., and Raymond A. Mohl. *The Paradox of Progressive Education: The Gary Plan and Urban Schooling*. Port Washington: Kennikat Press, 1979.

Cohen, Sol. *Education in the United States: A Documentary History*. 5 vols. New York: Random House, 1973.

Coker, Robert A. *Stephen Forbes and the Rise of American Ecology*. Washington: Smithsonian Institution Press, 2001.

Coleman, Sydney H. *Humane Society Leaders in America, with a Sketch of the Early History of the Humane Movement in England*. Albany: American Humane Association, 1924.

Colman, Gould. *Education and Agriculture: A History of the New York State College of Agriculture at Cornell University*. Ithaca: Cornell University Press, 1963.

Colton, Buel P. *An Elementary Course in Practical Zoology*. Boston: D. C. Heath, 1893.

Comstock, Anna Botsford. *The Comstocks of Cornell: John Henry Comstock and Anna Botsford Comstock*. Ithaca: Cornell University Press, 1953.

———. "Editorial." *Nature-Study Review* 14 (February 1918): 83.

———. "The Future of the Nature-Study Review." *Nature-Study Review* 19 (November 1923): 368–69.

———. *Handbook of Nature Study*. Ithaca: Comstock Publishing Co., 1911.

———. "In Memoriam: Ada E. Georgia." *Nature-Study Review* 17 (February 1921): 94.

———. "Nature-Study and Agriculture." *Nature-Study Review* 1 (July 1905): 143–44.

———. "Nature-Study as Servant." *Nature-Study Review* 8 (April 1912): 131–32.

———. "A New Venture in Nature Lore." *Nature-Study Review* 19 (March 1923): 136.

———. *The Ways of the Six-Footed*. New York: Ginn and Co., 1903.

———. "What Nature-Study Does for the Child and for the Teacher." *Nature-Study Review* 7 (May 1907): 134–35.

Comstock, John Henry, and Anna Botsford Comstock. *How to Know the Butterflies: A Manual of Those Which Exist in the Eastern United States*. New York: Appleton, 1904.

———. *Insect Life: An Introduction to Nature-Study and a Guide for Teachers, Students, and Others Interested in Out of Door Life*. New York: D. Appleton, 1897.

———. *A Manual for the Study of Insects*. Ithaca: Comstock Publishing Co., 1895.

Confino, Alon. *The Nation as a Local Metaphor: Wurttemberg, Imperial Germany, and National Memory, 1871–1918*. Chapel Hill: University of North Carolina Press, 1997.

Conklin, Keith Paul. *Peabody College: From a Frontier Academy to the Frontiers of Teaching and Learning*. Nashville: Vanderbilt University Press, 2002.

"Constitution [of the American Nature-Study Society]." *Nature-Study Review* 4 (1908): 5–7.

Cooke, Emmy Lou. "A Study of the Materials and Methods of Nature Study in Camps, as Shown by Literature on the Subject." M.A. thesis, Cornell University, 1940.

Corby, Betsy. "The Centralization of Educational Administration in the New York State Normal Schools, 1863–1917." Ph.D. diss., State University of New York at Buffalo, 1993.

Corcoran, Peter. "A Formative Evaluation of Science and Natural History Curriculum Materials and Their Influence on Maine Science Teaching, K–12." Ed.D. diss., University of Maine, 1985.

Cordier, Mary Hurlbut. *The Schoolwomen of the Prairies and Plains: Personal Narratives from Iowa, Kansas, and Nebraska, 1860s–1920s.* Albuquerque: University of New Mexico Press, 1992.

Cornell Nature-Study Leaflets, Being a Selection, with Revision, from the Teachers' Leaflets, Home Nature-Study Lessons, Junior Naturalist Monthlies and Other Publications from the College of Agriculture, Cornell University, Ithaca, N.Y., 1896–1904. Albany: J. B. Lyon, 1904.

Correspondence between J. Tod Kennedy and Edward F. Bigelow, Innis Arden House, Sound Beach (Old Greenwich) Connecticut. Privately printed and undated.

Coulter, John G. *The Dean [Stanley Coulter]: An Account of His Career and His Convictions.* Lafayette: Purdue Alumni Office, 1940.

Coulter, John M. "The Mission of Science in Education." *Science* 12 (1900): 281–93.

———. "Nature Study and Intellectual Culture." *Science* 4 (1896): 740–44.

———. "Principles of Nature-Study." *Nature-Study Review* 1 (January 1905): 57–60.

Coulter, John M., John G. Coulter, and Alice Jean Patterson. *Practical Nature Study and Elementary Agriculture: A Manual for the Use of Teachers and Normal Schools.* New York: D. Appleton and Co., 1909.

Coulter, Stanley. "Educational Values and the Aims of Nature-Study." *Nature-Study Review* 1 (1905): 50–53.

———. "Nature-Study in Indiana." *Nature-Study Review* 5 (May 1909): 33–35.

Counts, George. *School and Society in Chicago.* New York: Harcourt Brace and Co., 1928.

Craig, Gerald S. "Certain Techniques in Developing a Course of Study in Science for the Horace Mann Elementary School." Ph.D. diss., Columbia University, 1927.

———. *A New Science Program for Elementary Schools.* New York: Columbia Teachers College, Columbia University.

———. *Science for the Elementary School Teacher.* New York: Ginn and Co., 1940.

Cranz, Galen. *The Politics of Park Design: A History of Urban Parks in America.* Cambridge: MIT Press, 1982.

Cravens, Hamilton. *Before Head Start: The Iowa Station and America's Children.* Chapel Hill: University of North Carolina Press, 1993.

Crawford, Mattie Rose. *Guide to Nature-Study for the Use of Teachers.* Toronto: Copp, Clark Co., 1902.

Creed, Percy R. *Boston Society of Natural History, 1830–1930.* Boston: The Society, 1930.

Cremin, Lawrence. *Traditions of American Education.* New York: Basic Books, 1977.

———. *The Transformation of the Schools: Progressivism in American Education, 1876–1957.* New York: Knopf, 1961.

Cremin, Lawrence A., David A. Shannon, and Mary Evelyn Townsend. *A History of Teachers College, Columbia University.* New York: Columbia University Press, 1954.

Crist, Eileen. *Images of Animals: Anthropomorphism and Animal Mind.* Philadelphia: Temple University Press, 1999.

Crocker, Lucretia. *Methods of Teaching Geography.* Boston: Boston School Supply, 1883.

Croker, R. A. *Stephen A. Forbes and the Rise of American Ecology.* Washington, DC: Smithsonian Institution Press, 2001.

Crosby, Dick Jay. "A Few Good Books and Bulletins on Nature Study, School Gardening, and Elementary Agriculture for the Common Schools." U.S. Department of Agriculture, Office of Experiment Stations. *Circular* 52 (1904).

Cruikshank, Kathleen Anne. "The Rise and Fall of American Herbartianism: Dynamics of an Educational Reform Movement." Ph.D. diss., University of Wisconsin-Madison, 1993.

Cuban, Larry. *How Teachers Taught: Constancy and Change in American Classrooms, 1880–1990.* New York: Teachers College Press, 1993.

——. "The Persistence of Reform in American Schools." In *American Teachers: Histories of a Profession at Work*. Edited by Donald Warren. New York: MacMillan, 1989.

Cubberley, Ellwood P. *Changing Conceptions of Education*. Cambridge: Riverside Press, 1909.

Cummings, Horace H. *Nature Study by Grades: Teachers Book for Primary Grades*. New York: American Book Co., 1908.

Curtis, Henry S. "Vacation Schools, Playgrounds, and Settlements." In *Report of the U.S. Commissioner of Education for the Year 1903*. Washington, DC: Government Printing Office, 1905.

Cushing, Thomas. *Reminiscences of School Teachers in Dorchester and Boston*. [Hartford: Brown and Gross], c. 1884.

Cutts, Richard. *Index to Youth's Companion, 1871-1929*. 2 vols. Metuchen: Scarecrow Press, 1972.

Danbom, David B. *The Resisted Revolution: Urban America and the Industrialization of American Agriculture, 1900-1930*. Ames: Iowa State University Press, 1979.

Daniels, George H. *American Science in the Age of Jackson*. New York: Columbia University Press, 1968.

Dann, Kevin. *Across the Great Border Divide: The Naturalist Myth in America*. New Brunswick: Rutgers University Press, 2000.

Daum, Andreas W. *Wissenchaftpopularisierung im 19. Jahrhundert: Burgerliche Kultur naturwissenshchaftlickie Bildung und die deutchef Offenlichkeit, 1848-1914*. Munich: R Oldenbourg Verlag, 1998.

Davis, B. M. "Factors and Influences Now Contributing to the Progress of Nature Study in the California Schools." *Nature-Study Review* 2 (November 1906): 257-65.

Davis, Sheldon E. *Educational Periodicals during the Nineteenth Century*. Metuchen: Scarecrow Reprint Corp., [1919] 1970.

Deacon, Delsey. *Elsie Clews Parsons: Inventing Modern Life*. Chicago: University of Chicago Press, 1997.

DeAntoni, Edward Paul. "Coming-of-Age in the Industrial State: The Ideology and Implementation of Rural School Reform, 1893-1925." Ph.D. diss., Cornell University, 1971.

Dearborn, Ned Harland. *The Oswego Movement in American Education*. New York: Arno Press, [1925] 1969.

DeGarmo, Charles. "Ethical Training in the Public Schools." *Annals of the American Academy of Political and Social Science* 2 (Philadelphia, 1892): 1-23.

Densmore, Alice. "Nature-Study at Cornell." *Scientific American* 81 (August 12, 1899): 101.

DePencier, Ida B. *History of the Laboratory Schools: The University of Chicago, 1896-1965*. Chicago: Quadrangle Books, 1967.

Dewey, Evelyn. *New Schools for Old: The Regeneration of the Porter School*. New York: E. P. Dutton and Co., 1919.

Dexter, Ralph W. "An Early Movement to Promote Field Study in the Public Schools." *Science Education* 42 (October 1958): 344-46.

——. "The Annisquam Sea-side Laboratory of Alpheus Hyatt." *Scientific Monthly* 74 (1952): 112-16.

——. "Views of Alpheus Hyatt's Sea-side Laboratory and Excerpts from his Expeditionary Correspondence." *The Biologist* 39 (1956-57): 5-11.

"Discussion: Dr. Hornaday's 'The Weakness of Nature Study.'" *Nature-Study Review* 3 (January 1907): 30-32.

Donaldson, Elizabeth J. "Picturesque Scenes, Sentimental Creatures: The Rhetoric and Politics of American Nature Writing, 1890-1920." Ph.D. diss., State University of New York at Stony Brook, 1997.

Doolittle, Maude A. "The Rambler at Summer Camp." *Nature-Study Review* 19 (April 1923): 156-68.

Dorf, Philip. *Liberty Hyde Bailey: An Informal Biography*. Ithaca: Cornell University Press, 1956.

Doris, Ellen Elizabeth. "The Practice of Nature-Study: What Reformers Imagined and What Teachers Do." D.Ed. diss., Harvard University, 2002.

Downing, Elliot R. *Field and Laboratory Guide to Biological Nature Study*. New York: Longmans, Green and Co., 1918.

———. *Field and Laboratory Guide in Physical Nature Study*. Chicago: University of Chicago Press, 1924.

———. "Editorial." *Nature-Study Review* 9 (September 1913): 191–92.

———. "List of Instructors in Nature-Study." *Nature-Study Review* 12 (February 1916): 71.

———. *Our Physical World: A Source Book of Physical Nature Study . . . with a Chapter on Radio Communication by Fred G. Anibal*. Chicago: University of Chicago Press, [1920].

———. "Preparation of Teachers for Nature Study and Elementary Agriculture by the Normal Schools." *School Science and Mathematics* 17 (October 1917): 609–21.

———. "The Present Status of Nature-Study." *School and Society* 5 (June 23, 1917): 741–42.

———. *A Source Book of Biological Nature Study*. Chicago: University of Chicago Press, 1919.

———. "The Summer Outing." *Nature-Study Review* 12 (May 1916): 229–44.

———. *Yearbook of the American Nature-Study Society*. Toledo: Board of Education, 1925.

Downing, Elliot R., and Theodosia Hadley. *The Physiography of the Marquette Region with the Hypotheses of Its Origin*. Marquette: Northern Michigan State Normal School, 1909.

Dewey, John. "Science as Subject Matter and as Method." *Science* 31 (January 28, 1910): 123.

"Dr. Elliott Downing, Retiring Editor." *Nature-Study Review* 13 (February 1917): 41.

Dunkel, Harold B. *Herbart and Herbartianism: An Educational Ghost Story*. Chicago: University of Chicago Press, 1970.

Dunlap, Thomas R. *Saving America's Wildlife: Ecology and the American Mind, 1850–1990*. Princeton: Princeton University Press, 1988.

Dymond, T. S. *Suggestions for Rural Education, together with Some Specimen Courses of Nature Study, Gardening, and Rural Economy*. London: His Majesty's Stationery Office, 1908.

Eastman, Rebecca Hooper. *The Story of the Brooklyn Institute of Arts and Sciences, 1824–1924*. Brooklyn: [The Institute], 1924.

Eccles, Priscilla. "Nature Study Revisited." *Science and Children* 2 (November 1964): 18–21.

Eddy, Jacalyn. *Bookwomen: Creating an Empire in Children's Book Publishing, 1919–1935*. Madison: University of Wisconsin Press, 2006.

Edwards, Charles Lincoln. "The Los Angeles Nature-Study Exhibition." *Nature-Study Review* 10 (October 1914): 263–70.

———. "Nature Study Supervision in the Los Angeles City Schools." *Nature and Science Education Review* 1 (1928): 32.

———. *A Preliminary Plan for the Los Angeles Zoological Park and Aquarium*. [Proceedings of the Los Angeles Zoological Society]. Los Angeles, 1912.

Eikenbery, W. L. *The Teaching of General Science*. Chicago: University of Chicago, 1922.

Eliot, Charles W. "Dr. Hornaday's 'Weakness of Nature Study.'" *Nature-Study Review* 3 (February 1907): 52.

Emerson, Donald A., ed. *Nature Study Teachers Manual, Elementary Schools, State of Oregon*. Salem: Office of Superintendent of Public Instruction, 1937.

Eschenbacher, Herman F. *A History of Land Grant Education in Rhode Island*. New York: Appleton-Century-Crofts, 1967.

Fairbanks H. W. *Home Geography for the Primary Grades*. Boston: Education Publishing Co., 1902.

———. "Nature Study and Its Relation to Natural Science: A Symposium by H. W. Fairbanks, C. F. Hodge, T.H. McBride, F. L. Stevens, and M. A. Bigelow." *Nature-Study Review* 1 (January 1905): 3–6.

Farber, Paul. *Discovering Birds: The Emergence of Ornithology as a Discipline, 1960–1850*. Princeton: Princeton University Press, 1996.

Fear, Christena. "Nature and the Americanization as Allies." *Nature-Study Review* 19 (January 1923): 1–4.

Fielde, Adele M. "The Communal Life of Ants." *Nature-Study Review* 1 (November 1905): 239–51.

Findling, John E., ed. *Historical Dictionary of World's Fairs and Expositions, 1851–1988.* Westport: Greenwood Press, 1990.

Finklestein, Barbara. "Governing the Young: Teacher Behavior in American Primary Schools, 1820–1880." Ph.D. diss., Columbia University, 1970.

Finley, Kimberley D. "Cultural Monitors: Club Women and Public Art Instruction in Chicago, 1890–1920." Ph.D. diss., Ohio State University, 1989.

"First Directory of Members of the American Nature-Study Society." *Nature-Study Review* 5 (March 1909): 1–91.

Fisher, [George] Clyde, and Marion I. Langham. *Required Nature Study in the New York City Public Schools.* New York: Noble and Noble, 1934.

Fitzgerald, Deborah. *Every Farm a Factory: The Industrial Ideal in American Agriculture.* New Haven: Yale University Press, 2003.

Flexner, Abraham. *A Modern School.* New York: General Education Board, 1919.

Flexner, Abraham, and Frank B. Bachman. *The Gary Schools: A General Account.* New York: General Education Board, 1918.

Foote, William B. *The Teachers' Institute; or Familiar Hints to Young Teachers.* New York: A. S. Barnes, 1866.

Foght, Harold W. "The Country School." *Annals of the American Academy of Political and Social Science* 40 (March 1912): 149–57.

Franklin, Parker. *George Peabody: A Biography.* Nashville: Vanderbilt University Press, [1971] 1995.

Fraser, James W. *Preparing America's Teachers: A History.* New York: Teachers College Press, 2006.

Freeman, William G. *Nature Teaching Based upon the General Principles of Agriculture for the Use of the Schools.* New York: E. P. Dutton, 1904 [1901].

Fuller, Wayne E. *The Old Country School: The Story of Rural Education in the Middle West.* Chicago: University of Chicago Press, 1982.

———. "Making Better Farmers: The Study of Agriculture in Midwestern Country Schools, 1900–1923." *Agricultural History* 60 (Spring 1986): 154–68.

Gallaudet, T. H. *The Class Book of Natural Theology for Common Schools and Academies.* Hartford: Belknap and Hamersley, 1837.

Gallup, Anna Billings. "Children's Museum as an Educator." *Popular Science Monthly* 72 (April 1908): 371–79.

———. "Work of a Children's Museum." *Nature-Study Review* 2 (May 1906): 153–63.

Gates, Barbara T. *Kindred Nature: Victorian and Edwardian Women Embrace the Living World.* Chicago: University of Chicago Press, 1998.

Gates, Barbara T., and Ann B. Shteir, eds. *Natural Eloquence: Women Reinscribe Science.* Madison: University of Wisconsin Press, 1997.

Geddes, Patrick. *Introductory Course on Nature Study.* Cambridge: Cambridge University Press, n.d.

———. *City Development: A Study of Parks, Gardens, and Culture-Institutes; A Report to the Carnegie Dunfermline Trust.* Edinburgh: Geddes and Co., 1904.

Gehrs, John H. *Agricultural Nature Study.* New York: American Book Co., 1929.

Geiger, Roger L. *To Advance Knowledge: The Growth of American Research Universities, 1900–1940.* New York: Oxford University Press, 1986.

Geitz, Henry, Jurgen Heideking, and Jurgen Herbst, eds. *German Influences on Education in the United States to 1917.* Cambridge: Cambridge University Press, 1995.

Giordano, Gerald. *How Testing Came to Dominate American Schools: The History of Educational Assessment*. New York: P. Lang, 2005.

Goethe, C. M. "The California Nature Study League." *Nature-Study Review* 16 (May 1920): 204–7.

Goodrich, Charles L. *The First Book of Farming*. New York: Doubleday, Page, and Co. 1905.

Goodrich, Samuel. "About the Leaves of Trees." *Parley's Magazine* 3 (1835): 41–42.

Gordon, Eva L. "Cornell Nature-Study Leaflets, 1896–1956." *Cornell Rural School Leaflet* 50 (Fall 1956).

Gordon, Lynn D. *Gender and Higher Education in the Progressive Era*. New Haven: Yale University Press, 1990.

Gorelick, Sherry. *City College and the Jewish Poor: Education in New York, 1880–1924*. New Brunswick: Rutgers University Press, 1981.

Gould, A. A. "On the Introduction of Natural History as a Study to Common Schools." In *Lectures*. American Institute of Instruction. Boston: Carter, Hindee, and Co., 1835.

Graham, Patricia. "Expansion and Exclusion: A History of Women in Higher Education." *Signs* 3 (1978): 759–73.

Gray, Asa. *Botany for Young People and Common Schools: How Plants Grow, a Simple Introduction to Structural Botany*. New York: American Book Co., undated reprint of 1858 edition.

———. *Field, Forest, and Garden Botany: A Simple Introduction to the Common Plants East of the 100th Meridian, Both Wild and Cultivated*. Revised and extended by L. H. Bailey. New York: American Book Co., 1895.

Green, Amy. "'She Touched Fifty Million Lives': Gene Stratton-Porter and Nature Conservation." In *Seeing Nature Through Gender*. Edited by Virginia J. Scharff. Lawrence: University Press of Kansas, 2003, 221–41.

Green, Amy Susan. "Savage Childhood: The Scientific Construction of Girlhood and Boyhood in the Progressive Era." Ph.D. diss., Yale University, 1995.

Green, Norma Kidd. *A Forgotten Chapter in American Education: Jane Andrews of Newburyport*. Alumnae Association of the State College at Framingham, 1961.

———. "Lucretia Crocker." *Notable American Women, 1607–1950* (1971), 1:407–9.

Greene, John C. *American Science in the Age of Jefferson*. Ames: Iowa State University Press, 1984.

Greene, Kristen Jane. "The Macdonald Robertson Movement, 1899–1909." Ph.D. diss., University of British Columbia, 1992.

Gregg, F. A. "Hygiene as Nature Study." *Nature-Study Review* 8 (September and October 1912): 225–29.

Griffith, Nellie Lucy. "A History of the Origins of the Laboratory School at the University of Chicago." A.M. thesis, University of Chicago Press, 1927.

Grunfeld, Katherine Krop. "Purpose and Ambiguity: The Feminine World of Hunter College, 1869–1915." Ph.D. diss., Columbia University Teachers College, 1991.

Guyot, Arnold. *Elementary Geography for Primary Classes*. New York: C. Scribner, 1868.

Hadley, Theodosia. "The Relation of Nature-Study to Boys' and Girls' Club Work." *Nature-Study Review* 16 (February 1920): 54–61.

Hadsall, Leo Franklin. "The Extension Activities of Certain Publicly Supported Institutions in Assisting Teachers in Service in Elementary Science or Nature Studies." *Science Education* 20 (February 1936): 7–11.

Hagen, Joel. *An Entangled Bank: The Origins of Ecosystem Ecology*. New Brunswick: Rutgers University Press, 1992.

Hailmann, William. "German Views of American Education." Translated and published by the Department of the Interior, Bureau of Education. *Bulletin* 2, no. 361 (1907).

Hall, G. Stanley. "The Contents of Children's Minds." *Princeton Review* 11 (1883): 249–73.

———. "On Entering School." *Pedagogical Seminary* 11 (1891): 145.

Hanson, Elizabeth. *Animal Attractions: Nature on Display in American Zoos*. Princeton: Princeton University Press, 2002.

Harper, Charles A. *Development of the Teachers College in the United States, with Special Reference to the Illinois State Normal University*. Bloomington: McKnight and McKnight, 1935.

Harris, Thaddeus T. *How to Teach Natural Science in Public Schools*. Syracuse: C. W. Bardeen, 1887.

———. "The Study of Natural Science—Its Uses and Dangers." *Education* 10 (January 1890): 277–87.

Hatch, Luther A. "Why Many Fail in Teaching Nature-Study." *Nature-Study Review* 1 (May 1905): 97–100.

Hawkins, Layton S. *Methods of Nature Study*. Cortland: State Normal School, 1912.

Hays, Samuel P. *Conservation and the Gospel of Efficiency: The Progressive Conservation Movement, 1890–1920*. Cambridge: Harvard University Press, 1959.

Hayter, Earl W. *Education in Transition: The History of Northern Illinois University*. DeKalb: Northern Illinois University, 1974.

Heffron, John M. "The Knowledge Most Worth Having: Otis W. Caldwell (1869–1947) and the Rise of the General Science Course." *Science and Education* 4 (1995): 227–52.

———. "Nation Building for a Venerable South: Moral and Practical Uplift in the New Agricultural Education, 1900–1920." In *Essays in Twentieth Century Southern Education: Exceptionalism and Its Limits*. Edited by Wayne J. Urban. New York: Garland Pub. Co., 1999.

Helgren, Jennifer Hillman. "Inventing American Girlhood: Gender and Citizenship in the Twentieth-Century Camp Fire Girls." Ph.D. diss., Claremont Graduate University, 2005.

Henson, Pamela Marianne. "Evolution and Taxonomy: J. H. Comstock's Research School in Evolutionary Entomology at Cornell, 1874–1930." Ph.D. diss., University of Maryland, 1990.

———. "The Comstocks of Cornell: A Marriage of Interests." In *Creative Couples in the Sciences*. Edited by Helena M. Pycior, Nancy G. Slack, and Pnina G. Abir-Am. New Brunswick: Rutgers University Press, 1995.

Herbst, Jurgen. *And Sadly Teach: Teacher Education and the Professionalization in American Culture*. Madison: University of Wisconsin Press, 1989.

——— "Teacher Preparation in the Nineteenth Century: Institutions and Purposes." In *American Teachers: Histories of a Profession at Work*. Edited by Donald Warren. New York: Macmillan, 1989.

Hermand, Jost, and James Steakley, eds. *Heimat, Nation, and Fatherland: The German Sense of Belonging*. New York: Lang, 1997.

Herrick, Mary. *The Chicago Schools: A Social and Political History*. Beverly Hills: Sage Publications, 1971.

Herzig, Rebecca. "In the Name of Truth: Sacrificial Ideals and American Science, 1870–1930." Ph.D. diss., Massachusetts Institute of Technology, 1998.

Hevly, Bruce. "The Heroic Science of Glacier Motion." *Osiris* 11 (1996): 66–86.

Hill, Thomas. "The Truth Order of [Natural History] Studies." *American Journal of Education* 7 (1859): 273–84.

Hine, Lewis W. "The School Camera." *Elementary School Teacher* 6 (1905–6): 345.

Hodge, Clifton F. "The Established Principles of Nature Study." *Nature-Study Review* 3 (January 1907): 7–8.

———. "Foundations of Nature Study." *Pedagogical Seminary* 7 (July 1900): 206–28.

———. "Nature Study and the Bobolink." *Nature-Study Review* 6 (March 1910): 53–64.

———. *Nature Study and Life*. Boston: Ginn and Co., 1903.

———. "Passenger Pigeon Investigation." *Nature-Study Review* 6 (May 1910): 110–11; and 8 (November 1910): 248–51.

———. "Practical Work with Mosquitoes." *Nature-Study Review* 3 (1907): 33–36.

———. "Preparation of Teachers for National, State and Civic Biology." *Nature-Study Review* 10 (November 1914): 294–307.

Hodge, Clifton F., and Jean Dawson. *Civic Biology: A Textbook of Problems, Local and National, That Can Be Solved Only by Civic Cooperation*. Boston: Ginn and Co., 1918.

Hoffman, Nancy. *Women's True Profession: Voices from the History of Teaching*. Cambridge: Harvard University Press, 2003.

Holmes, John S. *Organization of Co-Operative Forest-Fire Protective Areas in North Carolina . . . Part of the Conference on Forestry and Nature Study in Montreat, North Carolina, July 8, 1915*. Raleigh: State Printers, 1915.

Holt, Henry. *Sixty Years as a Publisher*. London: George Allen and Unwin, [1923].

Holtz, Frederick L. "The Course of Nature-Study in the Elementary School." *Nature-Study Review* 6 (October 1910): 189–92.

———. *Nature-Study: A Manual for Teachers and for Students*. New York: Charles Scribner's Sons, 1908.

———. "Standardizing Nature-Study." *Nature-Study Review* 13 (February 1917): 52–54.

Hooker, Worthington. *The Child's Book of Nature*. New York: Harper and Brothers, 1885.

Hornaday, William. *The American Natural History: A Foundation of Useful Knowledge of the Higher Animals of North America*. New York: Charles Scribner's Sons, 1904.

———. "The Weakness in Teaching Nature-Study." *Nature Study Review* 2 (October 1906): 241–43.

Houghton, William. *Country Walks of a Naturalist with His Children*. London: Groombridge and Sons, 1869.

Howe, Edward Gardiner. *Systematic Science Teaching: A Manual of Inductive Elementary Work for All Instructors*. New York: D. Appleton and Co., [1893] 1895.

Hoyt, William A. "The Home of Nature at the Root of Teaching and Learning the Sciences." *Pedagogical Seminary* 3 (1894): 61.

Humphrey, Carpenter. *Secret Gardens: A Study of the Golden Age of Children's Literature*. Boston: Houghton Mifflin, 1985.

Hunter, George W. "Correlation between Nature-Study and High School Biology." *Nature-Study Review* 5 (May 1909): 127–31.

Husin, Torsden, and T. Neville Postlethwaite. *The International Encyclopedia of Education*. 2d ed. Oxford: Pergamon Press, 1994.

Hutchinson, Joseph C. *The Laws of Physiology, Hygiene, Stimulants, Narcotics*. New York: Clark and Maynard, 1891.

Huth, Hans. *Nature and the Americans*. Berkeley: University of California Press, 1955.

Huxley, Thomas H. "Natural History." *American Journal of Education* 3 (1872): 473.

Ilerbaig, Juan. "Allied Sciences and Fundamental Problems: C. C. Adams and the Search for Method in Early American Ecology." *Journal of the History of Biology* 32 (December 1999): 439–63.

———. "Pride of Place: Fieldwork, Geography, and American Field Zoology." Ph.D. diss., University of Minnesota, 2002.

Illinois Education Association. *Educational Papers*, 1889–1890.

"Inaccurate Nature Books." *Nature-Study Review* 3 (November 1907): 240–42.

[Irish Free State Department of Education]. *Teaching of Rural Science and Nature Study in Primary Schools: Regulations and Explanatory Notes for Teachers based on the Syllabuses in Rural Science and Nature Study contained in the Report and Programme presented by the National Programme Conference and adopted by the Department of Education*. Dublin, 1927.

Irwin, Manley E. "The Measurement of Nature Study in the Primary Grades in the Detroit Public Schools." *Science Education* 15 (November 1930): 23–32.

Isbell, Egbert R. *A History of Eastern Michigan University, 1848–1965*. Ypsilanti: Eastern Michigan University Press, 1971.

Jackman, Wilbur. *Field Work in Nature Study*. Chicago: A. Flanagan, 1894.

——. "Natural Science of the Common Schools." National Education Association. *Journal of the Proceedings* 30 (1891): 581.

——. *Nature Study.* Chicago: University of Chicago Press, 1904.

——. *Nature Study for the Common Schools.* New York: Holt and Co., 1896.

——. *Nature Study for the Grammar Grades: A Manual for Teachers and Pupils below the High School in the Study of Nature.* New York: Macmillan, [1898] 1909.

——. *Nature Study Record for the Common Schools.* Chicago, 1895.

——. *Nature Study and Related Subjects for Common Schools.* Chicago: F. Bartsch, 1896.

——. "Nature Study and Religious Training." *Educational Review* 30 (June 1905): 12–30.

——. *Number Work in Nature Study, Part I.* Published by the Author, 1893.

——. "Representative Expression in Nature Study." *Educational Review* 10 (October 1895): 248–61.

——. *Third Yearbook of the National Society for the Scientific Study of Education. Part 2: Nature-Study.* Chicago: University of Chicago Press, 1904.

Jardine, N., J. A. Secord, and E. C. Spary, eds. *Cultures of Natural History.* Cambridge: Cambridge University Press, 1996.

Jenkins, E. W. "Science, Sentimentalism, or Social Control? The Nature Study Movement in England and Wales, 1888–1914." *History of Education* 10 (1981): 33–43.

Jenkins, Oliver P. *Interesting Neighbors.* Philadelphia: P. Blakiston's Sons and Co., c. 1922.

——. *Lessons in Nature Study.* San Francisco: Whitaker and Ray Co., 1900.

——. "Nature Study." In *School Report of Oakland, California, 1897–1898.* Oakland: Kitchener, 1897, 6–10.

Jewell, James Ralph. *Agricultural Education, Including Nature-Study and School Gardens.* Department of Interior, U.S. Bureau of Education. *Bulletin* 2 (1907).

Johonnot, James. "The Story of a School." *Popular Science Monthly* 34 (1888): 496–508.

Joliffe, Ruby M. "The Palisades Interstate Park." *Nature-Study Review* 19 (April 1923): 148–51.

Joncich, Geraldine. *The Sane Positivist: A Biography of Edward L. Thorndike.* Middletown: Wesleyan University Press, 1986.

Jones, Howard Mumford. *O Strange New World: American Culture: The Formative Years.* New York: Viking Press, 1952.

Jordan, David Starr. "Nature Study and Intellectual Culture, and Moral Culture." *Science* 4 (1896): 149–56.

——. "Nature Study and Moral Life." *Annual Proceedings of the National Educational Association* (1896): 130.

——. *Scientific Sketches.* Chicago: A. C. McClurg and Co., 1888.

——. *The Story of Innumerable Company and Other Sketches.* San Francisco: Whitaker and Ray Co., 1896

Kaestle, Carl. "Literacy and Diversity: Themes from a Social History of the American Reading Public." *History of Education Quarterly* 28 (Winter, 1988): 523–49.

Katz, Michael B. "The 'New Departure' in Quincy 1873–1881: The Nature of the Nineteenth Century Educational Reform." *New England Quarterly* 40 (March 1967): 3–30.

Kaufman, Paula Welts. *Boston Women and City School Politics, 1872–1905.* New York: Garland Publishing, 1994.

——. *National Parks and the Woman's Voice.* Albuquerque: University of New Mexico Press, 1996.

Keeney, Elizabeth. *The Botanists: Amateur Scientists in Nineteenth-Century America.* Chapel Hill: University of North Carolina Press, 1992.

Kegley, Tracy Mitchell. "The Peabody Scholarships, 1877–1899." Ph.D. diss., George Peabody College for Teachers, 1949.

Kelly, Henry A. "Report." Superintendent of Public Schools, *Annual Report.* New York, 1899.

Kellogg, Vernon L. "American Nature Study Society." *Science* 57 (February 9, 1923): 185.

———. *Insect Stories*. New York: Henry Holt and Co., 1908.

———. *Nuova: The New Bee*. New York: Houghton Mifflin Co., 1921.

Kern, O. J. *Outline of Course of Instruction in Agricultural Nature Study for the Rural Schools of California*. College of Agriculture, Agricultural Experiment Station (December 1924).

Kessler, J. T. "Nature Study and Biology." Baylor University. *Bulletin* 9 (January 1906): 10.

Kingsland, Sharon E. *The Evolution of American Ecology, 1890-2000*. Baltimore: Johns Hopkins University Press, 2005.

Klassen, Kenneth G. "The School of Nature: An Annotated Index of Writings on Nature in *St. Nicolas Magazine* during the Editorship of Mary Mapes Dodge, 1873-1905." Ph.D. diss., University of Kansas, 1989.

Kliebard, Herbert M. *Forging the American Curriculum: Essays in Curriculum History and Theory*. London: Routledge, 1992.

Kloppenberg, James T. *Uncertain Victory: Social Democracy and Progressivism in European and American Thought, 1870-1920*. New York: Oxford University Press, 1988.

Klunk, Edward Timothy. "The Chicago Academy of Sciences: The Development and Method of Educational Work in Natural History." Ph.D. diss., Loyola University, 1996.

Kohler, Robert E. *All Creatures: Naturalists, Collectors, and Biodiversity, 1850-1950*. Princeton: Princeton University Press, 2006.

———. *Landscapes and Labscapes: Exploring the Lab-Field Border in Biology*. Chicago: University of Chicago Press, 2002.

Kohlstedt, Sally Gregory. "'A Better Crop of Boys and Girls': The School Gardening Movement, 1890-1920." *History of Education Quarterly* 48 (February 2008): 58-93

———. "Collections and Cabinets: Natural History Museums on Campus, to 1860." *Isis* 79 (Fall 1988): 405-26.

———. "Collections, Cabinets and Summer Camp: Natural History in the Public Life of Nineteenth-Century Worcester." *Museum Studies Journal* 2 (Fall 1985): 10-23.

———. "In from the Periphery: American Women in Science, 1830-1880." *Signs* 4 (Fall 1978): 81-96.

———. "Museums on Campus: A Tradition of Inquiry and Teaching." In *The American Development of Biology*. Edited by Ronald Rainger, Keith Benson, and Jane Maienschein. Philadelphia: University of Pennsylvania Press, 1988.

———. "Nature not Books: Scientists and the Origins of the Nature Study Movement in the 1890s." *Isis* 96 (September 2005): 324-52.

———. "Nature Study in North America and Australasia, 1890-1945." *Historical Records of Australian Science* 11 (June 1997): 439-54.

———. "Parlors, Primers, and Public Schooling: Education for Science in Nineteenth Century America." *Isis* 81 (Fall 1990): 424-45.

———. "*Science*: The Struggle for Survival, 1880-1894." *Science* 208 (July 4, 1980): 33-42.

———. "Single-sex Education and Leadership: The Early Years of Simmons College." In *Women and Educational Leadership: A Reader*. Edited by Sari Knopp Biklen and Marilyn Brannigan. Boston: Lexington Press, 1979.

Kohlstedt, Sally Gregory, Bruce V. Lewenstein, and Michael M. Sokal. *The Establishment of Science in America: 150 Years of the American Association for the Advancement of Science*. New Brunswick: Rutgers University Press, 1999.

Kraus, Helen. *Manual of Moral and Humane Education*. Chicago: R. R. Donnelly and Sons, 1910.

Kraus, Marcus. "Science Education in the National Parks of the United States: A Descriptive Study of the Development of Science Education Programs and Facilities by the National Park Service and the Relationship of these to the Advent of Nature Study and Conservation Education in America." Ph.D. diss., New York University, 1973.

Krug, Edward R. *The Shaping of the American High School, 1890-1920*. New York: Harper and Row, 1964.

Kuritz, Hyman. "The Popularization of Science in Nineteenth-Century America." *History of Education Quarterly* 21 (1981): 259-74.

Kuslan, Louis I. "Elementary Science in Connecticut, 1850-1900." *Science Education* 43 (1959): 286-89.

———. "Rensselaer and Bridgewater: A Footnote in the History of American Scientific Education." *Science Education* 50 (February 1966): 64-68.

LaBud, Sister Verona. "Liberty Hyde Bailey: His Impact on Science Education." Ph.D. diss., Syracuse University, 1963.

Lafollette, Marcel C. *Science on Air: Popularizers and Personalities on Radio and Early Television.* Chicago: University of Chicago Press, 2008.

———. "A Survey of Science Content in U.S. Radio Broadcasting, 1920s through 1940s: Scientist Speak in Their Own Voices." *Science Communication* 24 (2002): 4-33.

Lane, Robert Hill. *The Teacher in Modern Elementary Science.* New York: Houghton Mifflin Co., 1941.

Lange, Dietrich. *Course of Study in Nature Study, Grades One to Eight.* St. Paul: E. S. Ferry, 1907.

———. *Handbook of Nature Study.* New York: Macmillan, [1898] 1902.

———. "Nature Study in the Public Schools." National Education Association. *Addresses and Proceedings* (1900): 407-8.

———. *Our Native Birds: How to Protect Them and Attract Them to Our Homes.* New York: Macmillan, 1899.

Langemann, Ellen Condliffe. "Experimenting with Education: John Dewey and Ella Flagg Young at the University of Chicago." *American Journal of Education* 104 (May 1996): 171-85.

Lansbury, Coral. *Old Brown Dog: Women, Workers and Vivisection in Edwardian England.* Madison: University of Wisconsin Press, 1985.

Largent, Mark. "These Are Times of Scientific Ideals: Vernon Lyman Kellogg and Scientific Activism, 1890-1930." Ph.D. diss., University of Minnesota, 2000.

Larson, Edward. *Trial and Error: The American Controversy over Creation and Evolution.* New York: Oxford University Press, [1985] 1989.

Lavender, Linda. "A History of Nature Study in Texas." M.S. thesis, Texas Women's University, 1997.

Lawler, Thomas B. *Seventy Years of Textbook Publishing: A History of Ginn and Company, 1867-1937.* Boston: Ginn and Co., 1938.

Lazerson, Marvin. *Origins of the Urban School: Public Education in Massachusetts, 1870-1915.* Cambridge: Harvard University Press, 1971.

Lear, Linda. *Rachel Carson: Witness for Nature.* New York: Henry Holt and Co., 1997.

Lears, Jackson. *No Place of Grace: Antimodernism and the Transformation of American Culture, 1880-1920.* Chicago: University of Chicago Press, 1983.

"Lectures at Flathead Lake." University of Montana. *Bulletin* 5 (1902).

Lerman, Nina. "The Uses of Useful Knowledge: Science, Technology, and Social Boundaries in an Industrializing City." *Osiris* 12 (1997): 39-59.

Lightman, Bernard. "Marketing Knowledge for the General Reader: Victorian Popularizers of Science." *Endeavor* 24 (September 1, 2001): 100-106.

Lillie, Florence E. *Course in Nature-Study for Elementary Grades of Minnesota Public Schools.* Minneapolis: Syndicate Printing, 1909.

Lind, David. *Methods of Teaching in a Country School.* Danville: Normal Teacher Publishing House, 1880.

Lindsay, Deborah. "Intimate Inmates: Wives, Households, and Science in Nineteenth-Century America." *Isis* 68 (1998): 631-52.

Lindsey, Donald F. *Indians at Hampton Institute, 1877-1923.* Urbana: University of Illinois Press, 1995.

Link, William A. "The School That Built a Town: Public Education and the Southern Landscape,

1880–1930." In *Essays in Twentieth Century Southern Education: Exceptionalism and Its Limits.* Edited by Wayne J. Urban. New York: Garland Publishing Co., 1999.

Linn, Marcia C., Nancy B. Songer, and Bai-Sheva Eylon. "Shifts and Convergences in Science Learning and Instruction." In *Handbook of Educational Psychology.* New York: Macmillan, 1994.

Livingstone, David N. *Nathaniel Southgate Shaler and the Culture of American Science.* Tuscaloosa: University of Alabama Press, 1987.

Lloyd, Francis E., and Maurice Bigelow. "Courses in Biology in the Horace Mann High School." *Teachers College Record* 2 (1901).

"Local Sections." *Nature-Study Review* 9 (January 1913): 32.

Lochhead, W. "Canadian Department." *Nature-Study Review* 3 (January 1907): 17.

Long, Judith Reick. *Gene Stratton Porter: Novelist and Naturalist.* Indianapolis: Indiana Historical Society, 1990.

Lorrawairra, K. Tsianina. *They Called It Prairie Light: The Study of a Chilocco Indian School.* Lincoln: University of Nebraska Press, 1994.

Louisiana Nature Guardian Handbook for the Schools of the State. New Orleans: Bureau of Education for the Department of Conservation, 1931.

Louisiana Department of Education. *Course of Study for the Elementary Schools.* New Orleans, 1913.

Louv, Richard. *Last Child in the Woods: Saving Children from Nature-Deficit Disorder.* Chapel Hill: University of North Carolina Press, 2005.

Lovely, Robert A. "Mastering Nature's Harmony: Stephen Forbes and the Roots of American Ecology." Ph.D. diss., University of Wisconsin, 1995.

Lurie, Edward. *Louis Agassiz: A Life in Science.* Baltimore: Johns Hopkins University Press, 1988.

Lupfer, Eric. "Reading Nature Writing: Houghton Mifflin Company: The Ohio Teachers' Reading Circle, and *In American Fields and Forests* (1909)." *Harvard Library Bulletin* 13 (Spring 2002): 37–58.

Lutts, Ralph H. *The Nature Fakers: Wildlife, Science, and Sentiment.* Golden, CO: Fulcrum Press, 1990.

———, ed. *The Wild Animal Story.* Philadelphia: Temple University Press, 1998.

Lydston, G. Frank. *Sex Hygiene for the Male and What to Say to the Boy.* Chicago: Riverton Press, 1912.

MacCaughey, Vaughan. *The Natural History of Chautauqua.* New York: B. W. Huebsch, 1917.

MacKay, A. H. "Nature Study in the Schools of Nova Scotia." *Ottawa Naturalist* 13 (1905): 209–12.

MacPhail, Elizabeth C. *Kate Sessions: Pioneer Horticulturalist.* San Diego: San Diego Historical Society, 1976.

Macomber, Alice M., and Delia I. Griffin. *Outline of Nature Study for Primary and Grammar Grades.* Newton, MA, 1900.

McKee, Leila Gay Mitchell. "Voluntary Youth Organizations in Toronto, 1880–1930." Ph.D. diss., York University, 1983.

McLaughlin, Charles Hugh, Jr. "The Interpretation of the Environmental Movement within Manual Arts, Industrial Arts, Education, and Technology Education, 1875–1985." Ph.D. diss., University of Maryland, 1991.

McMurry, Charles A. *Special Method in Elementary Science for the Common School.* New York: Macmillan, 1904.

———. *Special Method in Natural Science for the First Four Grades.* Bloomington: Public School Publishing Co., 1896.

McMurry, Frank, and Henry E. Armstrong. "Some Recent Criticism of Nature-Study." *Nature-Study Review* 1 (1905): 22–26.

McMurry, Lida Brown. *Correlation of Studies with the Interests of the Child for the First and Second School Years*. DeKalb: Northern Illinois Normal School, 1907.

———. *Nature Study Lessons for Primary Grades*. New York: Macmillan, 1905.

McMurry, Lida Brown, and Agnes Spofford. *Songs of the Treetop and Meadow*. Bloomington, IL: Public School Publishing Co., 1899.

Maienschein, Jane. "Whitman at Chicago: Establishing a Chicago Style of Biology?" In *The American Development of Biology*. Edited by Ronald Rainger, Keith Benson, and Jane Maienschein. Philadelphia: University of Pennsylvania Press, 1988.

Mann, [Mary Peabody]. "Reminiscences of School Life and Teaching." *American Journal of Education* 32 (1882): 743–52.

Marshall, Helen E. *Grandest of Enterprises: Illinois State Normal University, 1857–1957*. Normal: Northern Illinois University Press, 1956.

Marshall, Ruth. "A Course in Nature-Study for Teachers." *Nature-Study Review* 15 (October 1909): 184–89.

Marx, Leo. *The Machine in the Garden: Technology and the Pastoral Ideal in America*. New York: Oxford University Press, 1964.

Mason, Jennifer. *Civilized Creatures: Urban Animals, Sentimental Culture, and Literature*. Baltimore: Johns Hopkins University Press, 2005.

Massachusetts State Assembly of the Agassiz Association. *Proceedings of the Meeting Held in Fitchburg, 30 May 1890*. The Assembly, c. 1891.

Mathien, Frances Joan. "Lucy L. W. Wilson, Ph.D.: An Eastern Educator and the Southwestern Pueblos." In *Philadelphia and the Development of Americanist Archaeology*. Edited by Don D. Fowler and David R. Wilcox. Tuscaloosa: University of Alabama Press, 2003.

Martusewicz, Rebecca A. "The Will to Reason: An Archeology of Womanhood and Education, 1880–1920." D.Ed. diss., University of Rochester, 1988.

Mattingly, Carol. *Well-Tempered Women: Nineteenth-Century Temperance Rhetoric*. Carbondale: Southern Illinois University Press, 1998.

Mattingly, Paul. *The Classless Profession: American Schoolmen in the Nineteenth Century*. New York: New York University Press, 1975.

Mayberry, B. D. *A Century of Agricultural Extension at 1890 Land-Grant Institutions and Tuskegee Institute*. New York: Vantage Press, 1992.

Mayhew, Katherine Camp, and Anna Camp Edwards. *The Dewey School: The Laboratory School of the University of Chicago*. New York: Atherton Press, 1966.

Mayman, Jacob Edward. *Teaching Elementary Science in Elementary Schools*. Publication 15. New York Department of Education, 1915.

———. "An Experimental Investigation of the Book Method, Lecture Method, and Experimental Method of Teaching Elementary Science in Elementary Schools." Ph.D. diss., New York University, 1912.

Mayo, Elizabeth. *Lessons on Shells as Given in a Pestalozzian School, at Cheam, Surrey*. New York: P. Hill, 1833.

Meier, William H. D. "Economic Biology in the Elementary Schools." Ph.D. diss., Harvard, 1919.

Meine, Curt. *Aldo Leopold: His Life and Work*. Madison: University of Wisconsin Press, 1991.

Merryman, John L. *The Indiana Story: Pennsylvania's First State University*. Clearfield, PA: Kurtz Brothers, 1976.

Mershon, Phillip. "The Brooklyn Children's Museum: The Story of a Pioneer." Typescript (d. 1959). Provided by the Brooklyn Institute Library.

Messerli, Jonathan. "Mary Peabody Mann." *Notable American Women* (1971), 2:488–90.

"Methods in Teaching Primary Science." *Outline of Methods for Georgia Teachers for the Summer Institutes for 1896*. Atlanta: Franklin Printing and Publishing, 1896.

Meyers, Ira B. "The Evolution of Aim and Method in the Teaching of Nature Study in the Common Schools of the United States." *Elementary School Teacher* 11 (December 1910): 205–13.

Miall, L. C. *House, Garden and Field*. New York: Longmans, 1904.

——. "Ready-Made Lessons in Nature Study." *Nature-Study Review* 1 (May 1905): 101–2.

Michigan Association for Childhood Education Research Committee. *Nature Study Units and Suggestions for the Early Elementary Grades*. The Association, [1930].

Mille, Marion. *An Out-of-Door Diary for Boys and Girls*. New York: Sturgis and Walton, 1910.

Miller, Edward Andrew. *Exercises with Plants and Animals for Southern Rural Schools*. Washington, DC: Department of Agriculture, 1915.

Miller, Janet A. "Urban Education and the New City: Cincinnati's Elementary Schools, 1870–1914." Ph.D. diss., Miami University, 1974.

Miller, Louise. *A Course in Nature Study for Use in the Public Schools*. Department of Agriculture, Commonwealth of Pennsylvania. *Bulletin* 63 (1900).

Miller, Olive Thorne. *The First Book of Birds*. Boston: Houghton Mifflin, 1899.

——. *Little Brothers of the Air*. Boston: Houghton Mifflin, 1892.

——. *True Bird Stories from My Note-books*. Boston: Houghton Mifflin, 1903.

Miller, Perry. *The Life of the Mind in America from the Revolution to the Civil War*. New York: Harcourt, Brace, and Co., 1965.

Miller, Susan. "Girls in Nature/the Nature of Girls: Transforming Female Adolescence at Summer Camp, 1900–1939." Ph.D. diss., University of Pennsylvania, 2001.

Mills, Enos A. "The Long's Peak Trail School and Nature Guiding." *Nature-Study Review* 17 (March 1921): 95–98.

Minton, Tyree G. "The History of the Nature-Study Movement and Its Role in the Development of Environmental Education." Ph.D. diss., University of Massachusetts, 1980.

Mitchell, Dora Otis. "A History of Nature-Study." *Nature-Study Review* 19 (September 1923): 258–74; (October 1923): 295–321.

Mitman, Gregg. *Reel Nature: America's Romance with Wildlife on Film*. Cambridge: Harvard University Press, 1999.

——. *The State of Nature: Ecology, Community, and American Social Thought, 1900–1950*. Chicago: University of Chicago Press, 1992.

Möbius, Karl. *Naturgeschichte in der Volksschule. Der Dorfteich als Lebensgemeinschaft*. Kiel: Lipsius and Tischer, 1885.

Mohl, Raymond A., and Neil Betten. *Steel City: Urban and Ethnic Patterns in Gary Indiana, 1906–1950*. New York: Holmes and Meier, 1986.

Monroe, Paul. *A Cyclopedia of Education*. New York: Macmillan, 1914.

Monroe, W. S. *History of the Pestalozzian Movement in the United States*. Syracuse: Bardeen, 1907.

Moran, Jeffrey P. *Teaching Sex: The Shaping of Adolescence in the 20th Century*. Cambridge: Harvard University Press, 2000.

Morrow, Prince Albert. *Report of the Special Committee on the Matter of Methods of Sex Education*. New York: American Federation for Sex Hygiene, 1913.

Muliak, Stanley B. "Nature Counselor Preparation in Relation to the Status of Nature Study in Camps." M.A. thesis, Cornell University, 1931.

Muldrew, Dean W. H. "Nature-Study and Elementary Agriculture in Canada." *Nature- Study Review* 1 (January 1905): 20–22.

Mulligan, Maureen Kay Harley. "Common Cares: Women and the Family Farm in the Midwest, 1870–1930." Ph.D. diss., University of Notre Dame, 1996.

Munson, John P. *Education through Nature Study: Foundations and Method*. New York: E. L. Kellogg and Co., 1903.

Nash, Roderick. *Wilderness and the American Mind*. New Haven: Yale University Press, 1967.

"National Council of Elementary Science." *Science Education* 31 (March 1947): 78–80.

The National Herbart Society. *Second Annual Yearbook*, 1896.

The Nature Almanac: A Handbook of Nature Education. Washington, DC: American Nature Association, 1930.

"The Nature Guide Service." *Yosemite* 39 (July 1960): 152–65.

"Nature Lore School." *Nature-Study Review* 19 (August 1923): 227–29.

Nature Study, A Course Designed Especially for Teachers and Students. Marine Biological Laboratory, 1900.

"Nature-study and Gardening for Indian Schools." *Nature-Study Review* 2 (April 1906): 141–43.

Nature Study for the Oakland Public Schools, Report Supplement, 1896–97.

"Nature-Study in Rural Schools." *Nature Magazine: A Weekly Illustrated Journal of Science* 61 (1899–1900): 553.

"Nebraska Teachers' Reading Circle." *Biennial Report of the [Nebraska] State Superintendent of Public Instruction for 1901–1903.* Lincoln, 1903: 125–27.

Needham, James G. "Agreement as to the Nature-Study Program." *Nature-Study Review* 13 (June 1917): 2–3.

———. *Outdoor Studies: A Reading Book of Nature Studies.* New York: American Book Co., 1926 [1898].

Neely, Wayne Caldwell. *The Agricultural Fair.* New York: Columbia University Press, 1935.

Nelson, Norma E. "Nature-Study in Our City Parks." *Nature-Study Review* 10 (September 1914): 210–13.

New York Humane Education Association. *A List of Books Recommended for Humane Reading and the Teaching of Humane Education and Nature Study.* New York: Association, 1912.

"News Notes." *Nature-Study Review* 2 (December 1906): 319–20.

"News Notes." *Nature-Study Review* 3 (September 1907): 190.

Nice, Margaret Morse. *Research is a Passion with Me: Autobiography of a Bird Lover.* Toronto: Consolidated Amethyst Communications, 1979.

Nietz, John A. *The Evolution of American Secondary Textbooks.* Pittsburgh: University of Pittsburgh Press, 1961.

Nordin, Dennis Sven. *Rich Harvest: A History of the Grange, 1867–1900.* Jackson: University Press of Mississippi, 1974.

Northrop, Alice Rich. *Through Field and Woodland: A Companion for Nature Students.* New York: G. P. Putnam's Sons, 1925.

Norwood, Vera. *Made from This Earth: American Women and Nature.* Chapel Hill: University of North Carolina Press, 1993.

Nutt, Patricia Margaret. "History and Development of Nature Education in California with Emphasis on the Period up to 1935." M.A. thesis, Cornell University, 1953.

Nyhart, Lynn K. "Economic and Civic Zoology in Late Nineteenth-Century Germany: The 'Living Communities' of Karl Möbius." *Isis* 89 (1998): 605–30.

———. *Modern Nature: The Rise of the Biological Perspective in Germany.* Chicago: University of Chicago Press, 2009.

Official Report of the Nature-Study Exhibition and Conference held in the Royal Botanic Society's Gardens, Regents Park, London. London: Blackie and Sons, 1903.

Ogren, Christine A. *The American Normal School: "An Instrument of Great Good."* New York: Palgrave Macmillan, 2005.

———. "Where Coeds were Coeducated: Normal Schools in Wisconsin, 1870–1920." *History of Education Quarterly* 35 (Spring 1995): 1–26.

Olmstead, Emma Jean. "Commentary." National Education Association. *Journal of Proceedings and Addresses* (1903): 418.

Olmsted, Richard R. "The Nature-Study Movement in American Education." Ed.D. diss., Indiana University, 1967.

Olsen, John W. "Nature Study: Some Physical Laws Necessary to the Study of Geography and Agriculture." In *Course of Study for the Common Schools of Minnesota* (n.p., 1908).

Osborn, Henry Fairfield, and George H. Sherwood. *The Museum and Nature Study in the Public Schools.* New York: American Museum of Natural History, 1913.

Overfield, Richard A. *Science with Practice: Charles E. Bessey and the Maturing of American Botany*. Ames: Iowa State University, 1993.

Overton, Frank, assisted by Mary E. Hill. *Nature Study: A Pupil's Text-book*. New York: American Book Co., 1905.

Packard, Alpheus S. *Zoology for Students and General Readers*. New York: Henry Holt and Co., 1879.

Page, Max. "'Uses of the Axe': Towards a Treeless New York." *American Studies* 40 (1999): 41–64.

Palmer, Ephraim Laurence. "The American Nature Study Society." *Science* 57 (February 9, 1923): 183.

———. *Fieldbook of Nature Study*. Ithaca: Comstock Publishing Co., 1927.

———. "Nature-study and the Scouting Movement or Vice Versa." *Nature-Study Review* 19 (April 1923): 168–76.

Pappas, Jeffrey Peter. "Forest Scholars: The Early History of Nature Guiding at Yosemite National Park, 1913–1925." Ph.D. diss., Arizona State University, 2003.

Park, David N., and Richard Ginsberg, eds. *Southern Cities, Southern Schools: Public Education in the Urban South*. New York: Greenwood, 1990.

Parker, Francis W. "The Child." National Education Association. *Addresses and Proceedings*, 1889.

———. "Syllabus of a Course of Lectures upon the Philosophy of Education." *The Course of Study* 1 (July 1900): 16.

Parker, Franklin. "Francis Wayland Parker, 1837–1902." *Paedagogica Historica* 1 (1961): 120–33.

Parker, Alison M. *Purifying America: Women, Cultural Reform, and Pro-censorship Activism, 1873–1933*. Urbana: University of Illinois Press, 1997.

Parker, Samuel C. *The History of Modern Elementary Education*. Totowa, NJ: Littlefield, Adams and Co., 1970.

Parker, Samuel Sheets. *A Textbook in the History of Modern Elementary Education*. Boston: Ginn and Co., 1912.

Paroni, Clelia A. *Course of Study Monographs: Nature Study*. Berkeley: Berkeley Public Schools, 1931.

Parsons, Fannie Griscom. *The First Children's Farm School in New York City*. New York: DeWitt Farm School, 1903.

———. "A Day in the Children's School Farm in New York City." *Nature-Study Review* 1 (November 1905): 255–61.

Patch, Edith M. *First Lessons in Nature Study*. New York: Macmillan, 1927.

———. *A Little Gateway to Science: Hexapod Stories*. Boston: Atlantic Press, 1920.

Patterson, Alice Jean. *Nature Study and Health Education for the First and Second Year*. Normal: McKnight and McKnight, 1928.

———. "Present Trends in the Teaching of Nature Study." *Nature and Science Education Review* 2 (1929): 1–5.

———. "Survey of Twenty Years Progress Made in the Courses of Nature Study." *Nature-Study Review* 17 (February 1921): 55–62.

Paulsen, Friedrich. *German Education: Past and Present*. Translated by T. Lorenz. London: T. Fisher Unwin, 1908.

Pauly, Philip J. *Biologists and the Promise of American Life: From Meriwether Lewis to Alfred Kinsey*. Princeton: Princeton University Press, 2000.

———. "The Development of High School Biology: New York City, 1900–1915." *Isis* 82 (1991): 662–88.

———. *Fruits and Plains: The Horticultural Transformation of America*. Cambridge: Harvard University Press, 2007.

Payne, Frank Owen. *Geographical Nature Studies for Primary Work in Home Geography*. New York: American Book Co., 1898.

——. *Lessons in Nature Study around My School*. New York: E. L. Kellogg, 1895.

Peabody Educational Fund. *Proceedings of the Trustees for 1894*. Cambridge: John Wilson and Sons, 1900.

Pearson, T. Gilbert. "Twenty-Five Thousand Dollars to Aid Teachers in Bird-Study." *Nature-Study Review* 11 (1915): 33–34.

Perez, Kimberly. "Fancy and Imagination: Cultivating Sympathy and Envisioning the Natural World for the Child." Ph.D. diss., University of Oklahoma, 2006.

Perkins, Linda M. "The History of Blacks in Teaching: Growth and Decline within the Profession." In *American Teachers: Histories of a Profession at Work*. Edited by Donald Warren. New York: Macmillan, 1989.

Persing, Ellis C., and Elizabeth K. Peeples, *Elementary Science by Grades: A Nature Study and Science Reader*. New York: D. Appleton, 1928.

Phelps, Almira Hart Lincoln. *Familiar Lectures on Botany*. Hartford: H. and F. J. Huntington, [1829] 1852.

Philippon, Daniel J. *Conserving Words: How American Nature Writers Shaped the Environmental Movement*. Athens: University of Georgia Press, 2004.

Phillips, Claude A. *Modern Methods and the Elementary Curriculum*. New York: Century Co., 1923.

Pond, Patricia. *Science in Nineteenth-Century Children's Books: An Exhibition Based on the Encyclopedia Britannica Historical Collection of Books for Children in the University of Chicago Library*. University of Chicago Library, Department of Special Collections, Exhibition Catalogues, vol. 25 (1966).

Porter, Charlotte M. *The Eagle's Nest: Natural History and American Ideas, 1812–1842*. Tuscaloosa: University of Alabama Press, 1985.

Porter, Gene Stratton. *A Girl of the Limberlost*. New York: Grosset and Dunlap, 1909.

Potter, Beatrix. *The Tale of Peter Rabbit*. New York, Saalfield Publishing Co., [1905] 1916.

Prindeville, Edith. "Children's Pets as Disease Carriers." *Nature-Study Review* 7 (December 1911): 260–63.

Program of Study and Syllabus in Physical Culture, Physiology and Hygiene, Nature Study at the Archdiocese of New York. New York: Catholic School Board, 1911. Found in the Andover Cambridge Divinity School Library, Cambridge, MA.

Pruitt, Clarence M. "Otis William Caldwell." *Science Education* 5 (December 1947): 285–86.

Pryon, Helen B. *Lou Henry Hoover, Gallant First Lady*. New York: Dodd, Mead, and Co., 1969.

Pycior, Helena M., Nancy G. Slack, and Pnina G. Abir-Am, eds. *Creative Couples in the Sciences*. New Brunswick: Rutgers University Press, 1996.

Raftery, Judith Rosenberg. *Land of Fair Promise: Politics and Reform in Los Angeles Schools, 1885–1941*. Palo Alto: Stanford University Press, 1992.

Rainger, Ronald, Keith Benson, and Jane Maienschein, eds. *The American Development of Biology*. Philadelphia: University of Pennsylvania Press, 1988.

Ravitch, Diane. *The Great School Wars, New York City, 1805–1973: A History of the Public Schools as Battlefield of Social Change*. New York: Basic Books, 1974.

——. *The Revisionists Revised: A Critique of the Radical Attack on the Schools*. New York: Basic Books, 1978.

Recchiuti, John L. "The Origins of American Progressivism: New York's Social Science Community, 1880–1917." Ph.D. diss., Columbia University, 1992.

"Recent Publications." *Garden and Forest* 5 (September 14, 1892): 443.

Reese, William J. "Between Home and Family: Organized Parents, Club Women, and Urban Education in the Progressive Era." *School Review* 87 (1978): 3–28.

——. *Power and the Promise of School Reform: Grassroots Movements during the Progressive Era*. London: Routledge & Kegan Paul, 1986.

———. "The School Health Movement." In *Power and the Promise of School Reform: Grassroots Movements during the Progressive Era*. London: Routledge & Kegan Paul, 1986: 209–37.

"A Report of the 1920 Annual Meeting." *Nature-Study Review* 17 (February 1921): 47.

"A Report of the 1921 Annual Meeting." *Nature-Study Review* 18 (January/February 1922): 62–63.

Report of the Teachers Training Schools of Minnesota. Minneapolis: School Education Company, 1895.

Rennie, John. *The Aims and Methods of Nature Study: A Guide for Teachers*. Warwick: University Tutorial Press, [1910].

Richardson, Cynthia Watkins. "Picturing Nature: Education, Ornithology and Photography in the Life of Cordelia Stanwood: 1865–1958." Ph.D. diss., University of Maine, 2002.

Rice, Joseph M. "The Public Schools of Minneapolis and Others." *Forum* 15 (May 1893): 376.

———. *The Public School System of the United States*. New York: Century Co., 1914.

Rice, William North. *Science-Teaching in the Schools: An Address Delivered before the American Society of Naturalists*. Boston: D. C. Heath, 1889.

Ricks, George. *Natural History Object Lessons: A Manual for Teachers*. London: Isbister, n.d.

Riley, Glenda. *Women and Nature: Saving the "Wild" West*. Lincoln: University of Nebraska Press, 1999.

Rindge, Debora Anne. "The Painted Desert: Images of the American West from the Geological and Geographical Surveys of the Western Territories, 1867–1879." Ph.D. diss., University of Maryland College Park, 1993.

Riney, Scott. *The Rapid City Indian School 1898–1933*. Norman: University of Oklahoma Press, 1999.

Ripley, Ellor Carlisle. *Provisional Course of Study in Elementary Science for the Elementary Schools*. Boston School Committee, School Document no. 5 (1911).

Ritvo, Harriet. "Learning from Animals: Natural History for Children in the Eighteenth and Nineteenth Centuries." *Children's Literature* 113 (1985): 72–93.

Roberts, I. P. *Nature-Study Quarterly*. New York: Agricultural Experiment Station, 1899–1900.

Robin, Regina Spires. "Science for Children and the Untutored in Eighteenth-Century England." Ph.D. diss., City University of New York, 1998.

Rogers, Dorothy. *Oswego: Fountainhead of Teacher Education*. New York: Appleton, Century, Crofts, 1961.

Rogers, Julia Ellen. *A Key to the Nature Library, the Open Book of Nature; with Practical Suggestions on the Every-day Use of the Volumes by Children and Grown People in Homes, Schools, and Nature Clubs*. New York: Doubleday, Page, and Co., 1909.

———. *The Shell Book: A Popular Guide to a Knowledge of the Families of Living Mollusks, and an Aid to the Identification of Shells Native and Foreign*. New York: Doubleday, Page, and Co., 1908.

———. *The Tree Book: A Popular Guide to a Knowledge of the Trees of North American and to their Uses and Cultivation*. New York: Doubleday, Page, and Co., 1905.

Rogers, Margaret Jennings. "From True to New Womanhood: The Rise of the Girl Scouts, 1912–1930." Ph.D. diss., Stanford University, 1992.

Rome, Adam. "'Political Hermaphrodites': Gender and Environmental Reform in Progressive America." *Environmental History* 11 (July 2006): 440–63.

Rose, Flora, Esther H. Stocks, and Michael W. Whittier. *A Growing College: Home Economics at Cornell University*. Ithaca: Cornell University Press, 1969.

Rosenberg, Rosalind. *Beyond Separate Spheres: Intellectual Roots of Modern Feminism*. New Haven: Yale University Press, 1982.

Ross, Dorothy. *G. Stanley Hall: The Psychologist as Prophet*. Chicago: University of Chicago Press, 1972.

Ross, Kirstie. "The 'Two Lucy's': Collaborative Work of Lucy Moore and Lucy Cranwell (1928–1938)." *New Zealand Science Review* 28:4 (2001): 138–42.

Rossiter, Margaret. *The Emergence of Agricultural Science: Justus Liebig and the Americans.* New Haven: Yale University Press, 1975.

———. "Sexual Segregation in the Sciences: Some Data and a Model." *Signs* 4 (1978): 146–51.

———. *Women Scientists in America: Struggles and Strategies to 1940.* Baltimore: Johns Hopkins University Press, 1982.

———. "Women's Work in Science, 1880–1910." *Isis* 71 (1980): 381–98.

Royston, H. R. *The Unity of Life: A Book of Nature Study for Parents and Teachers.* London: George G. Harrap and Co., 1925.

Rudolph, Emanuel D. "Women in Nineteenth Century American Botany: A Generally Unrecognized Constituency." *American Journal of Botany* 69 (September 1982): 1346–55.

Ruediger, William Carl. *Agencies for the Improvement of Teachers in Service.* U.S. Department of Education. *Bulletin* 449 (1911).

Ruggles, C. O. *Semi-Centennial Historical Sketch and Notes: Winona State Normal School, 1860–1910.* Winona: Jones and Kroeger, 1910.

Rury, John L. "Who Became Teachers? The Social Characteristics of Teachers in American History." In *American Teachers: Histories of a Profession at Work.* Edited by Donald Warren. New York: Macmillan, 1989.

Russell, B. B. "Nature Study and Intellectual Culture in Our Schools." *Education: A Monthly Magazine Devoted to the Science, Art, Philosophy and Literature of Education* 12 (1891–92): 345.

Russell, James Earl. *Founding Teachers: Reminiscences of the Dean Emeritus.* New York: Teachers College, 1937.

———. "The Function of the University in the Training of Teachers." *Teachers College Record* 1 (1900): 1–6.

"Ruth Marshall, 1869–1955." *Transactions of the American Microscopical Society* 75 (January 1956): 144.

Sandweiss, Martha. *Print the Legend: Photography and the American West.* New Haven: Yale University Press, 2002.

Sarver, Stephanie L. *Uneven Land: Nature and Agriculture in American Writing.* Lincoln: University of Nebraska Press, 1999.

Satterthwait, Alfred. "A Survey of Twenty Years' Progress in Nature Study in Extension Work." *Nature-Study Review* 17 (February 1921): 71–80.

Scharff, Virginia J., ed. *Seeing Nature through Gender.* Lawrence: University Press of Kansas, 2003.

Scheele, Irmtraut. *Von Lüeben bis Schmeil: Die Entwicklung von der Schulnaturgeschichte zum Biologieunterricht zwischen 1830 und 1933.* Berlin: Dietrich Reimer Verlag, 1981.

Schiltz, Mary Pieroni. "The Dunne School Board: Reform in Chicago, 1905–1908." Ph.D. diss., Loyola University, 1993.

Schmidt, John P. *Nature Study and Agriculture.* Boston: D. C. Heath, 1920.

Schmitt, Peter. *Back to Nature: The Arcadian Myth in Urban America.* New York: Oxford University Press, 1969.

Schmucker, Samuel Christian. "Science and Nature Study." *Nature-Study Review* 14 (February 1918): 48–52.

———. *The Study of Nature.* Philadelphia: J. B. Lippincott Co., 1908.

Schorger, A. W. *The Passenger Pigeon: Its Natural History and Extinction.* Madison: University of Wisconsin Press, 1955.

Schulten, Susan. *Geographical Imagination in America, 1880–1950.* Chicago: University of Chicago Press, 2001.

Schuyler, David. *The New Urban Landscape: The Redefinition of City Form in Nineteenth-Century America.* Baltimore: Johns Hopkins University Press, 1986.

Scott, Charles B. *Nature Study and the Child.* Boston: D. C. Heath, [1900] 1910.

———. *One Hundred Lessons in Nature Study Around My School*. New York: A. S. Barnes, 1895.

Secord, James. *Victorian Sensation: The Extraordinary Publication, Reception, and Secret Authorship of Vestiges of the Natural History of Creation*. Chicago: University of Chicago Press, 2000.

Sedlak, Michael W. "'Let Us Go and Buy a School Master:' Historical Perspectives on the Hiring of Teachers in the United States." In *American Teachers: Histories of a Profession at Work*. Edited by Donald Warren. New York: Macmillan, 1989.

Selle, Erwin S., ed. *The Winona State Teachers College Historical Notes, 1910–1935*. N.p., 1935.

Selleck, R. S. W. *The New Education: The English Background, 1870–1914*. Melbourne, Australia: Pitman and Co., 1968.

Seller, Maxine Schwartz. "G. Stanley Hall and Edward Thorndike on the Education of Women: Theory and Policy in the Progressive Era." *History of Education Quarterly* 29 (Spring 1989): 95–107.

Selzer, Catherine R. "A History of the Chicago Normal School, 1855–1905." M.Ed. thesis, Chicago Teachers College, 1940.

Sequel, Mary L. *The Curriculum Field: Its Formative Years*. New York: Teachers College Press, 1966.

Shapin, Steven. *The Scientific Life: A Moral History of a Late Modern Vocation*. Chicago: University of Chicago Press, 2008.

Shapiro, Adam R. "Civic Biology and the Origin of the School Antievolution Movement." *Journal of the History of Biology* 41 (September 2008): 409–33.

Shapiro, Michael S. *Child's Garden: The Kindergarten Movement from Froebel to Dewey*. University Park: Pennsylvania State University, 1983.

Shaw, Ellen Eddy. "The American Nature-Study Society." In *Nature Almanac: A Handbook of Nature Education*. Washington, DC: American Nature Association, 1930.

———. "The Place of Children's Gardens." *Nature-Study Review* 6 (February 1910): 43–45.

Sheehan, Nancy M. "The WCTU and Educational Strategies on the Canadian Prairie." *History of Education Quarterly* 24 (Spring 1984): 101–19.

Sheldon, Dorothy Dean. *The Blind Child in the World of Nature*. New York: American Foundation for the Blind, 1929.

Sheldon, Edward. *Lessons on Objects*. New York: Ivison, Blakeman & Co., 1863.

Sheldon, Jennie Arms. *Observation Lessons on Animals . . . for the Use of Teachers of Primary and Grammar Schools*. Deerfield: n.p., 1931.

Sherwood, Herbert Francis. "Children of the Land: The Story of the Macdonald Movement in Canada." *The Outlook* 10 (April 1910): 891–99.

Sherwood, George H. *Free Nature Education by the American Museum of Natural History in Public Schools and Colleges*. New York: American Museum of Natural History, 1920.

Shi, David. *The Simple Life: Plain Living and High Thinking in American Culture*. New York: Oxford University Press, 1985.

Shteir, Ann C. *Cultivating Women, Cultivating Science*. Baltimore: Johns Hopkins University Press, 1996.

Sinclair, S. B. "The Time Required for Nature-Study." *Nature-Study Review* 1 (May 1905): 122–24.

Sizer, Theodore R. *Secondary Schools at the Turn of the Century*. New Haven: Yale University Press, 1964.

Sloan, Joan Butin. "The Founding of the Naples Table Association for Promoting Scientific Research by Women." *Signs* 4 (Autumn 1978): 208–16.

Smallwood, James. *And Gladly Teach: Reminiscences of Teachers from Frontier Dugout to Modern Module*. Norman: University of Oklahoma Press, 1976.

Smith, Ruby Green. *The People's Colleges: A History of the New York State Extension Service in Cornell University and the State, 1876–1948*. Ithaca: Cornell University Press, 1949.

Solberg, Winton U. *The University of Illinois, 1894–1904: The Shaping of a University*. Urbana: University of Illinois Press, 2002.

Spearman, Melissa. "The Peripatetic Normal School: Teachers' Institutes in Five Southwestern Cities (1880–1920)." Ph.D. diss., University of Texas at Austin, 2006.

Spencer, John Walton. "The Diplomacy of the Good Teacher." *Nature-Study Review* 13 (1917): 22–23.

Stage, Sarah, and Virginia B. Vincenti, eds. *Rethinking Home Economics: Women and the History of the Professions*. Ithaca: Cornell University Press, 1997.

State of New York, Department of Education. *First Annual Report, Supplemental Volume*. Albany: Department of Education, 1905.

Stearns, Raymond P. *Science in the British Colonies of America*. Urbana: University of Illinois Press, 1970.

Steele, J. Dorman. *Fourteen Weeks in Zoology*. A. S. Barnes and Co., 1876.

Stein, Barbara R. *On Her Own Terms: Annie Montague Alexander and the Rise of Science in the American West*. Berkeley: University of California Press, 2001.

Stemmons, Walter. *Connecticut Agricultural College: A History*. Storrs: Connecticut Agricultural College, 1931.

Stenhouse, Ernest. *An Introduction to Nature-Study*. London: Macmillan, 1903.

Stratton-Porter, Gene. *A Girl of the Limberlost*. New York: Grosset and Dunlap, 1909.

Sullivan, E. N. *Friends of the Fields*. Boston: Educational Publishing Co., 1898.

"The Summer School—for What?" *School Education* 14 (December 1895): 5–6.

Syllabus for Nature Study: Humaneness, Elementary Agriculture, and Homemaking. State University of New York. Bulletin 675 (November 15, 1918).

Swiss Cross: A Monthly Magazine of Popular Science. Vols. 1–5, 1887–89.

Talmage, James E. *First Book of Nature*. Salt Lake City: George Cannon and Sons, 1892.

Tebbel, John. *A History of Publishing in the United States: The Expansion of the Industry, 1865–1919*. New York: R. R. Bowker Co., [1972] 1981.

"Theodosia Hadley." *Western Michigan College News Magazine* 5 (1947): 8–9.

Thorndike, Edward Lee. "The Feminization of American Education." Vermont State Teachers Association. *Proceedings and Addresses* (October 1909): 48–62.

——. "Reading as Nature Study." *Education* 19 (1899): 386–91.

——. "Sentimentality in Science Teaching." *Educational Review* 17 (1899): 56–64.

Tichi, Cecelia. *Shifting Gears: Technology, Literature, and Culture in Modernist America*. Chapel Hill: University of North Carolina Press, 1987.

Tobey, Ronald C. *Saving the Prairies: The Life Cycle of the Founding School of Plant Ecology, 1895–1955*. Berkeley: University of California Press, 1981.

Tolley, Kimberley F. Higgins. *The Science Education of American Girls: An American Perspective*. New York: Routledge Falmer, 2003.

——. "The Science Education of American Girls, 1784–1932." Ph.D. diss., University of California at Berkeley, 1996.

Tomes, Nancy. *The Gospel of Germs: Men, Women and the Microbe in American Life*. Cambridge: Harvard University Press, 1998.

Tovel, A. "Plea for the Study of Nature in School." *Education: A Monthly Magazine Devoted to the Science, Art, Philosophy and Literature of Education* 8 (1887–1888): 375.

Townsend, Manley. "What Do I Expect That Nature-Study Should Do for My Child?" *Nature-Study Review* 19 (February 1923): 61–64.

Trafton, Gilbert H. *Nature Study and Science for Intermediate Grades*. New York: Macmillan, 1927.

——. "Outline of Nature-Study for Use in the Elementary School of the State Normal School, Mankato, Minnesota." *Nature-Study Review* 11 (March 1915): 92–167.

——. *The Teaching of Science in the Elementary School*. Boston: Houghton Mifflin, 1918.

Trennert, Robert A., Jr. *The Phoenix Indian School: Forced Assimilation in Arizona, 1891–1935*. Norman: University of Oklahoma Press, 1988.

———. "Selling Indian Education at World's Fairs and Expositions, 1893–1904." *American Indian Quarterly* 11 (1987): 203–22.

Troeger, John. *Harold's Rambles*. New York: Appleton and Co., 1898.

Troen, Selwyn K. *The Public and the Schools: Shaping the St. Louis System, 1838–1920*. Columbia: University of Missouri Press, 1975.

Tyack, David B. *The One Best System: A History of American Urban Education*. Cambridge: Harvard University Press, 1974.

Tyack, David, and Larry Cuban. *Tinkering toward Utopia: A Century of Public School Reform*. Cambridge: Harvard University Press, 1995.

Tyack, David, and Elizabeth Hansot. *Managers of Virtue: Public School Leadership in America, 1820–1980*. New York: Basic Books, 1982.

Underhill, Orra Ervin. *The Origin and Development of Elementary School Science*. Chicago: Scott, Freeman, and Co., 1941.

United States Bureau of Education. *Report of the Committee of Ten on Secondary Studies*. Washington, DC, 1893.

"University and Education News." *Science* 38 (October 17, 1913): 544–45.

Urban, Wayne. "Historical Studies of Teacher Education." In *Handbook of Research on Teacher Education*. Edited by W. Robert Houston. New York: Macmillan, 1990.

Urban, Wayne J. and Jennings L. Wagoner, Jr. *American Education: A History*. New York: McGraw Hill, 2004.

Van Buskirk, Kate. "Introduction." *Course of Education for Elementary Schools of Arizona in Nature Study*. Phoenix: State Department of Education, 1936.

Van De Voort, Alice M. *The Teaching of Science in Normal Schools and Teachers Colleges in Contributions to Education*, no. 287. New York: Teachers College, 1927.

Van Slyck, Abigail A. *Carnegie Libraries and American Culture, 1890–1920*. Chicago: University of Chicago Press, 1998.

———. *A Manufactured Wilderness: Summer Camps and the Shaping of American Youth, 1890–1960*. Minneapolis: University Minnesota Press, 2006.

Verrill, Addison E. *True Nature Stories*. Boston: Gorham Press, 1929.

Vinal, William Gould. "The First Nature Lore School." *Nature-Study Review* 17 (April 1921): 145–46.

———. *Nature Guiding*. Ithaca: Comstock Publishing Co., 1926.

———. *Nature Recreation: Group Guidance for the Out-of-Doors*. New York: Dover, 1963 [1940].

———. "Summer Camp and Nature-Study." *Nature-Study Review* 18 (April 1922): 113–18.

Wagner, R. E. "On the Training of Teachers of Nature-Study." *Nature-Study Review* 12 (February 1916): 47–55.

Walker, Jerome. *Anatomy, Physiology, and Hygiene: A Manual for the Use of Colleges, Schools, and General Readers*. New York: A. Lovell and Company, 1887.

Warner, Deborah Jean. "Science Education for Women in Antebellum America." *Isis* 69 (1978): 58–67.

Warren, Donald, ed. *American Teachers: Histories of a Profession at Work*. New York: Macmillan, 1989.

Warren, Minetta L. *From September to June with Nature*. Boston: Heath, 1898.

Watterson, Ada. "Guide to Periodical Literature." *Nature-Study Review* 1 (March 1905): 90–96.

Webb, Edith J. *Boys' and Girls' 4-H Club Work in the United States: A Selected List*. U.S. Department of Agriculture Extension Service, 1932.

Weiler, Kathleen. *Country Schoolwomen: Teaching in Rural California, 1850–1950*. Stanford: Stanford University Press, 1998.

Welch, Margaret. *The Book of Nature, 1825–1875*. Boston: Northeastern University Press, 1998.

Wells, H. G. *A Graded Course of Instruction for the Graded School*. New York: A. S. Barnes, 1862.

Welker, Robert. *Birds and Men: American Birds in Science, Art, Literature and Conservation, 1800–1900*. Cambridge: Harvard University Press, 1955.

Weneck, Bette. "Social and Cultural Stratification in Women's Higher Education: Barnard College and Teachers College, 1898–1912." *History of Education Quarterly* 31 (1991): 1–25.

Wessel, Thomas, and Marilyn Wessel. *4-H, An American Idea, 1900–1980: A History of 4-H*. Chevy Chase, MD: National 4-H Council, 1982.

Westell, W. Percival. *Nature Stalking for Boys through Field Glasses, Stereoscopes and Camera with an Introduction by R. S. S. Baden-Powell*. London: J. M. Dart, 1909.

Whitney, E. R. "Nature Study as an Aid to Advanced Work in Science." National Education Association. *Addresses and Proceedings* (1904): 889–896.

Whitney, Evangeline E. "Discussion." National Education Association. *Journal of Proceedings and Addresses* (1903): 892.

———. "Vacation Schools, Playgrounds, and Recreation Centers." National Education Association. *Journal of Proceedings and Addresses* (1904): 298–303.

Wilkins, Thurman. *Clarence King: A Biography*. New York: Macmillan, 1958.

Wills, Harrington. *The Teaching of Nature Study and the Biological Sciences*. Boston: Christopher Publishing Co., 1936.

Wilson, Charles B. "How Can Advanced Science in the College and University and Nature Study Be Rendered Mutually Beneficial?" National Education Association. *Addresses and Proceedings* (1900).

Wilson, Charles R., and William Ferris, eds. *Encyclopedia of Southern Culture*. Chapel Hill: University of North Carolina Press, 1989.

Wilson, John D. *How to Study Nature in the Schools: A Flexible Manual for Teachers*. Syracuse: C. W. Bardeen, 1900.

Wilson, Lucy Langdon Williams. *Handbook of Domestic Science and Household Arts for Use in Elementary Schools*. New York: Macmillan, 1900.

———. *Nature Study in Elementary Schools: First Reader*. New York: Macmillan, 1900.

———. *Nature Study in Elementary Schools: A Manual for Teachers*. New York: Macmillan, [1897] 1900.

Winterburn, Rosa V. *Methodology in Teaching, Being the Stockton Methods in Elementary Schools*. New York: Macmillan, 1907.

Wisconsin Alumni Directory. Madison: University of Wisconsin, 1921.

Withers, Charles, and Diarmid Finnegan. "Natural History Societies, Fieldwork, and Local Knowledge in Nineteenth-Century Scotland: Towards a Historical Geography of Civic Science." *Cultural Geographies* 10 (2003): 334–53.

Wolcott, Robert H. "Aim and Method in Nature Study." *University [of Nebraska] Journal* 5 (November 1908): 24–25.

———. "What Nature Study Should Be." *University [of Nebraska] Journal* 1 (March 1905): 7–8.

Wood, Carolyn D. *Animals: A Nature Study Textbook*. Boston: Ginn and Co., 1912.

Woodhull, John F. "Physical Nature Study." *Nature-Study Review* 1 (January 1905): 18–19.

Woodward, Laura E. "The Cape May Summer School." *Nature-Study Review* 10 (March 1914): 99–102.

Woonspe Wankantu. Santee Normal Training School, 1897–1898. Santee Agency, Nebraska, c. 1898.

Works, George A. *Rural School Survey of New York State: A Report to Patrons, by the Joint Committee on Rural Schools*. Ithaca: n.p., 1922.

Worster, Donald. *Nature's Economy: A History of Environmental Ideas*. Cambridge: Cambridge University Press, 1977.

Wright, Julia McNair. *Seaside and Wayside*. Boston: D. C. Heath, 1885–1903.

Wright, Mabel Osgood. *Gray Lady and the Birds: Stories of the Bird Year for Home and School*. New York: Macmillan, 1907.

Yearbook of the American Nature-Study Society. Toledo: Board of Education, 1925.

Zenz, W. "Naturgeschichte." In *Enzyklopadisches handbuch der Erziehungskunde*. Edited by Joseph Loos. Vienna and Liepzig: A. Pichlers Witwe & Sohn, 1908, 2:103–6.

Index